7th High Performance Yacht Design Conference 2021

Auckland, New Zealand
11 – 12 March 2021

ISBN: 978-1-7138-2919-5

Printed from e-media with permission by:

Curran Associates, Inc.
57 Morehouse Lane
Red Hook, NY 12571

Some format issues inherent in the e-media version may also appear in this print version.

High Performance Yacht Design Conference (HPYD)
c/o David Le Pelley
University of Auckland, Mechanical Engineering Department
Private Bag 92019
Auckland, New Zealand

Phone: +64 9 527 5086

info@hpyd.org.nz

Additional copies of this publication are available from:

Curran Associates, Inc.
57 Morehouse Lane
Red Hook, NY 12571 USA
Phone: 845-758-0400
Fax: 845-758-2633
Email: curran@proceedings.com
Web: www.proceedings.com

TABLE OF CONTENTS

CHARACTERIZATION OF A FLOW SEPARATION SENSOR ON A HYDROFOIL: THE HYDRO E-TELLTALE 1
P. Bot, R. Alas, D. Voisin, A. Soulier, C. Braud, P. Marcillat, B. Pezeril, F. D'Arco

REVIEW OF UNDEWATER FLUID-STRUCTURE INTERACTION MEASURING TECHNIQUES 7
L. Giovannetti, I. Stenius

EXPERIMENTAL INVESTIGATIONS ON CONVENTIONAL AND UNCONVENTIONAL BOOST SYSTEMS TO ASSIST TAKE-OFF OF HYDROFOIL CRAFTS 15
L. Minerva, F. Montero

SPLIT-FLAPS FOR INCREASED HEEL STABILITY OF T-FOIL CONFIGURATIONS 25
N. Prabahar, L. Larsson, A. Persson

ANALYTICAL STUDY OF A FOILING CATAMARAN IN WAVES 39
I. Sabadell, M. Drummond, E. Copeland, T. Simhony

F50 WING 2.0: DEVELOPING A NEXT-GENERATION WING SAIL CONTROL SYSTEM 49
E. Hawthorne, J. Nicols

PERFORMANCE PREDICTIONS OF LONG-DISTANCE SAILING VESSELS 59
S. Werner, J. Kuttenkeuler, A. Horteborn, O. Lundback, M. Razola, U. Dhome

VARIATION OF UNDERWATER APPENDAGES AND THEIR EFFECTS ON WIND POWERED SHIPS 67
L. Giovannetti, K. Ljungqvist, S. Werner

EFFICIENT JIB-MAINSAIL FLUID-STRUCTURE INTERACTION MODELLING BASED ON AN EXTENDED LIFTING-LINE THEORY – VALIDATIONS WITH SEMI-RIGID SAILS EXPERIMENTS 76
A. Morvan, M. Sacher, A. Neme, J. Leroux, C. Jochum, N. Abiven

ON THE EFFECT OF THE LEADING-EDGE SEPARATION BUBBLE ON THE AERODYNAMICS OF SPINNAKERS 87
J. Souppez, P. Bot, I. Viola

FORCE GENERATION MECHANISMS OF DOWNWIND SAILS 98
A. Arredondo-Galeana, I. Viola

SELECTION OF EXISTING SAIL DESIGNS FOR MULTI-FIDELITY SURROGATE MODELS 108
T. Peart, N. Aubin, S. Nava, J. Cater, S. Norris

AERODYNAMIC PERFORMANCE - TRANSFERRING LOAD BETWEEN CABLE AND SAIL 118
J. Braun

QUANTITATIVE DYNAMIC ENERGY ABSORPTION OF SANDWICH CORE MATERIAL 129
T. Basset, M. Battley, R. Gerard, T. Allen

LOCALISED CORE SHEAR STRESS AT IN SANDWICHES WITH FACESHEET THICKNESS VARIATIONS 136
T. Allen, E. Grant

Author Index

CHARACTERIZATION OF A FLOW SEPARATION SENSOR ON A HYDROFOIL: THE HYDRO E-TELLTALE

Patrick Bot[1], patrick.bot@ecole-navale.fr
Rémi Alas[1], remi.alas@gmail.com
Dimitri Voisin[2], dvoisin@meragitee.com
Antoine Soulier[2], asoulier@meragitee.com
Caroline Braud[3], caroline.braud@ec-nantes.fr
Paul Marcillat[1], marcillat.eleve@ecole-navale.fr
Baudoin Pezeril[1], pezeril.eleve@ecole-navale.fr
François d'Arco[1], darco2.eleve@ecole-navale.fr

Abstract. The performance of lifting bodies such as hydrofoils is determined by the flow state and particular attention should be paid to flow separation, as this greatly affects the generated lift and drag. Sailors are used to look at telltales (woolies) to trim their sails or steer their yacht. A French company developed an electronic telltale for sails based on a strain gauge activated by a silicon strand, with the appropriate signal processing to deliver the same information as a classical wool-made telltale, basically attached or separated flow. This new sensor proved useful when woolies are not visible or to deliver a signal to feed a control system, such as the autopilot for example. It was also applied to wind turbines to control the blade pitch. Mer Agitée is now developing an equivalent hydrodynamic e-Telltale to be used on hydrofoils and rudders to help trimming and controlling. The present work presents the investigation of a foil section fitted with this new sensor in a water tunnel, combining force and PIV measurements with the sensor signal, on a wide range of angle of attack. Results show that the hydro e-Telltale enables detecting the flow separation and anticipate stall, and possibly allows for detecting the boundary layer transition to turbulence. The feedback from this new sensor could be used for example in a control loop to make an "Anti-Stall System" on a rudder or a foil.

1. NOMENCLATURE

α	=	*angle of attack (°)*
c	=	chord length (m)
C_L	=	lift coefficient ( )
C_D	=	drag coefficient ( )
$Re=U_0\, c/\nu$	=	Reynolds number ( )
U_0	=	upstream velocity (m/s)
ρ	=	fluid density (kg/m^3)
ν	=	fluid kinematic viscosity (m^2/s)
s	=	profile span (m)
L	=	lift force (N)
D	=	drag force (N)

2. INTRODUCTION

On foiling yachts, dynamic trimming of the foils may be key to the flight control and some feedback on the flow state around the foils would be very helpful. Similarly on any yacht, monitoring the flow on the rudder may help anticipate and prevent stall. More generally in many cases of fluid flow over a lifting body, it is interesting to get some feedback from the flow state in order to help optimizing performance or controlling the system. On yacht sails, sailors are used to using "woollies" to detect flow separation and help trimming the sails and steering the yacht. A limiting factor is that these common telltales are to be seen by a human who can take action according to the information from the telltale: basically whether the flow is attached or separated. The company Mer Agitée has developed an electronic telltale based on a strain gauge embedded in a silicon ribbon [1]. A basic processing of the output voltage is made to deliver a signal equivalent to a classical telltale, namely attached or separated flow, that can be transmitted to a controlling system to trigger an alarm, to automatically trim the sail, or to the autopilot for example. This aero version of the electronic telltale has been extensively characterized and investigated for the application to wind turbines [2-5], and is routinely used on yachts (IMOCA for example). More recently, Mer Agitee has developed a similar sensor for hydrodynamic applications, based also on a strain gauge fitted on a thin metal sheet, to be flush-mounted on a hydrofoil. This paper reports the investigation of this sensor, in a well-controlled water flow, with simultaneous measurements of fluid forces, fluid velocity field and e-telltale response. Some preliminary tests and settings adjustments can be found in [6].

The flow over a wing or foil is mostly attached at low incidence, with a lift coefficient almost linearly increasing with the angle of attack. At higher incidence, flow separation becomes larger, generally reducing lift, and resulting in flow unsteadiness and varying loads related to vortex shedding. In some applications, it is interesting to monitor the flow state in real-time in order to feed a closed-loop controlling system (e.g. [7,8]). Hot-film measurements have already been used to detect transition and stall in the aeronautical sector [9], but seem difficult to be implemented on the appendages of a high-performance sailing vessel.

[1] IRENav, Ecole Navale, Naval Academy Research Institute, France
[2] Mer Agitee, France
[3] LHEEA Ecole Centrale de Nantes, France

3. EXPERIMENTAL SET-UP

Experiments were carried out in the IRENav hydrodynamic tunnel (Fig. 1) to measure the e-telltale output, the forces and the velocity fields on a two-dimensional foil section NACA 66(mod)-312 as in [10,11]. A recess is machined in the stainless steel foil to accommodate the e-telltale and the wire to carry the voltage signal. The e-telltale is mounted flush on the suction side of the foil, positioned between x/c=0.7 and x/c=0.8 from the leading edge (Fig. 2). The sensor consists in a strain gauge fitted on a 0.05mm-thick metal sheet. The output signal is conditioned locally to a 0-5V voltage and carried by a wire through the foil support, out of the tunnel test section to the acquisition system. The sensor system and wire are covered by a resin in order to achieve a smooth surface.

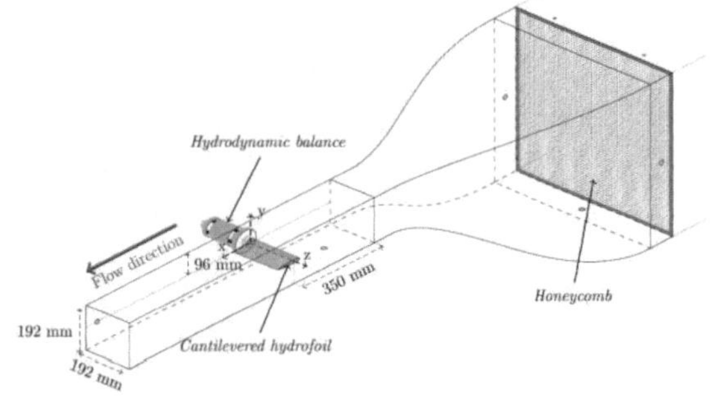

Figure 1: Water tunnel test section with the hydrofoil cantilevered on the force balance.

The test section measures 192 × 192 mm² and is 1 m long, located downstream of honeycombs and a 1/9 contraction. The measured turbulence intensity is 1.8%. The foil model spans almost all of the channel width, except for a small gap on each end to avoid contacts with the walls that would alter force measurements. The profile is mounted at mid-height of the test section, cantilevered on a strain gauge hydrodynamic balance to measure the forces. The balance is fixed on a rotating frame mounted on bearings and driven in rotation by a stepper motor with 6 10^5 impulsions per revolution. The axis of rotation was located at 25% of the chord from the leading edge. The angle of attack is controlled by a home-made Labview interface with accuracy better than 0.1°. The e-telltale and balance strain gauges are controlled by an IOtech Strainbook 616 and forces are determined thanks to a previous calibration with accuracy better than 1.7N for the lift and 0.5N for the drag. The upstream velocity U_0 is 6m/s resulting in a Reynolds number Re = 5.17 10^5. The force coefficient uncertainties at the investigated velocity are 0.006 for C_L and 0.002 for C_D. The units of the e-telltale output are arbitrary and only relative variations are considered.

Figure 2: Foil section fitted with the e-telltale sensor. The sensitive moving part is shown by the red circle. The strain gauge is glued at the root (upper end on the picture) of the moving part.

Forces and the e-telltale signal are recorded at 1 kHz, during 30 seconds when recorded without particle image velocimetry (PIV), and during 1 or 10s when time-resolved particle image velocimetry (TR-PIV) measurements are simultaneously recorded. The lift and drag coefficients are defined as:

$$C_L = \frac{L}{\frac{1}{2}.\rho.c.s.U_0^2} \quad (1)$$

$$C_D = \frac{D}{\frac{1}{2}.\rho.c.s.U_0^2} \quad (2)$$

Velocity field measurements are performed around the foil section using time-resolved 2D particle image velocimetry (TR-PIV). The flow is illuminated from above with a laser sheet normal to the spanwise direction and located at mid span. Two mirrors are placed below the tunnel test section to illuminate the other side of the foil, which allows for the whole velocity field around the obstacle to be measured at once. The light sheet thickness is around 2mm in the recorded flow field. The water flow is seeded with µm-sized polyamide particles and visualized by a 1280*800 pixel² speedsense 9600 camera at a sampling frequency of 1 kHz in double frame. For each value of α, either 1,000 or 10,000 image pairs are recorded and processed with Dantec's DynamicStudio™ in order to compute the snapshots of the instantaneous flow field. Raw images are pre-processed the following way: the time-averaged image is subtracted to eliminate the light flare on the section walls, and the resulting images are normalized by the time-averaged image to mitigate illumination inhomogeneity. Pre-processed image pairs are then processed with the AdaptivePIV algorithm [12]. The final interrogation window is 32 × 32 pixels² with a 50% overlap, yielding a spatial resolution of 1.75mm (0.017c), and typically 3 correlation passes are processed. Considering the whole measurement process, the measurement uncertainty on velocity is estimated to be around 1% of the free stream velocity.

4. RESULTS

4.1. Time-averaged force coefficients and e-telltale voltage

Figure 3 shows the time-averaged lift and drag coefficients and the e-telltale output voltage recorded during several runs on a wide range of angle of attack, from -4° up to 21°. The lift coefficient is increasing almost linearly with a slope slightly higher than the theoretical 2π/rad from 0° to 6°. It shows a small plateau at $C_L \approx 0.8$ from 6° to 8°, and increases again up to a maximum of $C_L \approx 1.15$ at 17° and then smoothly decreases. A change of slope can be noticed at 13°. The zero-lift angle of attack is -2°.

The drag coefficient is minimal at -1°, increases gently almost linearly up to $C_D \approx 0.018$ at 12°, and then increases much more sharply. This shape of the C_L curve is characteristic of NACA 6-series foils operating at moderate Reynolds number. These results are consistent with previous findings on the same hydrofoil at a similar Reynolds number [10,11] where the transition in the boundary layers was shown to occur via a Laminar Separation Bubble. On the suction side, the transition location was found to migrate from near the trailing edge to the leading edge around 6° angle of attack, associated to a plateau in the evolution of C_L between 5° and 7°. Here, this plateau is found from 6° to 8°, but the Reynold number is slightly lower (Re = 5.17 10^5 in the present study, Re = 7.5 10^5 in [10]). The elimination of the Laminar Separation Bubble on the pressure side may also contribute to the C_L plateau, as explained in [10,13]. It should be noted that these studies are in the critical Reynolds number range, where the separation behaviour and related forces evolution change drastically.

The e-telltale output is minimal and almost constant around 0.27V from -5° to 4° angle of attack It then shows a bell curve between 5° and 12° with a maximum around 0.66V at 9°. Then, it increases sharply for AoA ≥ 14° up to a maximum mean value close to 1.5V at 17° before dropping again. It is noticeable that the fluctuations are very small (RMS < 3% of mean value) for angles of attack below 13°, and increase for higher angles to reach RMS ≈ 70% of mean value at 17°, in the stall regime.

4.2. Velocity fields

Time-averaged velocity fields around the hydrofoil are shown on Figure 4 at four characteristic values of the angle of attack. A small area of low velocity appears near the trailing edge at α=7° (Fig. 4.b) and increases in size for increasing angle of attack. At 13°, another low velocity area shows up near the leading edge (Fig. 4.c), and for higher angles, the flow is fully separated right from the leading edge (Fig. 4.d). A closer inspection in the velocity vectors along the wall enables detecting flow separation where a negative tangential velocity component is visible. On the time-averaged velocity fields, the following is found: no separation for α ≤ 7°; flow separation near the trailing edge for 9° ≤ α ≤ 11°; flow separation at the leading edge for α ≥ 15°. Interestingly, no separation is visible at α = 13°.

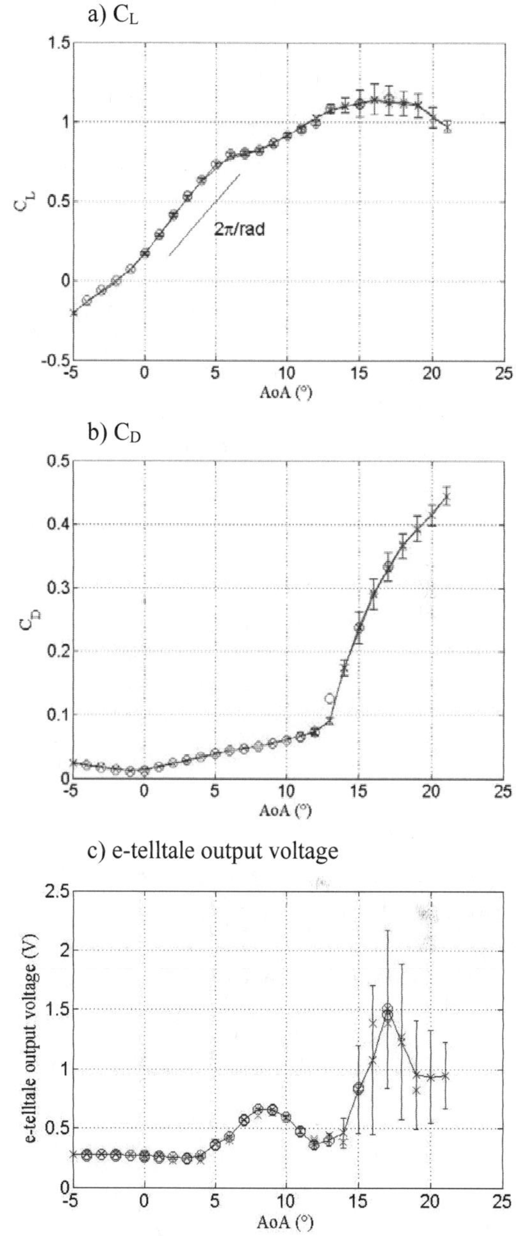

Figure 3: Time-averaged lift coefficient (a), drag coefficient (b) and e-telltale output voltage (c) as a function of angle of attack. Errorbars show the standard deviation of fluctuations in the time series. Different symbols and colours correspond to different experimental runs.

Flow separation is also detected into the instantaneous velocity fields. As these velocity fields are nosier than the time-averaged fields, and as demonstrated by [14], detection of actual separation was retained if at least three consecutive vectors along the suction wall showed a reverse tangential velocity component, and the same criterion was used to detect flow reattachment. This looked consistent with what would be considered separated / reattached when looking at the vector fields.

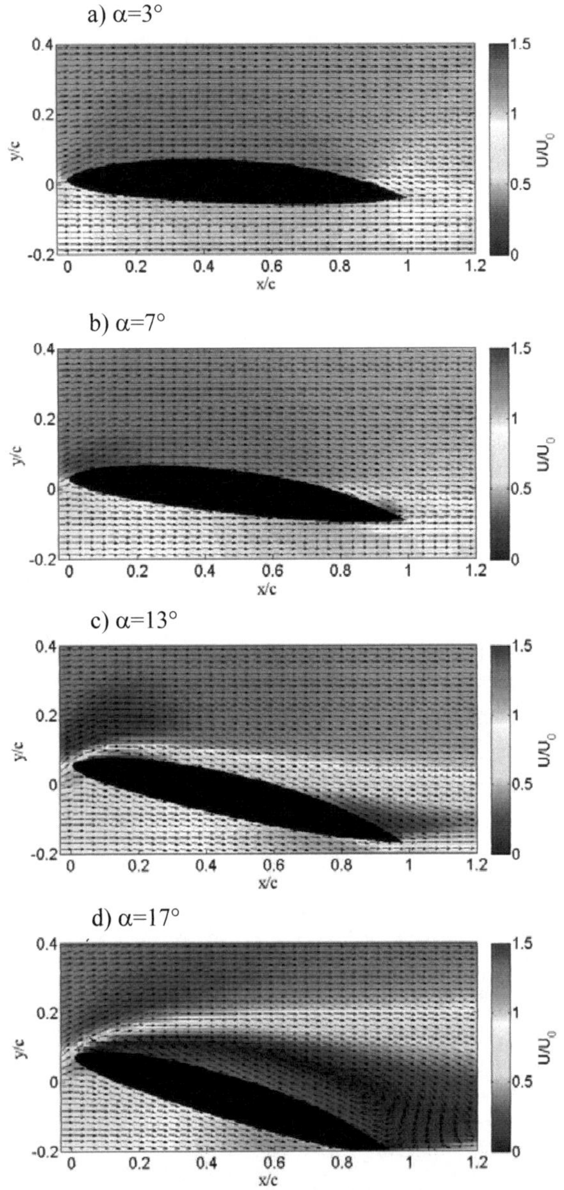

Figure 4: Time-averaged velocity field for angle of attacks 3°, 7°, 13° and 17°. All vectors are of unit length and the velocity magnitude is shown by the colormap. Only every second vector is shown for clarity.

Below 7° angle of attack, no separation is detected on any of the instantaneous velocity fields. At α=7°, very rare events of flow separation are detected near the trailing edge (0.5% of the snapshots), while frequent separation is observed near the trailing edge for higher angles: 62% at α=9°, and 95.8% at α=11°. At α=13°, the flow remains fully attached in 33.4% of the snapshots, while the flow separates near the trailing edge in 13.3%, and near the leading edge in 53.3% of the snapshots. When the flow separates at the leading edge, it either reattaches (25.1% of total snapshots), reattaches and separates again near the trailing edge (24.5% of total) or more seldom remains fully separated (3.7% of total).

Figure 5 shows an example of instantaneous velocity field where the flow separates at the leading edge, reattaches further downstream and separates again near the trailing edge.

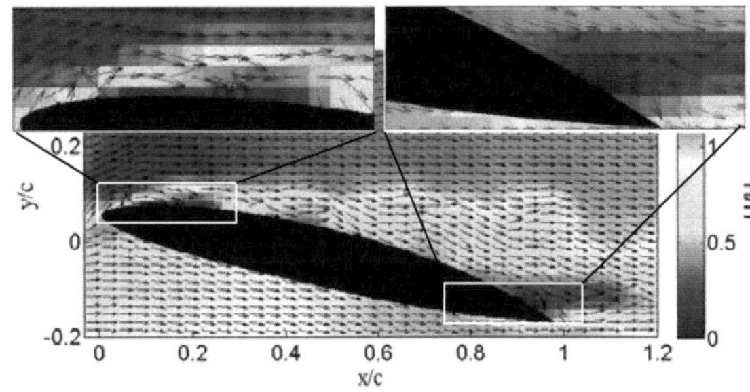

Figure 5: Snapshot of the instantaneous velocity field for angle of attack α = 13°. All vectors are of unit length and the velocity magnitude is shown by the colourmap. This snapshot shows a separation near the leading edge, flow reattachment, then separation again near the trailing edge.

For higher angles, the flow separates in more than 99% of the snapshots, most often close to the leading edge: at x/c < 0.3 in 96% of cases for α = 15° and 85% of cases for α = 17°. Reattachment and subsequent separation may occur occasionally, but the flow being turbulent, the detection is not always reliable. These results are summarized in Figure 6 where the distribution of separation events along the suction side of the foil is shown for different values of angle of attack.

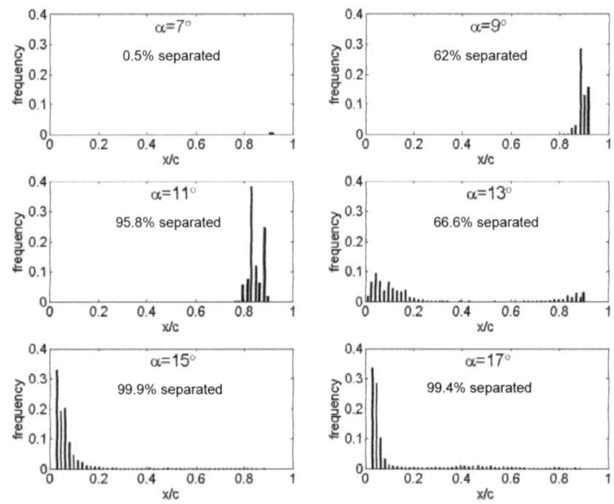

Figure 6: Distribution along the suction side of flow separation for α from 7° up to 17°. The vertical coordinate is the frequency of occurrence of each class, i.e. the number of snapshots where separations is detected at each location along the chord, divided by the total number of velocity snapshots (10,000).

4.3. E-telltale visualization

On the PIV raw images, the tip of the e-tellale can be seen, and the deformation can be estimated when it is large enough. No deformation is visible (less than 1 pixel ≈

0.1mm) for angles of attack lower than 14°. For $\alpha \geq 14°$, the metal sheet oscillates between its position at rest (flush on the surface) and the maximum deformation, as shown for example on Figure 7 at $\alpha = 17°$, where the maximum deformation corresponds to a maximum deflection of 3.3mm upwards.

a) t=0.919s: minimum deflection

b) t=0.950s: maximum deflection

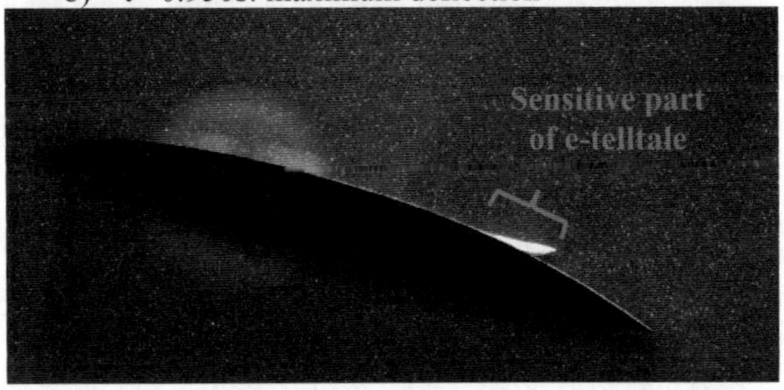

Figure 7: Raw image from the high-speed PIV camera at $\alpha = 17°$ at two instants showing e-telltale sensor with the minimum deflection (a) and maximum deflection (b). The minimum deflection corresponds to the position at rest with the e-telltale flush on the wall: it is the reflexion of the laser light and pixel saturation that make it erroneously appear higher than the surrounding wall.

5. DISCUSSION

From these observations, we can conclude that the hydro e-telltale effectively responds to the flow state. Indeed, the sensor is intermittently deformed and bent upward when the flow is massively separated, and the output signal is strongly oscillating with an increased mean value. Similarly to the aero version of the e-telltale [1-5], the output signal, and particularly the fluctuations intensity can be used to detect flow separation and anticipate stall. More surprisingly, the sensor output is significantly increased in the range $5° \leq \alpha \leq 12°$ with a local maximum for α = 8-9°, even if no sensor deformation is visible on the visualization images. Moreover, the flow is still attached at the location of the sensor in this range of angle of attack. As explained in Section 3 and supported by the work of Delafin et al. [10], it is speculated that this range of α corresponds to the migration of the boundary layer transition point from near the trailing edge (downstream of the sensor) to near the leading edge (upstream of the sensor). Hence, this means that the e-telltale is sensible to boundary layer transition.

Figure 8 shows time series of the e-telltale output for four different characteristic angles of attack, 3°, 9°, 13° and 15°. When the flow is attached, fluctuations are very small, even at 9° where the sensor response is at a local maximum (see Figure 3.c). At 15°, fluctuations are huge with very strong events associated to the dynamic vortex shedding due to flow separation upstream of the sensor. In the intermediate situation of $\alpha = 13°$ where separation intermittently occurs at the leading edge or the trailing edge, fluctuations become significant but the mean value remains smaller than at $\alpha = 9°$. Then, it seems that the turbulent nature of the boundary layer over the sensor results in an increased output value compared to the laminar boundary layer, but with no significant fluctuations, whereas flow separation –particularly ahead of the sensor– gives rise to violent fluctuations of the output. This signal could be used as a real-time feedback on the flow state which could be very useful in the perspective of closed-loop flow control. In Fig. 8, the typical time scale of fluctuations seems to be of order 20ms, which is 1.2 c/U. At full scale for a speed of 30 knots and a chord length of 0.5m, this corresponds to a typical time scale of order 40ms. This would probably be too short to enable human reaction, but a real-time signal about the system state would anyway be useful for a closed-loop controlling system, based for example on a reduced-order model. Further work is underway to better characterize the dynamic response of the sensor and relate it to the dynamics of the flow such as shedding events.

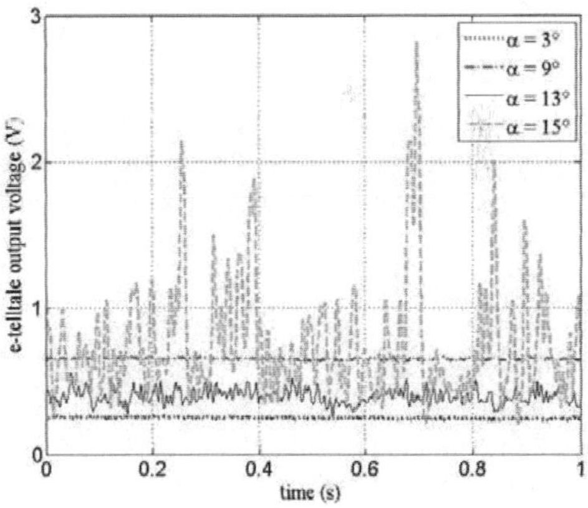

Figure 8: Time series of the e-telltale output voltage for $\alpha = 3°$, 9° and 15°.

6. CONCLUSIONS

A new flow sensor inspired by telltales has been built and was used on a hydrofoil to detect the flow regime. It is the hydro equivalent of the e-telltale developed by Mer Agitee for aero applications and already in use on yacht sails (the IMOCA-Class yacht SMA for example). The hydro e-telltale has been mounted on a hydrofoil model and tested in a water tunnel with force measurements and flow visualisation. The sensor response clearly shows the occurrence of flow separation and can be used to anticipate stall. A prototype has been fitted on an IMOCA-class yacht rudder and the feasibility to record useful information while sailing has been demonstrated, even if no specific testing campaign is available yet. We believe that this signal may be useful to estimate the loading on a rudder or a foil to avoid over-loading and risk of stall. This is a successful extension of the TrimControl concept [1] to hydrofoils and rudders. Moreover, the output signal may prove useful as a real-time feedback to help dynamic control of foils or rudders.

Acknowledgements

The authors are grateful to the IRENav technical staff (SCEFER) for their help in the experimental work, and to Tristan Thibaut for his participation in preliminary tests.

References

1. Patent claim : FRA N°07.06196, EUR N° 08 851 441.9, US N° 12/675 565, The wireless system fitted with the TrimControl App is commercialized by Mer Agitee (https://www.meragitee.com/trimcontrol).
2. Soulier, A., Braud, C., Voisin, D., and Podvin, B., 2020, "Ability of the e-TellTale sensor to detect flow features over wind turbine blades: flow stall/reattachment dynamics", Wind Energ. Sci. Discuss., https://doi.org/10.5194/wes-2020-13, in review.
3. Soulier, A., Voisin, D., Danbon, F., Braud, C., 2019, "Electronic TellTale (E - penon) sensor to detect flow separation on wind turbine blades". Wind Energy Science Conference, Jun 2019, Cork, Ireland. ⟨hal-02888568⟩
4. Soulier, A., Braud, C., Voisin, D., Danbon, F., Jaunet, V., 2018, "Electronic Tell-Tale, a sensor to detect flow separation on wind turbine blades". SMARTEOLE colloquium, Aug 2018, Orléans, France. ⟨hal-02888598⟩
5. Soulier, A., 2020, « Caractérisation du fonctionnement aérodynamique du e-Penon, capteur de décollement pour les pales d'éoliennes ». Milieux fluides et réactifs. PhD, Ecole Centrale Nantes, Français. ⟨hal-02572664⟩
6. Thibault, T., 2020, The "e-Telltale": characterization of an electronic sensor to detect flow phenomena, IRENav, Ecole Navale, MSc project.
7. Shaqarin, T., Braud, C., Coudert, S., Stanislas, M., 2013, "Open and closed-loop experiments to identify the separated flow dynamics of a thick turbulent boundary layer. Exp. Fluids 54, 1448
8. Becker, R., King, R., Petz, R., Nitsche, W., 2007. Adaptive closed-loop separation control on a high-lift configuration using extremum seeking. AIAA J. 45 (6), 1382–1392.
9. F. Feyzi, M. Kornberger, N. Rachor and B. Ilk, 1989, Development of two multisensor hot-film measuring techniques for free-flight experiments, International Congress on Instrumentation in Aerospace Simulation Facilities, Göttingen, West Germany, pp. 443-449, doi: 10.1109/ICIASF.1989.77699
10. Delafin, P.L., Deniset, F., Astolfi, J.A., 2014, "Effect of the laminar separation bubble induced transition on the hydrodynamic performance of a hydrofoil", Eur. J. Mech. B/Fluids, 46, pp. 190-200, DOI: 10.1016/j.euromechflu.2014.03.013
11. A. Ducoin, J. Astolfi, M. Gobert, 2012, An experimental study of boundary-layer transition induced vibrations on a hydrofoil, J. Fluids Struct. 32 (0), 37–51
12. DynamicStudio user's manual, Dantec.
13. Timmer, W., 2008, "Two-dimensional low-Reynolds number wind tunnel results for airfoil NACA0018, Wind Eng., 32(6), pp. 525-537, DOI: 10.1260/030952408787548848
14. De Gregorio, F. Albano, F. and Lupo, M., 2007, Flow separation investigation by PIV technique. In 7th International Symposium on Particle Image Velocimetry, Rome, Italy

REVIEW OF UNDEWATER FLUID-STRUCTURE INTERACTION MEASURING TECHNIQUES

Laura Marimon Giovannetti[1], Laura.Marimon@sspa.se
Ivan Stenius[2], stenius@kth.se

Abstract. This research aims to provide a feasibility study for an underwater Fluid Structure Interaction (FSI) experimental methodology. The possibility of measuring a structural deflection under fluid load and the flow behaviour around the structure is particularly interesting with the rise of composite materials, due to their inherent flexibility that permits them to be designed to a range of loading conditions. A review study on the experimental methodology available is carried out so that, in the future, a FSI measurement system can be installed in an underwater environment such as a towing tank or a marine basin. A robust and repeatable experimental methodology will also provide researchers with a validation case for numerical FSI simulations. The feasibility study starts with a thorough investigation of the existing measuring systems with an assessment of what could be used in SSPA Sweden underwater facilities (i.e. towing tank and marine dynamics laboratory). Currently there is a lack of FSI experimental methods, especially for underwater environment. The large majority of the current studies does not account for the complexity of the FSI coupled problems, where the structural response affects the flow behaviour and vice-versa. Bringing different techniques together will allow simultaneous measurements to be taken in a dynamic underwater environment. This would be especially important for performance prediction and numerical codes validations of foiling structures to assess the effects of deflections and twist when phenomena like cavitation and ventilation occur.

1. NOMENCLATURE

a	Distance to load [m]
c	Chord length [m]
E	Young's modulus [Pa]
G	Modulus of rigidity [Pa]
h	Maximum camber [m]
J	Sectional moment of inertia [m^4]
K_I	Proportionality coefficient [-]
I	Second moment of area [m^4]
l	Span [m]
P	Applied load [N]
T	Torque [Nm]
t	Thickness [m]
γ	Shear strain []
δ	Displacement [m]
ε	Strain []
θ	Twist angle [radians]
σ	Uniaxial stress [Pa]
τ	Shear stress [Pa]

2. INTRODUCTION

Recent developments in composite materials have led to a wide variety of new application areas, ranging from the aircraft industry, renewable energy devices, high performance sports applications, to marine transport and components. The key enabler is the ability to custom tailor the composite materials to produce unique mechanical properties that optimise the performance of the structure being designed. Composite materials typically have very high specific stiffness and strength ratios with excellent resistance to corrosion, ultraviolet (UV) light and fatigue loads making these materials very attractive in marine applications [1, 2].

To fully utilize their anisotropic properties, composite structures may be designed with increasing strength in a particular direction, thereby tailoring the material properties of a laminate to achieve a desired deformation response. Due to the interactions between the bending-twist and extension-twist within the stiffness matrices of the composite laminates, it is possible to maximise the performance of a given structure, leading to the possibility of exploring design techniques that account for the internal flexibility of the composite laminates for a specific loading condition, introducing the concept of Passive Adaptive Composites (PAC).

A large number of studies have been performed on lightweight aero-elastic structures during the past twenty years, investigating the possibility of designing composite structures tailoring the properties for a desired purpose. However, the large majority of these studies still does not fully account for the complexity of the FSI coupled problem, where the structural response affects the flow behaviour and vice-versa, as most researches were based on numerical simulations with for example very accurate flow models and simplified structural response or alternatively highly detailed structural models with medium fidelity flow solvers. Without a two-way coupled FSI model, the modelling of the structural behaviour or of flow features cannot be realistic, e.g. an unsteady aerofoil deformation affects the aerodynamic efficiency and the noise associated with it [3].

Analytical solutions do not exist for most FSI problems [3, 4]. Likewise, it has proved challenging to acquire experimental measurements of dynamic coupling between applied fluid loading and a structure response. As a result, research in this area has mainly focused on coupled Computational Fluid Dynamics (CFD) and structural Finite Element Analysis (FEA) simulations. Even though numerical studies have been extensive, there is a lack of experimental validation cases for FSI problems and, for

[1] Senior Researcher, SSPA Sweden AB, SWE
[2] Associate professor, Department of Engineering Mechanics, KTH, SWE

the limited cases that there are, the uncertainty in measurements is often unknown.

Moreover, even though some studies are now present in the aeronautics sector, the marine industry and therefore the underwater response to hydrodynamic loading is still an area that needs to be investigated. The harsh marine environment requires specific consideration of the durability, performance fatigue characteristic and potential failure mechanism of the material and structure when subject to long-term salt immersion, marine growth and reliability issues [1]. The low maturity of the experiments, low level of available technology and the lack of data and analysis tools for maritime applications are significant technological barriers.

In maritime applications, loading conditions can vary significantly due to interactions with the free surface, waves, currents, and adjacent bodies. In addition, cavitation and ventilation can develop for lifting bodies such as hydrofoils in water operating at high speeds and near the free surface. Added mass effects are posing additional challenges to take properly into account during the design process [5, 6]. Cavitation involves the formation and collapse of vapor bubbles caused by local pressure fluctuations [7]. Ventilation involves the entrainment of air into the flow around the body, and can occur simultaneously with cavitation [8, 9, 10]. The dynamic collapse of cavitation and/or ventilation bubbles can lead to large load fluctuations, vibrations, noise and instability issues [7]. The formation of small surface air bubbles can lead, with the use of composite materials, to resin peeling and fibre and matrix failure of laminated composites [11, 12]. A flexible hydrofoil, operating in water rather than in air will undergo an increased flow-induced deformation due to the higher fluid disturbing force.

In terms of experimental methodologies, especially looking at foil and vortex vibrations for both rigid and flexible hydrofoils, vibrations under hydrodynamic loading were investigated and simulated in [13]. Laminar to turbulent boundary layer transition induced vibrations were measured in [14, 15]. Recently, a strong coupling between the von Karman vortex-shedding and the structural response is measured in [16] with an interesting experimental setup for a morphing hydrofoil structure. However, all those presented methods don't account for the complexity of FSI experimental methodologies in a dynamic underwater environment such as a towing tank or a marine basin. In the last year several research in France have brought forward the advances in hydrofoil measurements [17, 18] both investigating a hydrofoil behaviour in cavitation tunnel and in flume tunnel.

For light weight composite structures, where the modelling is more complex than for quasi-isotropic materials, to assess the validity of numerical FSI simulations we require the ability to measure the influence of fluid load on the structural response. As also shown in [6], added mass effects may be even more influential and difficult to predict for light weight composite materials than for traditional aluminium and steel structures. The possibility of using full-field non-contact measurement techniques such as Digital Image Correlation (DIC) and tomographic flow measurement (e.g. shake the box) allow the measurements to be taken on large investigated areas, without modifying the properties of either the structure or the flow.

With the increase of foiling boats in the high-performance sailing, leisure crafts and passenger ferries industries, the performances, failure modes and vibration characteristics of foils need to be accurately tested [19, 20].

This research project builds on previous work, where a fully coupled FSI methodology was developed in a wind tunnel to assess the response of a PAC aerofoil [20, 21, 22]. DIC and Particle Image Velocimetry (PIV), were used in-air to obtain high-speed full-field wing deformation and flow field velocity data in synchronous with forces and moments. The tests presented in [20, 21, 22] show that it is possible to tailor a structure to a desired specification, opening the possibility of designing adaptable structures for new applications encompassing aeronautics, mechanical and marine engineering.

The aim of the current research is to acquire the background knowledge necessary to provide the possibility of testing in an underwater controlled environment a number of hydrofoils. Those foils could be optimised not only for high performance sailing, but could also be fitted in fully electric vessels, to reduce both the CO_2 emissions and the resistance of the wetted surface area.

3. BACKGROUND ON MEASUREMENT TECHNIQUES

Digital Image Correlation (DIC), Particle Image Velocimetry (PIV) and Shake the Box (STB) are all full field, non-contact, light based optical techniques and have been used and developed for a variety of experiments separately since the 1980s [23, 24].

DIC is generally used to measure the strain or displacement of a structure subjected to load. It is a white light full-field noncontact measurement technique [25]. It involves the use of Charged-Coupled Device (CCD) digital cameras that register a series of images of a surface on which a randomised speckle pattern is applied. From these images, the displacement of the specimen can be calculated using a correlation algorithm to determine the motion of the speckle pattern within the field of view [26]. This technique has been used at different scales, from high magnification [27] to large scale structures [28]. Within the DIC software, the speckle pattern is mapped to calculate its deformation, allowing the displacements and strains of the underlying structure to be measured [26].

DIC analysis of a foil structure within the working section of a wind tunnel was developed at the University of Southampton [29]. In order to correctly measure the structural response of the flexible aerofoil, high-speed three-dimensional DIC (3D DIC) is used in a stereo configuration. The stereo angle between the two cameras controls the measurement accuracy of the out of plane deformation [25, 30]. This was very important in the wind tunnel study as the major mode of deformation was out-of-plane and the stereo angle was carefully considered with respect to the dimensional constraints of the tunnel [31, 32]. As demonstrated in [27], the variations in speckle pattern density and sizes influence the measurement accuracy. Therefore, it is possible to design a speckle pattern tailored for each experimental test. Moreover, to guarantee accurate measurements, the illumination intensity should remain stable during the experiments. A number of studies were performed on the accuracy of DIC measurement of specimens subject to high temperatures and extreme lighting conditions. In these cases, it is necessary to combine bandpass filters with monochromatic light. Those findings were brought together in the wind tunnel research to be able to accurately measure the deflections of a specimen with a fixed speckle size and density under varying lighting conditions. The DIC uncertainty was measured in-air during wind tunnel experiments [21].

DIC has been recently used in MARIN both in their cavitation tunnel to assess both uniform and non-uniform flows [33] and in the Depressurised Wave Basin [34]. Those techniques are focused on propeller deflection and use cameras looking through glass-like structures to assess the propeller deflection underwater.

Tomographic volume measurement with STB approach was initially developed in 2013 as a Lagrangian particle tracking [35]. The technique has been recently adapted to be used in underwater measurement with LED flashlights rather than with laser [36]. STB finds particles previously inserted in the water, with the same seeding technique as for underwater PIV, via 3D triangulation and "shakes" each particle in space to find the best possible match with the recorded camera images [37]. Then, for each detected particle, the particle images on all camera sensors are found and subtracted from the camera images. This yields residual camera images only showing the signal from the remaining particles, revealing previously obscured particle images. This Lagraingian approach is not possible with PIV which is an Eulerian method to measure the flow field.

Due to image aberrations and optical distortions, as for example shown in Figure 1, caused by the camera lenses and blurred particles, the same particle in different positions in the volume can generate differently shaped particle images on a camera sensor. This knowledge is stored in the Optical Transfer Function (OTF) optimized for the current recording [38].

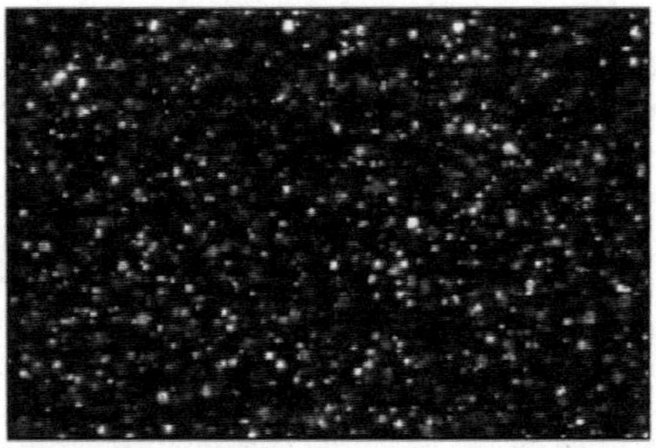

Figure 1: Unprocessed camera image of a TOMO-PIV experiment in air [38].

Having the cameras in the same position with respect to each other (Figure 2) aids with the recognition of distortions and spurious areas.

Figure 2: STB underwater arrangement from LaVision [39]. The 4 cameras are positioned in the bottom right corner of the torpedo-shaped housing. Their relative distance is always kept constant, aiding in particle tracking and calibration techniques.

The OTF serves as a basis for the STB processing. This ensures that even minor image aberrations are taken into account, which optimizes 3D particle position detection. The OTF also provides the appropriate particle images for computing residual camera images maximizing resolvable particle densities. In time-resolved STB a second order polynomial function is fit to each detected particle track. This yields a prediction of the particle position in the following step.

In the next time step, the algorithm prioritizes the search for particles in the vicinity of each predicted 3D particle position by 'shaking' the 3D particles into positions that maximize the match with the camera images. Then, the particle images are subtracted from the camera images. The resulting residual images contain only the particles that have yet to be tracked making the algorithm computationally efficient.

As shown by Jacobi for PIV arrangements, but applicable for most moving-underwater full-field measuring techniques, there are some critical areas to be considered [40, 41]. One essential problem, that in a laboratory environment, as well as in the wind-tunnel, could be addressed easily with changes in camera arrangements are stand-off distances. The distance to the specimen, that for

FSI cases include both the structural specimen and the flow around it, should be chosen in underwater measurements, not only for the ideal field of view size, but also to ensure non-intrusiveness and avoid flow disturbances from the torpedoes. When considering a towing tank environment, the seeding particles should also be investigated not only for the quality of seeds and dispersion level within the tank for flow measurements, but also to avoid disturbances to the structural DIC measurements. Within an air-environment in the wind tunnel (Figure 3) optical isolation between the structural and the flow measurements was ensured using magenta filters on LED lights [42]. As seen in Figure 4, the red bandwidth filters, ensure the highest transmittance level of light, approximately 90% further increased in the magenta colour, cutting-off the laser green-waveband that was captured by the PIV cameras.

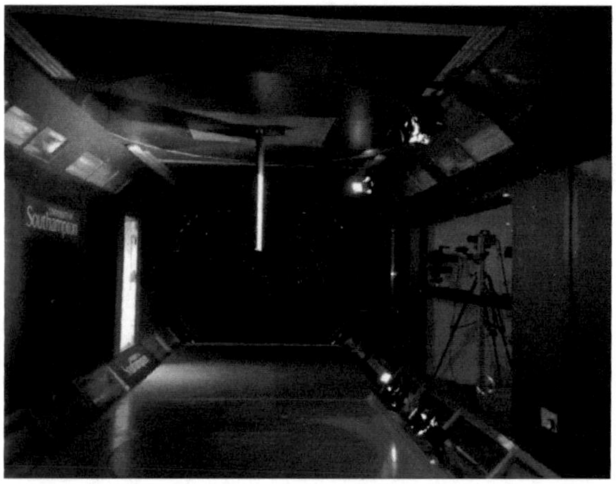

Figure 3: Wind tunnel setup as presented in [21, 42] to ensure optical isolation between the PIV and the DIC systems.

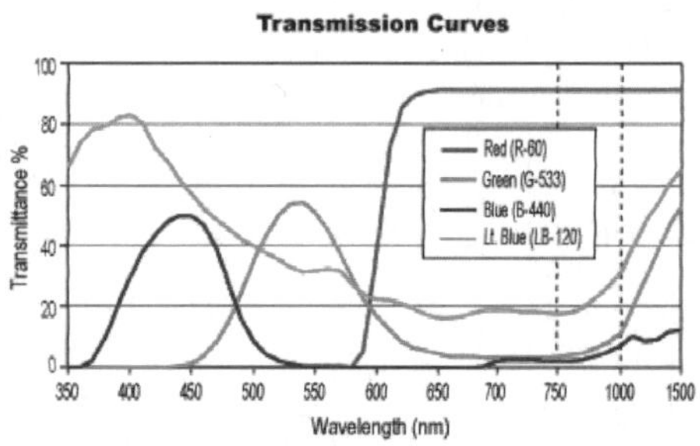

Figure 4: Transmittance curve for camera-filters [42].

When considering an underwater application where laser illumination cannot be used for safety purposes, the possibility of using white-light or blue high-power LED illumination needs to be considered. The selection of the different light source colour is linked to the choice of seeding particles. The application of fluorescent particles is known to prevent reflections from models and bubbles and the use of blue-flashlights with the corresponding camera filters will result in high-quality flow-measurements. If the FSI problem is investigated over a large area, with for example the structural deflection measured upstream and the flow wake measured downstream, a combination of white-light DIC set alongside the structure with blue-light STB positioned where the necessary flow measurements are taken can be considered as an option. The abovementioned set-up would be beneficial for flow measurements as it would avoid reflections from the specimen surface. While testing a hydrofoil, however, if events like cavitation or ventilation occur, light could be reflected not only near the free-surface but also by the air cavity bubbles formed on the specimen surface and shed downstream. The high light intensities from their reflections may result in exceedance of the saturation level on certain locations of the image sensor, especially for coupled charged device (CCD) cameras. Those problems are less affecting the newer complementary metal-oxide-semiconductor (CMOS) sensors [40]. The use of fluorescent seeding particles would reduce those reflection problems, at the expense of higher costs and difficulties in using them in large basins.

When testing high-speed hydrofoils, reaching the high-end limits of towing tank carriages to ensure Reynolds number and hence flow-features similarities, the main encountered issues are related to structural vibrations of both the carriage and the attached measuring devices. Most carriages are built on two rails that are carefully aligned with a laser and in case of long towing tank facilities are aligned accounting for the curvature of the earth. This ensures that the position of the towed model is not changed with respect to the free-surface. The carriages would move on metal wheels positioned on metal rails. The smallest imperfection in the rail-wheel system would therefore result in vibrations that are then transmitted to the carriage and specimen, with a change in frequency to the model itself and the underwater torpedoes. As both described systems are optical measurement techniques they rely on the perfect alignment of the cameras and, in case of vibrations, on their synchronous movement to avoid misalignments. Furthermore, the position of the cameras with respect to themselves and the specimens according to the initial calibrations will influence the overall accuracy. LaVision has worked on some of those issues, optimising the shape and the strength of the vertical strut attached to the torpedo to an Eppler E838 profile. Compared to the cylindrical shaped support tubes shown in Figure 2, lift, drag and overall deformation of the system are reduced to a minimum. Those results are however valid up to 7m/s, therefore further investigations on an optimised profile and structural integrity might be necessary for higher towing speeds, as the ones required to ensure Reynolds number similarity for hydrofoil testing.

Even for fully-coupled FSI experiment in an underwater environment in the near future, the recommended procedures detailed by the ITTC can be adapted to the new measuring systems to ensure that the criteria for testing procedures are followed and understood [44].

4. STRUCTURAL MEASUREMENTS

Hydrofoil systems used in high-performance sailing boats mostly rely on different types of manufacturing methods:

1. specialist one-off design and manufacturing of both the profile and lay-up used mainly by well-funded sailing teams for events such as the America's Cup or the Vendée Globe.
2. One-off or batch manufacturing for large development fleets such as the International Moth.
3. Batch manufacturing for one-design fleets such as the NACRA 17, IQfoil, Kiteboards etc.

Those different types of manufacturing methods will have dissimilar levels of accuracy, inversely proportional to their overall costs. The former type will have the highest accuracy and the finished foil would be as close to the design as possible with the building processes. The hydrofoils would be mainly manufactured with aeronautical-grade carbon fibres pre-preg and cured in autoclaves. Their internal lay-up would also be laid accurately to give the expected mechanical properties, tailored for the encountered sailing static and dynamic loads.

The second type of foils will range from pre-preg to resin infusion to hand lay-up, depending on the builder and on the sailor requests. This type of hydrofoils can be built with inputs from sailors and on-the-water testing, tailoring most of the types of foils to specific crew-weights and/or requests.

The latter type, being one-design will not allow the sailors to be involved in the design or manufacturing process. The building would range from pre-preg to resin infusion and the performance profiles (i.e. shape and thickness) and the mechanical properties will only be known to the approved manufacturers.

Composite materials indeed can have drastic changes to their mechanical properties with changes in the internal lay-up of the fibres. With fibres oriented at [0°, 90°, ±45°] composites specimens are considered quasi-isotropic with the same in-plane mechanical properties in all directions. However, the bending and torsional stiffness for the macroscopic structure will change when fibres are oriented at different angles, as the relationship of the stresses and strains must take into consideration the complete stiffness matrix as the stresses (σ and τ) and strains (ε and γ) are coupled to σ_1 and/or σ_2 in the two principal directions. This leads, under fluid loads, to a relationship between the forces and moments with the strains at laminate levels [18].

When analysing one-design foils, or indeed specimens where the material properties are unknown, it is possible to estimate the Young's modulus and the torsional stiffness of a structure, when subject to bending loads, following equations (1-3):

$$\delta_{max} = \frac{Pa^2}{6EI}(3l - a) \quad (1)$$

$$I = K_I ct(t^2 + h^2) \quad (2)$$

$$\theta = \frac{Tl}{JG} \quad (3)$$

An example of a hydrofoil structurally tested in a laboratory environment can be seen in Figure 5.

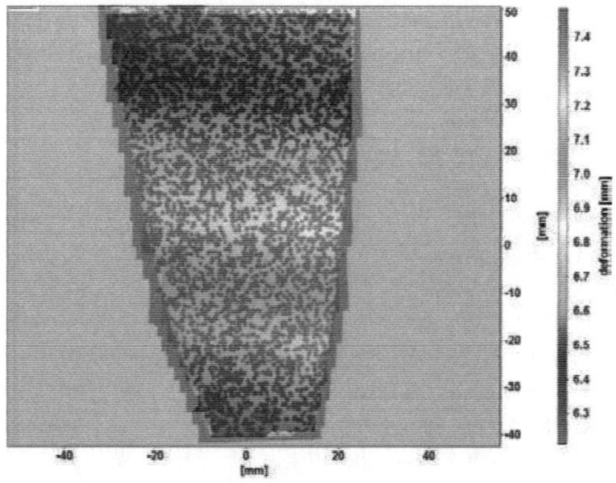

Figure 5: DIC deformation field for 178N load applied 14mm behind the leading edge of an International Moth horizontal foil [45].

Assessing the material properties and the shear centre location of hydrofoil systems, it is possible to investigate the expected deflections whilst sailing.

Having the possibility of testing a large number of specimens allows to understand the differences between them and the standard deviations in their manufacturing methods. Following those investigations, it is therefore possible to appreciate the differences in overall performances of a range of NACRA 17 hydrofoils, port and starboard, tested in bending load. The tests were replicating the boundary conditions of the daggerboard case, with fixed-fixed at the top case and fixed in displacement and free to rotate at the bottom hull-interface.

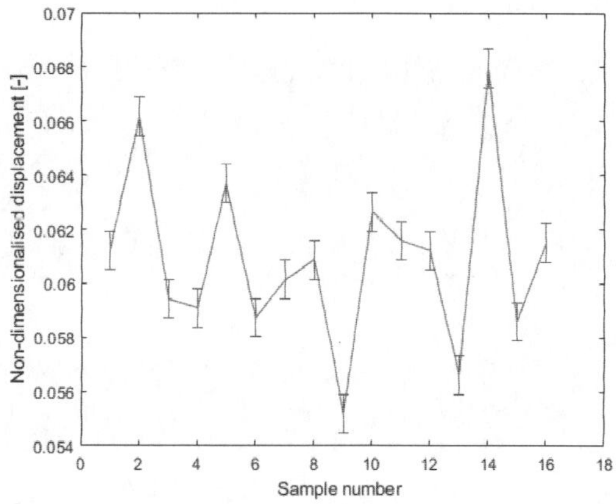

Figure 6: Non-dimensionalised displacement resulting from structural testing of a random sample size of one-design hydrofoils. Error bars represent the variation of the unloaded initial and final position of the specimens in the jig.

The load is statically applied at mid-span and at 25% of the chord with the displacements measured with a laser pointer. As seen in Figure 6 a random number of specimens has a large variation in bending response, changing the effective profile shape whilst sailing.

Being able to appreciate the complexity of fluid-structure interaction allows not only designers to tailor structures for specific loads, but also sailors to choose the best set-up for their specific sets of skills.

5. TESTING PLATFORM

In addition to the development of more accurate measurement techniques in laboratory settings such as the SSPA towing tank and Marine Dynamics Lab, testing and demonstrator platforms are important complementary tools when verifying new hydrofoiling technologies. With a test platform it is possible to test and verify full system concepts including control strategies and algorithms for un-stable hydrofoil configurations such as the KTH hydrofoil test rig for new electric motor-pods and roll stabilization techniques (Figure 7). Testing platforms are further very useful in verifying real scenario performance in waves, wind, with boat motions, etc.

Another important aspect is to minimise the risks of losing a hydrofoil during a real-event scenario (i.e. hitting an unidentified submerged object (USO) like a container or a marine wildlife that could result in abrupt breakage of the foil). Using a testing platform can help in developing foil casing and systems that are able to break in a controlled manner, without hindering the whole boat. This aspect of testing at model scale possible real-case scenarios could become attractive after the unfortunate race abandonment during the 2020 Vendée Globe edition. Many skippers in the two latest races were forced to retire or adapt to slower speeds after hitting USOs that damaged part or all of the foil structure and casing [46]. When considering the scalability of structural testing and the flow behaviour, it would be possible to develop such strategies as dynamic FSI experiments.

Figure 7: KTH test rig for new hydrofoiling concept with patented electric motor pods mounted on the wings and a unique roll stabilization configuration based on a mast rudder.

6. CONCLUSIONS

The paper presents a detailed review of the experimental methods able to describe fluid structure interaction events in an underwater dynamic environment. A feasibility study is carried out to understand the implications, limitations and advantages of being able to measure with full-field techniques in a towing tank. The necessity of experimental tests that are able to accurately replicate hydrofoils Reynolds number whilst measuring deflections and forces are becoming prominent to avoid full-scale or testing on a finished product, reducing the final cost and being able to assess the structural deformations at a design stage, taking advantage of those flexibilities. Moreover, the benefits of aligning underwater FSI tests, with structural measures and with testing platforms are explored for future assessment of hydrofoils within SSPA. Foiling boats are becoming more attractive to the general public, however issues such as reliability of the manufacturing methods and safety issues still need to be tackled. This feasibility study addresses some of those topics analysing them together with the advantages given by composite materials that allow designing for the expected loads, benefitting from hydroelasticity.

Acknowledgements

The authors would like to acknowledge the EPSRC for funding the research carried out in the past at the University of Southampton under the grant number EP/I009876/1. The authors would also thank the colleagues from the University of Southampton.

References

1 Y. L. Young, M. R. Motley, R. Barber, E. J. Chae and N. Garg, "Adaptive Composite Marine Propulsors and Turbines: Progress and Challenges," *Applied Mechanics Reviews,* vol. 68, 2016.

2 E. Greene, Marine Composites, Annapolis, MD: Eric Greene Associates, 1999.

3 R. de Borst, P. Nithiarasu, T. Tezduyar, G. Yagawa and T. Zohdi, Computational Fluid-Structure Interaction, John Wiley & Sons Ltd., 2013.

4 G. Hou, J. Wang and A. Layton, "Numerical methods for Fluid-Structure Interaction - A review," *Communications in Computational Physics,* vol. 12, no. 2, pp. 337-377, 2012.

5 I. Stenius, L. Fagerberg and J. Kuttenkeuler, "Experimental eigenfrequency study of dry and fully wetted rectangular composite and metallic plates by forced vibrations," *Ocean Engineering,* vol. 111, pp. 95-103, 2016.

6 I. Stenius, L. Fagerberg and A. Säther, "Predicting the natural frequency of submerged structures using coupled solid-acoustic finite element simulations," *Ocean Engineering,* vol. 159, pp. 37-46, 2018.

7 C. E. Brennen, Cavitation and Bubble Dynamics, New York: Oxford University Press, 1995.

8 Y. Young and S. Brizzolara, "Numerical and Physical Investigation of a Surface Piercing Hydrofoil," in *Third International Symposium on Marine Propulsors (SMP13)*, Launceston, Tasmania, 2013.

9 C. Harwood, Y. Young and S. Ceccio, "Ventilated Cavities on a Surface Piercing Hydrofoil at Moderate Froude Numbers: Cavity Formation, Elimination, and Stability," *Journal of Fluid Mechanics,* vol. 800, pp. 5-56, 2016.

10 Y. Young, C. Harwood, F. Montero, J. Ward and S. Ceccio, "Ventilation of Lifting Bodies: Review of the Physics and Discussion of Scaling Effects," *Applied Mechanics Review,* vol. 69, 2017.

11 D. Hammond, M. Amateau and R. Queeney, "Cavitation Erosion Performance of Fibre Reinforced Composites," *Journal of Composite Materials,* vol. 27, no. 16, pp. 1522-1544, 1993.

12 T. Yamatogi, H. Maruyama, K. Uzawa, K. Kageyama and N. Watanabe, "Study on Cavitation Erosion of Composite Materials for Marine Propeller," in *Seventeenth International Conference on Composite Materials (ICCM17)*, Edinburgh, UK, 2009.

13 D. Akcabay, E. J. Chae, Y. L. Young, A. Ducoin and J.-A. Astolfi, "Cavity induced vibration of flexible hydrofoils," *Journal of Fluids and Structures,* vol. 49, pp. 463-484, 2014.

14 A. Ducoin, J.-A. Astolfi and J. F. Sigrist, "An experimental analysis of fluid structure interaction on a flexible hydrofoil in various flow regimes including cavitating flow," *European journal of mechanics B/Fluids,* vol. 36, pp. 63-74, 2012.

15 A. Ducoin, J.-A. Astolfi and M.-L. Gobert, "An experimental study of boundary layer transition induced vibrations on a hydrofoil," *Journal of Fluids and Structures,* vol. 32, pp. 37-51, 2012.

16 L. Pernod, A. Ducoin, H. Le Sourne, J.-A. Astolfi and P. Casari, "Experimental and numerical investigation of the fluid-structure interaction on a flexible composite hydrofoil under viscous flows," *Ocean Engineering,* vol. 194, 2019.

17 M. A. Fatiha, B. Augier, F. Deniset, P. Casari and J. A. Astolfi, "orphing Hydrofoil Model Driven by Compliant Composite Structure and Internal Pressure," *Journal of Marine Science and Engineering*, vol. 7, 2019.

18 T. T. Vanilla, A. Benoit and P. Benoit, "Hydro-elastic response of composite hydrofoil with FSI", *Ocean Engineering*, vol. 221, 2021.

19 Y. Liao, N. Garg, J. R. Martins and Y. L. Young, "Viscous fluid-structure interaction response of composite hydrofoils," *Composite structures,* vol. 212, pp. 571-585, 2019.

20 L. Marimon Giovannetti, J. Banks, M. Ledri, S. R. Turnock and S. W. Boyd, "Toward the development of a hydrofoil tailored to passively reduce its lift response to fluid load," *Ocean Engineering,* vol. 167, pp. 1-10, 2018.

21 L. Marimon Giovannetti, J. Banks, S. R. Turnock and S. W. Boyd, "Uncertaintly assessment of coupled Digital Image Correlation and Particle Image Velocimetry for fluid-structure interaction wind tunnel experiments," *Journal of Fluids and Structures,* vol. 68, pp. 125-140, 2017.

22 L. Marimon Giovannetti, O. Charalampopoulos, J. Banks, S. W. Boyd and S. R. Turnock, "Developing fluid structure interaction experimental methodologies for dynamic foil measurements," Innov'sail 2020, Gothenburg, 2020.

23 M. Raffel, C. Willert, S. Wereley and J. Kompenhans, Particle Image Velocimetry - A Practical Guide, Springer, 2007.

24 F. Chen, X. Chen, X. Xie, X. Feng and L. Yang, "Full-field 3D measurement using multi-camera digital image correlation system," *Opt. Lasers Engineering,* vol. 51, pp. 1044-1052, 2013.

25 Z. Tang, J. Liang, Z. Xiao and C. Guo, "Large deformation measurement scheme for 3D digital image correlation method," *Opt. Lasers Engineering,* vol. 50, pp. 122-130, 2012.

26 P. K. Rastogi and H. Erwin, Optical Methods for Solids Mechanics: A Full-Field Approach, Wiley, 2012.

27 G. Crammond, S. W. Boyd and J. Dulieu-Barton, "Speckle pattern quality assessment for digital image correlation," *Opt. Lasers Engineering,* vol. 51, pp. 1368-1378, 2013.

28 N. Mccormick and J. Lord, "Digital image correlation for structural measurements," *Proceedings of the Institution of Civil Engineers,* vol. 165, pp. 185-190, 2012.

29 J. Banks, L. Marimon Giovannetti, X. Soubeyran, A. M. Wright, S. R. Turnock and S. W. Boyd, "Assessment of digital image correlation as a method of obtaining deformations of a structure under fluid load," *Journal of Fluids and Structures,* vol. 58, pp. 173-187, 2015.

30 H. Schreier, J.-J. Orteu and M. A. Sutton, Image Correlation for Shape, Motion and Deformation Measurements: Basic Concepts, Theory and Applications, Boston, MA, USA: Springer, 2009.

31 P. Reu, "Stereo-rig design: creating the stereo-rig layout – Part 1," *Exp. Tech.,* vol. 36, pp. 3-4, 2012.

32 Y. Q. Wang, M. A. Sutton, X. D. Ke, H. W. Schreier, P. L. Reu and T. J. Miller, "On error assessment in stereo-based deformation measurements," *Exp. Mech.,* vol. 51, pp. 405-422, 2011.

33 P. Maljaars, "Hydro-elastic analysis of flexible marine propellers," TU Delft, Delft, 2019.

34 G.-J. Zondervan , N. Grasso and W. Lafeber, "Hydrodynamic design and model testing techniques for composite ship propellers," in *Fifth International Symposium on Marine Propulsors, smp'17*, Espoo, Finland, 2017.

35 D. Schanz, S. Gesemann and A. Schroder, "Shake-the-box: Lagrangian particle tracking at high particle image densities," *Experiments in Fluids,* vol. 57, no. 70, 2016.

36 F. Roettig, K. Wennemar, K. Hoyer, R. Beslac, C. Hesseling, T. Beck and B. Friedhoff, "Volumetric flow measurement in the wake of a ducted propeller," in *Sixth International Conference on Advanced Model Measurement Technology for Maritime Industry*, Rome, Italy, 2019.

37 B. Wieneke, "Iterative reconstruction of volumetric particle distribution," *Measurement Science and Technology,* vol. 24, 2013.

38 D. Schanz, "Non-uniform optical transfer functions in particle imaging: calibration and application to tomographic reconstruction," *Measurement Science and Technology,* vol. 24, 2013.

39 LaVision, "MiniShaker Underwater," Germany, 2020, technical manual.

40 G. Jacobi, "The application of Particle Image Velocimetry for the analysis of high-speed craft hydrodynamics," TU Delft, Delft, 2020.

41 G. Jacobi, C. Thill and R. Huijsmans, "The application of Particle Image Velocimetry for the analysis of high-speed craft hydrodynamics," in *12th Intarnational Conference of Hydrodynamics - ICHD 2016*, 2016.

42 L. Marimon Giovannetti, "Fluid structure interaction testing, modelling and development of Passive Adaptive Composite foils," University of Southampton, Southampton, PhD thesis, UK, 2017.

43 Edmund Optics, "Edmund optics," 2013. [Online]. Available: https://www.edmundoptics.com/knowledge-center/application-notes/imaging/filtering-in-machine-vision/.

44 ITTC, "ITTC - Recommended Procedures and Guidelines: Guideline on Best Practices for the Applications of PIV/SPIV in Towing Tanks and Cavitation Tunnels," ITTC: International Towing Tank Conference, 2014.

45 J. Banks, L. Marimon Giovannetti, J. C. Taylor and S. R. Turnock, "Assessing human-fluid-structure interaction for the International Moth," *Procedia Engineering* , vol. 147, pp. 311-316, 2016.

46 Vendée Globe, "The abandoning trio in Cape Town evoke the origin of the Vendée Globe," 5 December 2020. [Online]. Available: https://www.vendeeglobe.org/en/news/20826/an-abandoning-trio-in-cape-town-evoke-the-origin-of-the-vendee-globe. [Accessed 2020].

EXPERIMENTAL INVESTIGATIONS ON CONVENTIONAL AND UNCONVENTIONAL BOOST SYSTEMS TO ASSIST TAKE-OFF OF HYDROFOIL CRAFTS

Luigi Francesco Minerva[1], l.minerva@marin.nl
Francisco Miguel Montero[2], f.miguelmontero@marin.nl

Abstract. In 2020, a dedicated research regarding the take-off was granted. The objective was to investigate experimentally unconventional solutions to overcome the take-off hump; this challenge implies the risk to identify also not successful solutions but, at the same time, to dare looking beyond the current state of art. Inspired by nature, an impulsive boost could be used to overcome the take-off hump of the hydrofoil craft's resistance: the modelled mechanics were inspired to the impulsive boost thrust generated by the ink fishes or by a rapidly flapping hydrofoil. Together with the experiments of physical unconventional boost systems, a systematic investigation on the required boost thrust, energy and power was also carried out by means of a dedicated model test set-up. Objective was to answer questions such as: What kind and how much of boost is needed to let the craft take off? For how long is this boost needed? Which solutions are more promising for further development?

1. INTRODUCTION AND OBJECTIVES

When thinking about high speed crafts and application of hydrofoils in the maritime world, the imagination of most goes to the sailing boats of the Americas cup AC75 class or high performance passenger or recreation vessels. These applications are typically characterised by light displacements. Foils for larger displacements are challenging to apply due to limitations such as the excessive power demand, in first instance, at the take-off. However the possible economic and emissions-related advantages to operate in foil-borne mode triggered the interest to explore new ways to assist the take-off phase, also for relatively "heavy" payloads.

In 2020, a research project was granted with the objective to investigate experimentally unconventional and innovative solutions to assist the take-off. This challenge implied the risk to identify also not successful solutions but, at the same time, to dare looking beyond the current state of art. The investigation of "out of the box" solutions implied that the required level of detail at this stage was low: the primary interest was to evaluate multiple promising solutions rather than focusing only on one case, with the knowledge that success was not guaranteed for any of the tested alternatives.

2. APPROACH

The main focus of the research was on solutions to develop an impulsive boost thrust to assist the take-off in calm water. An example of typical (but not general) relation between steady resistance in calm water and speed for a craft with hydrofoils is presented in Figure 1. Prior to take-off, the resistance has a peak ("take-off hump") to overcome. In that respect, from an ideal point of view, there are two options: one is an "oversized" standard propulsion system in order to generate the entire required thrust; another one is the integration of a boost system that provides only part of the total required thrust.

The last alternative appears to be more promising if:

- The standard propulsion system is capable to generate propulsive thrust for the speeds before and, especially, after the take-off hump (otherwise the craft will revert to the less efficient hull borne condition)
- The system to generate the required boost thrust should not take a too large percentage of the total displacement (at cost of the available payload on board)
- It should be practical to integrate such a boost system in the ship design.
- The overall life-time cost (in economic and environmental terms) should be lower than an equivalent vessel with a larger main propulsion unit.

The first priority of the experimental campaign was to answer questions such as: What kind and how much of boost thrust is needed to let the craft take off? For how long is this boost needed?

The energy/power demands to provide this required boost thrust were only briefly investigated in the this research campaign. Their role is absolutely fundamental to further advance the feasibility of solutions and to evaluate the impact on the total weight of the craft.

2.1. Take-off hump

Figure 1 shows a resistance curve with a pronounced take-off hump. If this hump were not very high, almost flat, the boost force required to take-off would be very small and, in general, the application of a dedicated boost system might be not so advantageous.

The take-off hump depends on multiple factors:

- Hull displacement
- Hull form

[1] Senior Project Manager, Maritime Institute Research Netherlands (MARIN), Wageningen, The Netherlands
[2] Senior Project Manager, Maritime Institute Research Netherlands (MARIN), Wageningen, The Netherlands

- Hydrofoils and their eventual surface control (such as flaps) design
- Sea state
- Design take-off and cruising speed

Once all of these factors are evaluated for a specific design, further studies can be performed to finalize a boost system for the take-off. As a consequence of this, the results of this research cannot be generalized to all hydrofoil crafts but only used as a first reference for feasibility studies in early design phase.

3. MODEL SET-UP

The following key elements are discussed in the next sub-sections with respect to the model test set-up:

- Hull form and hydrofoils
- Captive and semi-free sailing set-up
- Boost systems (virtual and physical)

3.1. Hull form and hydrofoils

The objective behind the selection of the hull form and hydrofoil was to have a representative demonstrator for the application of the different boost systems. Thus, there was no need for an in-depth optimisation as long as the complete package (hull plus hydrofoil) had a clear take-off resistance hump and the hydrofoils were delivering enough lift for take-off (and steady foil-borne condition).

The shape and main dimensions of the hull were chosen based on considering a selection from MARIN's database of planing crafts with (and without) hydrofoils. Figure 2 shows the body plan: a typical form for high speed hull (and foil)-borne applications. Figure 3 shows the main particulars of the selected hull compared with database values. Table 1 present the main characteristics of the hull on model and full scale (for the chosen scaling factor of 10).

The hull was equipped with two sets of hydrofoils: the front set consisted of surface piercing J-hydrofoils while the aft set consisted of one deeply submerged inverted π hydrofoil. Figure 4 shows an overview of the hull model with the hydrofoils.

The configuration with surface piercing hydrofoils was chosen in order to have a self-stabilizing platform, with no need for a dynamic lift control system.

The main dimensions and location of the hydrofoils were designed for stability in foil-borne condition. The design alternatives were evaluated by means of MARIN's in-house codes Hydsim and Hydres (based on simple lifting line theory and vortex lattice method). The resulting configuration was the well-known "airplane" with the largest percentage of lift allocated to the front foil (while the aft foil working mainly as a pitch stabiliser).

The design angle of attack of the foils was estimated in the design phase and further fine-tuned manually in the tank, to allow for scale effects compensation.

Figure 4 shows also anti-ventilation plates applied to the front foil. This was necessary due to frequent instability caused by random ventilation events.

3.2. Captive and semi-free sailing set-up

Two model test set-ups were used for the experimental campaign: captive and semi-free sailing.
The captive set-up consisted of a traditional system for towing tests: the hull was towed by the carriage via a six-component force frame; pitch and heave were free but not roll and yaw. The towing force, pitch and heave were measured. This set-up allowed to measure the steady resistance curve of the craft as well as static sink and trim as a function of speed, which represented the basic input information for the campaign with the second test set-up. In fact, recalling Figure 1, with the knowledge of the total resistance at the take-off hump it was possible:

- to identify the speeds in hull-borne and foil-borne condition and their correspondent resistance demands;
- to identify the thrust force needed for take-off and for steady foil-borne cruising;
- to identify the minimum boost force in order to reach the take-off speed.

The semi-free sailing set-up was meant to allow the craft to be accelerated by the combined action of main and boost thrust forces, to reach the take-off speed and overcome the take-off hump (if the boost impulse is sufficient) and then to keep flying (by means of only the main thrust force). Normally a completely free sailing and self-propelled model could have been also considered, despite for example the additional complications of the installation of a dynamic control for the course. For this experimental campaign a different approach was chosen:

- to simplify the model test set-up from the dynamic control aspects;
- to have a controllable towing force resembling the thrust action of the propulsors;
- to have a controllable boost force (in magnitude and release time) to superimpose to the aforementioned force;

Figure 5 shows photos of the semi-free sailing set-up and its schematic working principle. The hull was connected to the carriage via two longitudinal lines controlled by electrically driven winches and four transversal lines connected directly to the carriage with springs. The role of the transversal lines was to restrain the model from uncontrolled yaw motions. The longitudinal lines had two main tasks: to apply a controlled propulsive (and eventually boost) force and to control the position of the model. In fact the model was not towed by the carriage but freely accelerated thanks to the forces exerted by the winches. The carriage control was all the time following the virtually self-propelled model based on the input given by the winch position (and consequently) the relative position of the model respect to the carriage. Force transducers were applied to the front and aft winch lines in order to monitor the applied longitudinal force. The semi-free sailing set-up was also used in "towing" mode to measure the towing force. However in this mode the accuracy of the measured resistance was lower because the model was relatively free to yaw (implying extra additional drag) when advancing.

3.3. Boost systems

In this section, only the description of the boost systems is presented together with their working principle. In the discussion of the results, a more detailed analysis is given about their effectiveness and eventual improvements.

The previously described semi-free sailing set-up allowed to perform systematic investigations on the required boost thrust and release time in order to attain the take-off speed and then steadily remain in foil-borne condition. Therefore this was considered as a first "virtual booster" (also named Booster 1). The advantage was the possibility to derive some preliminary indication of the required boost energy and power for the take-off. The input parameters for the control of the winch force were defined as follows:

Name	Description
F_main	Main propulsion force to drive the boat to a higher constant speed once foil-borne
F_boost	Additional boost force exerted to reach the take-off speed
t_up	Time to sum the F_boost to F_main
t_boost	Release time of the boost force
t_down	Time to remove the boost force

This type of set-up could be generally applicable for any hull-hydrofoil design case, where the demand of thrust, energy and power could be investigated in order to identify the most suitable boost system .

In parallel, physical boost systems were developed based on nature concepts, such as the one generated by the ink fishes or by a rapidly flapping fin. The complexity of the engineered systems was kept simple, since the first objective was to evaluate how promising such concepts could be. Further design loops and refinements would be required for actual implementation.
The physical booster systems were tested with the semi-free sailing set-up. In this case only the main propulsion force (F_main) was constantly applied to steadily sail the boat before and, eventually, after the take-off.

The first physical system considered was the Flapping aft foil-Booster 2. Figure 6 shows a photo of this system: the aft foil was connected to a pneumatic linear actuator (manually driven for multiple actuations) pushing the foil vertically downwards (from a higher position to the one designed for the steady flight). The speed of actuation could be varied by changing the air pressure in the actuator, while the angle of attack of the aft foil remained fixed.

Impulsive lifting vertical and propulsive horizontal force were produced when the foil was pushed vertically downwards and advancing at speed, thanks to the instantaneously variation of the inflow conditions (effective angle of attack and velocity) to the foil.

The second physical system considered was the "Waterjet-Booster 3". Figure 7 shows an overview of the system and its installation on board of the model. The inspiration comes from the squid ability to generate an impulsive force by jettison water at high speed. The physical principle is the change in water momentum thanks to the conversion of pressure energy into kinetic energy. This principle is common in the maritime sector when thinking, for example, of the propulsion waterjets. However, unlike a normal waterjet, the adopted one was a closed system with a limited capacity of stored water. Quantity of water in the cylinder, water pressure and nozzle diameter were varied to control intensity and duration of the boost force.

In line with spirit of this research campaign for high risk (of failure) innovative solutions, an additional system, still inspired by the squid, was tested with a dedicated set-up in bollard pull condition. Figure 8 shows an overview of this concept booster, named "Squid concept – Booster 5": two elastic and strong bags were in contact with each other; one bag was filled with (variable) pressurised air and pressing the second bag (filled with water and connected to a nozzle variable in diameter). This booster represented a further step into the biomimetic, attempting to reproduce closely the squid mechanism (a water-inflated elastic membrane capable to release quickly water at high speed and consequent large developed boost thrust). The controllable parameters were the pressure (in this case only the air bag), quantity of water and nozzle diameter.

The last physical booster system mounted in the model (precisely on top of the hull deck at the stern) consisted of one high power air turbine (named "Air turbine jet – Booster 4). Figure 9 shows a photo of the system. By varying the revolution rate of the turbine different horizontal thrust force (or eventually a combination of vertical and horizontal) could be generated to provide the boost and to reach the foil-borne condition.

4. DISCUSSION OF RESULTS

The results are first discussed on model scale followed by considerations on full scale (when applicable). For simplicity, only tables of extrapolated results are presented in this paper.

4.1. Model scale

4.1(a) Captive set-up

The first step of the model test campaign was to determine the steady resistance curve, shown in Figure 10, in order to:

- Identify the take-off speed (around 5 m/s);

- Fix the minimum main propulsive force (around 101 N) to have the boat steadily sail in foil-borne condition at speeds beyond the hump;
- Identify the minimum boost force (about 4 N) to reach the take-off.

With the minimum boost force, time and distance for take-off were on the limit of the testing conditions (maximum length of the basin). Therefore the choice of the boost force input for the virtual system was set for values at least three times the minimum one so that take off could be achieved within the available basin length. For the physical boosters, it was not known in advance how much boost force and release time would have been deployed. It was directly found out during the experiments with the maximum available basin length.

From the captive set-up, the boost force and release time developed by the Flapping aft foil-Booster 2 and the Waterjet-Booster 3 were also preliminarily measured for some combinations of their input parameters.

In terms of deployed horizontal boost force, the "Flapping aft foil-Booster 2" was capable to generate enough force to push the craft beyond the take-off hump but the release time was very short (less than 1 second). In this experimental campaign a simplified set-up was chosen in the spirit of a preliminary investigation: both parameters, speed of actuation and angle of attack, were kept constant. The action of boost force (and consequent acceleration when in semi-free sailing set-up), obtained when kicking downwards, were unfortunately cancelled by the motion upwards. Both aspects are shown in Figure 11.

4.1(b) Semi-free sailing set-up

The second step of the model test campaign, in semi-free sailing set-up, investigated which systems (and their combination of parameters) could provide enough boost for the take-off and maintain a steady foil-borne condition. For all the tests, the testing sequence was as follows:

- A constant propulsive force "F_main" of 101 N was applied mimicking the action of a main propulsion system. Given the steady relation between resistance and speed, this force was capable to maintain a speed of about 4.6 m/s before take-off (and about 5.5 m/s once in foil-borne condition);
- Initial acceleration up to 4.47 m/s by means of towing action of the carriage via the winch system;
- Switch from "towing mode" to "following mode", where the model is free to sail and accelerate in response to the applied propulsive forces (F_main and F_boost);
- End of the run with deceleration controlled in "towing mode".

The tested physical systems, excluding the flapping aft foil-Booster 2, gave enough boost force to reach a stable flight, provided that the combination of booster force and release time was sufficient. The cases where take-off was not successfully reached are specifically mentioned in the tables introduced along the discussion.

4.1(c) Virtual booster

The first tested booster was the virtual one. Figure 12 shows the applied boost profiles in terms of force and release time.

A successful boost action was identified by take-off and steady sailing at speeds above 5 m/s (which corresponded to a full foiling condition). In that respect, only one combination of booster input (15 N force applied only for 1 sec) resulted insufficient to reach a steady flight.

The results for the virtual booster clearly indicated that, with the same booster force, release times shorter than 2 seconds were likely not enough to reach a steady full foiling condition.

4.1(d) Physical boosters

For the Flapping aft foil-Booster 2 the release time was very short (about half second): a longer actuation (and acceleration) time was needed by means of multiple kicks. However to be efficient and effective, these kicks demand necessarily a variable vertical speed of actuation combined eventually with a variable angle of attack so that no increase in the resistance is experience during the upwards motion. As said before both parameters were kept constant with the consequent continuous cancelling of the propulsive action (kicking downwards) by the upward actuation of the foil. Figure 13 shows an example of the model speed behaviour combined with the multiple kicks. Further development and optimisation in the biomimetic aspects of this mechanism could improve the effectiveness of the system.

For the waterjet-booster 3 some of the conditions (pressure and nozzle dimeter) tested in captive set-up were investigated to evaluate whether the take-off and steady flight were effectively reached or not.

The Air turbine jet – Booster 4 was tested only with the orientation to generate only horizontal boost force (measured at the location of the turbine).

For "Squid concept – Booster 5", Figure 14 indicates the possibility to eventually generate comparable booster force although the challenge (similarly to the Booster 2) lays in the release time: for Booster 2 the release time depends on factors such as the vertical length of the actuation motion and the possibility to perform multiple kicks to generate efficiently enough vertical and horizontal boost force; for Booster 4, the capacity of the "water bag" can represent the biggest challenge (with the same input pressure and nozzle diameter).

The working principle of the Waterjet-Booster 3 and Squid concept – Booster 5 are the same (impulsive force generated by change in momentum of water ejected at high speed). The mechanism is very different but both systems on model scale were capable to develop similar boost force. This comparison is however approximated and not entirely fair because the Squid concept – Booster 5 was tested only in bollard pull condition. Nonetheless it remains an interesting exercise to explore alternative ways to generate a boost force.

4.2. Full scale considerations

Table 3, Table 4and Table 5 shows the relevant parameters for the take-off extrapolated to full scale (especially concerning the boost force, release time, attained speeds, take-off distance). For the test with the virtual booster, there are additional output values discussed in the next subsection. Some considerations about the extrapolation procedure and possible scale effects are treated in sub-section 4.2(b).

The ensemble of investigated booster forces and release times gave a reference for the testing of the physical systems in terms of required conditions for successful take-off. In that respect Figure 15 shows a correlation between these two factors (booster thrust and release time) and the successful (or not) take-off (which includes a reached stable flight). It is important to realize that such a figure is not general applicable because the trends can differ in relation to different hull form, hydrofoils, displacement etc. However, as initially said about the representability of the hull model, this investigation can set some initial boundaries about feasibility and promising level of new ideas.

4.2(a) Power/energy requirements and impact on displacement

An additional analysis of the tests with the virtual booster was performed to evaluate the boost demands in terms of energy/power and the impact on the additional weight when existing systems are integrated to the standard propulsion system (battery or supercapacitors packs).

For each test, the booster thrust was constantly applied for the chosen release time. The speed before and after the boost is constant, steadily increasing during the release of the boost force. The total and boost power can be derived as the product of respectively the total thrust force and boost force with the instantaneous speed. Furthermore the integration of the power over time gives respectively the total and boost energy. As an example of this analysis on model scale, Figure 16 shows the model speed and the total power in function of time.

It was also considered to use modern electric drive systems to provide the boost power. Systems like batteries and supercapacitors are now used in various applications to provide high peak powers during short timeframes. These systems have higher power densities than conventional combustion engines, and thus are interesting to consider for the boost power. Table 3 shows the extrapolated booster power and energy together with the estimated weight of battery or supercapacitor packs.

The weight estimation of battery and supercapacitors is derived from the Ragone chart plotted in Figure 17: in correspondence of release times close to the measured ones (10 or 20 seconds) there are different effective power density values for batteries and supercapacitors. From the power density, energy density and the boost power, the system weight is calculated. In case of the battery package the impact of the discharge rating on the weight is also taken into account (batteries need to be of a certain size to be able to discharge an amount of power greater than their energy capacity). In the computation of the weight a conservative factor of 5 kg/kW was considered to include the additional weight of all the other systems integrated, like electric drives, converters. These considerations about weight are clearly preliminary and they can change based on the required power and actuation time. For example, longer actuations (in the order of 2-5 minutes) can invert the rank of the lightest solution between battery and supercapacitors.

The importance of the weight estimation of the booster systems is crucial to judge their feasibility (and worthiness): for a given total displacement, an excessive weight might lead to a not economically acceptable reduction of transported payload; furthermore the initial investment cost for a more complex design might not be justified economically.

Figure 18 shows some historical data [1] about the typical weight distribution of hydrofoil craft. The source of data is quite old (1954). It is reasonable to expect that with the current technology some weight components (such as hull and hydrofoil structures or machinery) could be reduced, but not drastically. However this figure provides a reasonable picture to derive initial feasibility considerations. In a range of displacement between 30 and 100 tons, the available payload (in percentage of total displacement) is about 30% and the presence of the boost systems (battery or supercapacitors) would reduce if by a range between 2 and 10%, depending on the take-off requirements (power, boost thrust, release time and acceleration).

For the tested physical boost systems, an estimation of their impact on the weight would require a more in depth analysis and design refinement.

4.2(b) Extrapolation and scale effects

The results of the measurements were scaled up to full size values according to Froude's law of similitude. The scale considered was 10 and in Table 2 the scaling factors as applied are shown.

No dedicated extrapolation procedures were adopted considering that this research belongs to a very preliminary design phase, where the focus was to derive trends and orders of magnitudes.

In a more advanced design stage, the boost system mechanisms described in Section 3.3 should be subject of additional in depth investigations. For their experimental assessment, scale effects can affect the performance as follows:

Boost system	Factors scale affected
Flapping aft foil-Booster 2	Viscous effects on angle of attack Viscous drag Actuation speed
Waterjet-Booster 3	Weight of the system Local headloss at nozzle
Air turbine jet–Booster 4	Turbine mechanisms and weight
Squid concept–Booster 5	Elastic response of the air and water bags Viscous flow in the bags Local headloss at nozzle

The above table is meant as an overview of the main scaling issues but, once identified a promising system, more in-depth studies are necessary.

Beyond the scale effects on the single booster mechanisms, it is worthy to remind other relevant aspects to take into considerations when testing a full foiling craft:

- Scale effects on the free surface (especially in relation to ventilation issues in case of surface piercing hydrofoils)
- Position of the towing point (unless a full self-propelled free sailing set-up is used).

5. CONCLUSIONS

An experimental model test campaign was carried out to investigate unconventional and innovative solutions to assist the take-off of hydrofoil crafts. The primary objective was to evaluate new "out of the box" ideas rather than performing a complete design loop for one system. This challenge implied the risk to identify also not successful solutions and/or to realize the steps to be taken for further development and optimisation.

The hull form was meant to be as representative as possible with respect to the main dimensions and displacement. The hydrofoil configuration was treated only as a "lift generator" for the take-off and not optimised.

One "virtual booster" system allowed to explore systematically the correlation between the boost force and release time to successfully take-off and to reach a steady foil-borne condition.

Four physical booster systems were experimentally tested. They were inspired to the impulsive boost thrust found in nature. Among them, the waterjet-Booster 3 proved to provide enough boost thrust and release time; similarly the Air turbine jet – Booster 4. The Flapping aft foil – Booster 2 did not succeed to reach take-off but it could be promising with further development and optimisation into a more elaborate motion pattern during the kick (including variable angle of attack).

A correlation between inputs (booster thrust and release time) and the successful (or not) take-off was estimated to provide a preliminary feasibility tool on the feasibility and on the promising level of these unconventional possible installations.
The impact on the total weight of the craft was estimated when using other new boost systems (such as battery or supercapacitors drive systems).

The investigations and results presented in this paper represent only a first step after which a more in depth research and design are required from the industry for those systems considered most promising (and suitable for any given strategic and design objectives).

Acknowledgements
This research was accomplished thanks to the support of the Dutch Ministry of Defence.

Reference

1. Hoerner, S.F., Michel W.H. et. al., (1954), *Hydrofoil Handbook*, Volume 1 Design of hydrofoil craft Unclassified document from Gibbs&Cox, Inc., US.

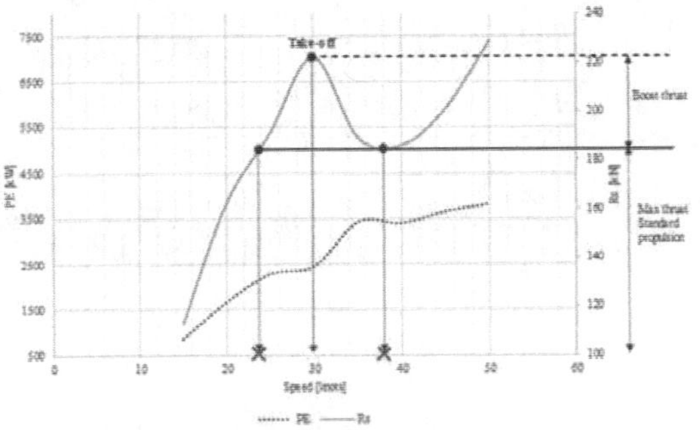

Figure 1. Example of resistance (and power) curve for a craft with hydrofoils

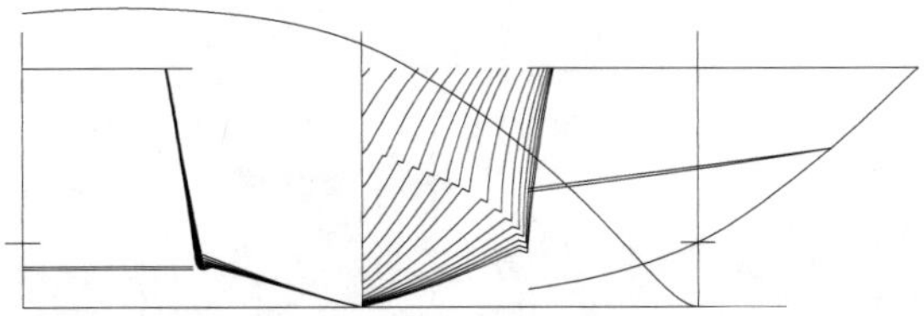

Figure 2. Body plan of hull M10185A

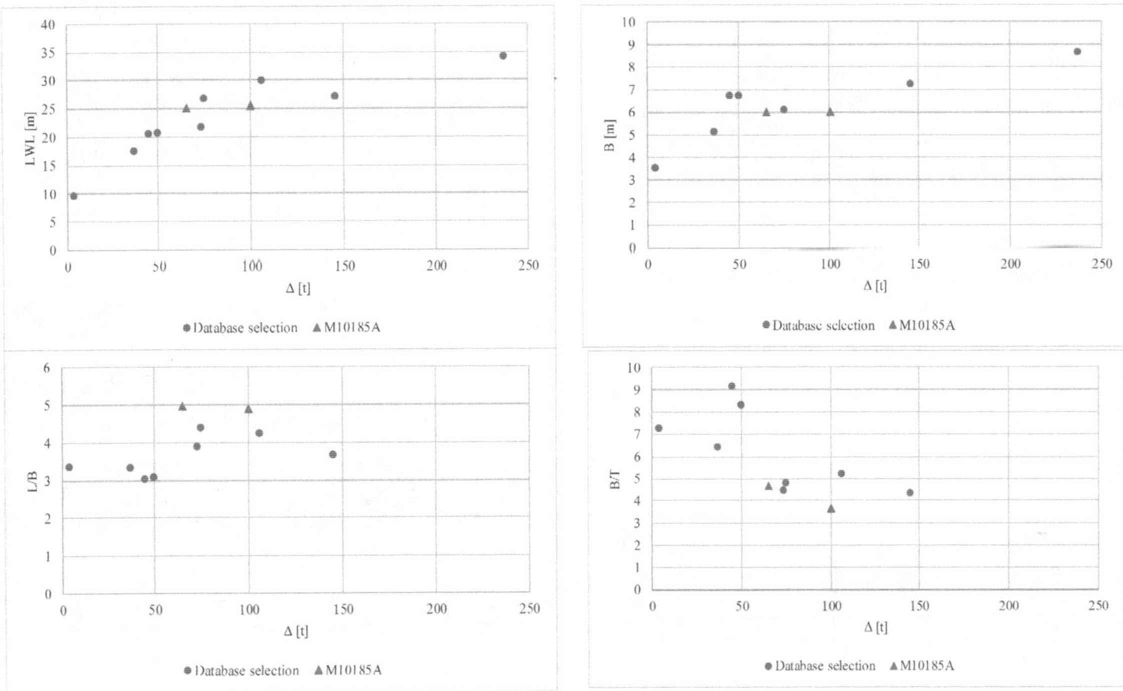

Figure 3. Parameters values of hull M10185A and database selection

Figure 4. Photo of hull model M10185A equipped with designed hydrofoils (before application of anti-ventilation plates) (top); details of front hydrofoils and anti-ventilation plates mounted on the front foils (bottom)

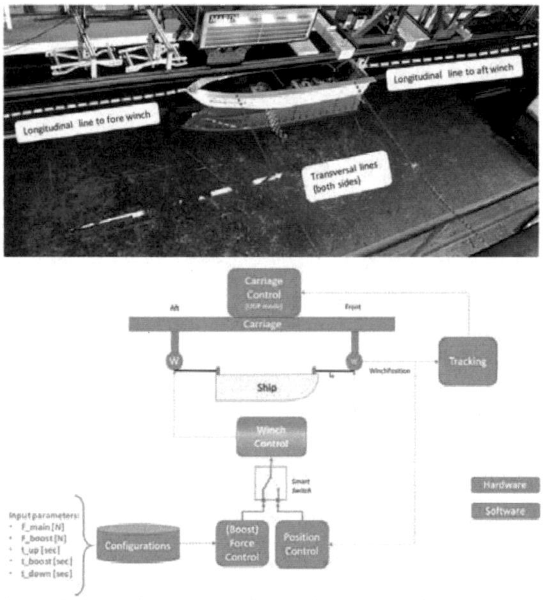

Figure 5. Overview (top) and working schematic (bottom) of semi-free sailing set-up

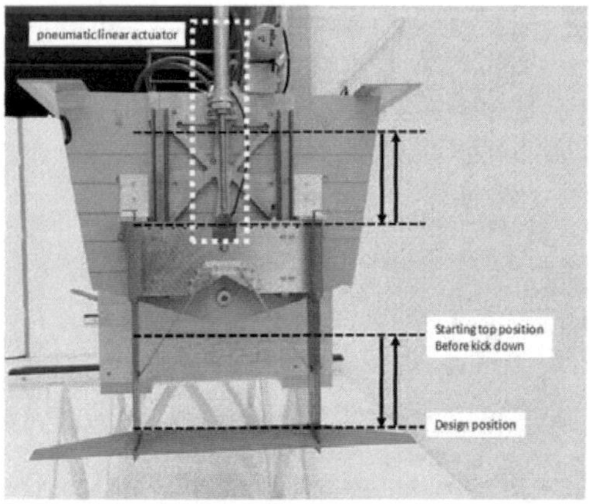

Figure 6. Flapping aft foil-Booster 2

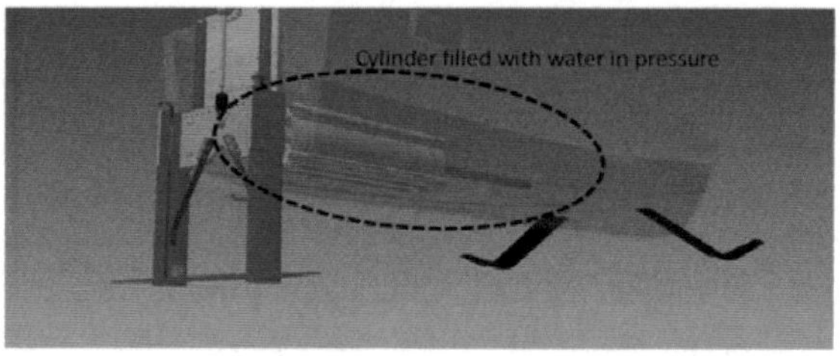

Figure 7. Waterjet-Booster 3

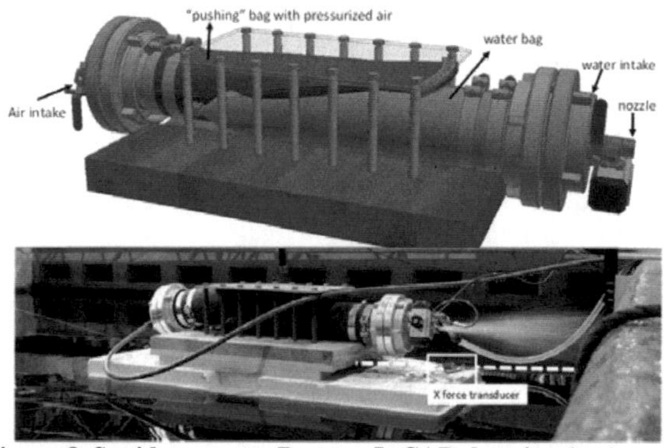

Figure 8. Squid concept – Booster 5; CAD drawing (top) and set-up photo (bottom)

Figure 9. Air turbine jet-Booster 4;

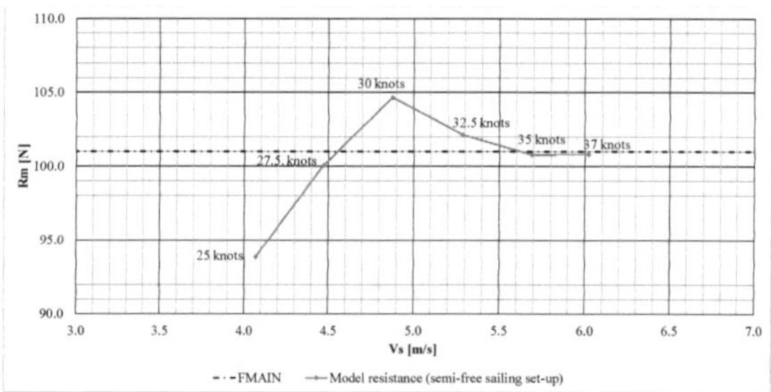

Figure 10. Resistance curve (model scale) with indication of speeds (in knots) at full scale

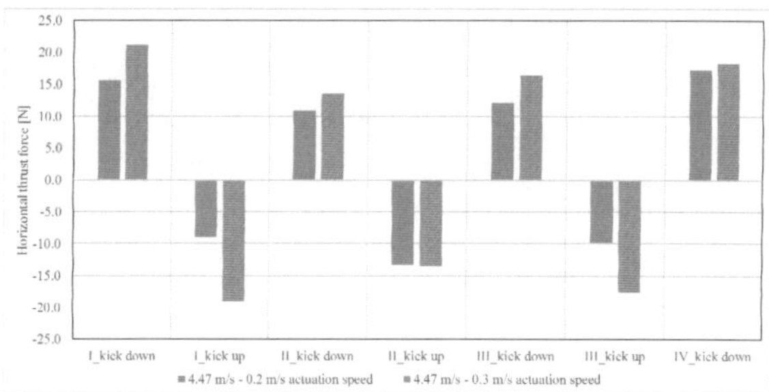

Figure 11. Flapping aft foil-Booster 2 (captive set-up) – Horizontal boost force during actuation

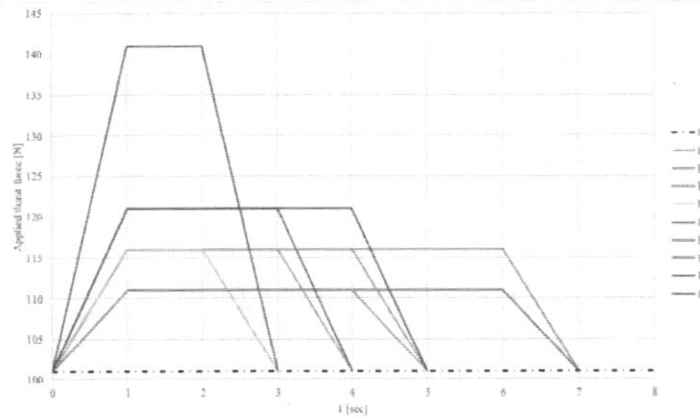

Figure 12. Boost profiles applied for virtual booster tests

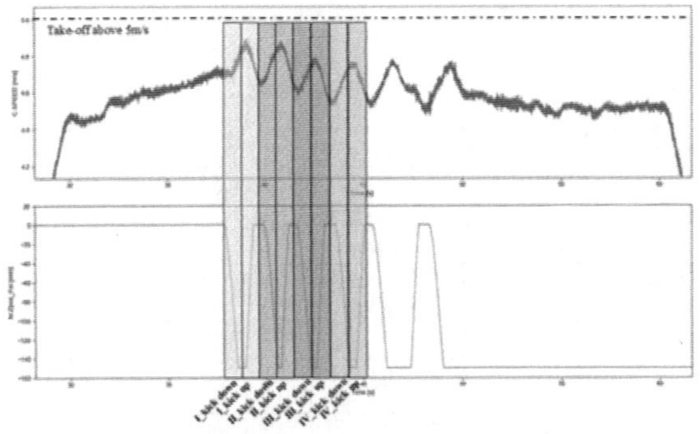

Figure 13. Flapping aft foil-Booster 2 (semi-free sailing set-up); example of model speed combined with indication of actuated kicks (down and up)

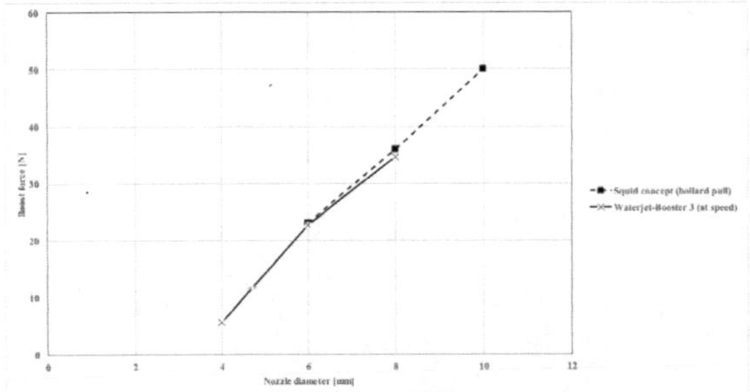

Figure 14. Comparison of boost force generated by Squid Concept-Booster 5 (in bollard pull) and Waterjet-Booster 3 (at speed)

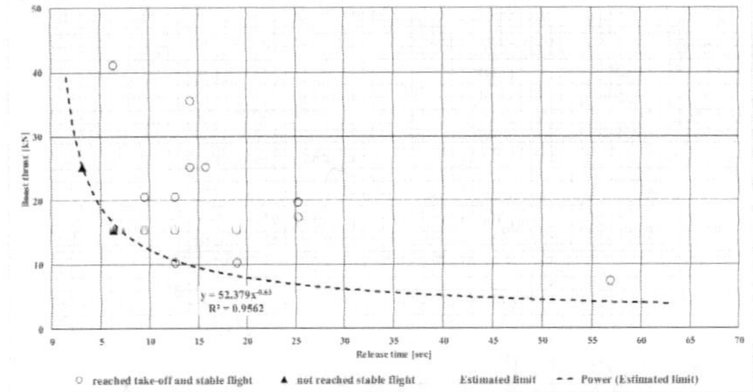

Figure 15. Correlation between booster thrust and release time for successful take-off

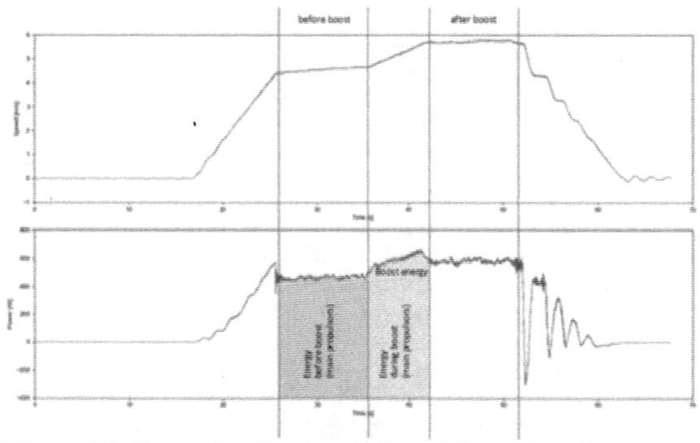

Figure 16. Example of computed model speed and instantaneous total power in function of time

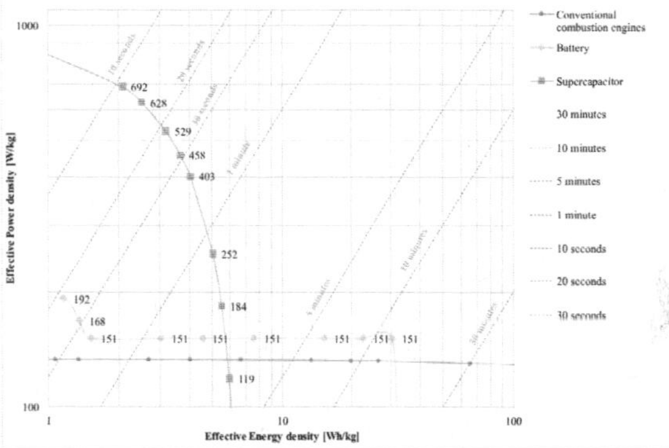

Figure 17. Effective Power and energy density for different systems and release times

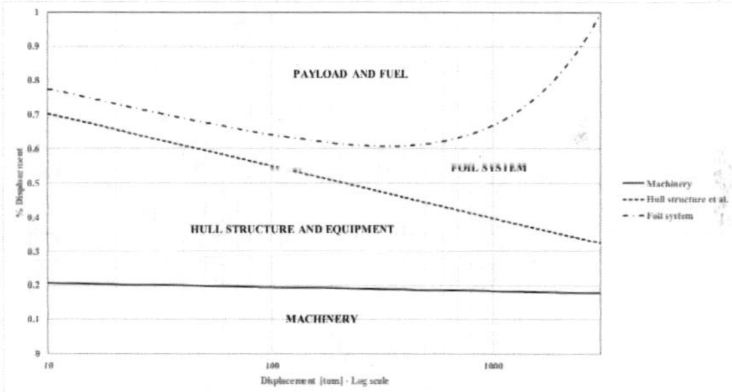

Figure 18. Weight distribution of hydrofoil craft (graph derived from information of [1])

Table 1: Hull form main characteristics

Symbol	Value	Unit	Description
L_{PP}	25.000	[m]	Length between perpendiculars
L_{WL}	24.995	[m]	Length on waterline
B	6.000	[m]	Breadth max. moulded
B_{WL}	5.013	[m]	Breadth max. moulded on waterline
TF	1.074	[m]	Draught moulded fore
TA	1.074	[m]	Draught moulded aft
DISV	63.401	[m³]	Displacement volume moulded
Δ	65.050	[t]	Displacement mass [water density = 1026.0 kg/m³]
FB	15.380	[m]	Position centre of buoyancy aft of FP
CB	0.471	[-]	Block coefficient
CP	0.719	[-]	Prismatic coefficient
L/B	4.987	[-]	Length-Breadth ratio
B/T	4.667	[-]	Breadth-Draught ratio

Table 2: Data scaling table

Quantity	Scaling factor
Linear	$\lambda = 10$
Volume	$\lambda^3 = 1000$
Force	$\lambda^3\gamma = 1027$
Pressure	$\lambda\gamma = 10.27$
Linear velocity	$\lambda^{0.5} = 3.162$
Time	$\lambda^{0.5} = 3.162$
Power	$\lambda^{3.5}\gamma = 3247$
Energy	$\lambda^4\gamma = 10269$
Note: γ is the ratio of the specific mass of seawater to that of the fresh water in the basin, with $\gamma = 1.027$	

Table 3: Virtual Booster tests input and output (full scale)

Test No.	Run No.	F_main	F_boost = Boost thrust	t_up	t_boost	t_down	Average speed before boost	Average speed after boost	Distance to take-off
		[kN]	[kN]	[sec]	[sec]	[sec]	[knots]	[knots]	[m]
0104015	011-03	104	15	3	16	3	28.3	35.2	310
0104015	011-04	104	15	3	9	3	28.3	33.8	202
0104015	011-05	104	15	3	6	3	28.3	31.3	146
0104015	011-06	104	15	3	3	3	28.3	30.1*	95
0104015	013-02	104	10	3	16	3	28.3	32.6	297
0104015	013-03	104	10	3	9	3	28.3	32.0	196
0104015	015-01	104	21	3	9	3	28.3	34.4	204
0104015	015-02	104	21	3	6	3	28.3	33.8	152
0104015	017-02	104	41	3	3	3	28.3	33.8	101

*Take-off not successful

Test No.	Run No.	Average speed after boost	Average boost power	Average boost energy	Estimated booster weight: Battery drive system		Estimated booster weight: Supercapacitors drive system	
		[knots]	[kW]	[kJ]	[kg]	%Δ	[kg]	%Δ
0104015	011-03	35.2	251	4755	2212	3.4	1727	2.7
0104015	011-04	33.8	202	2557	1785	2.7	1393	2.1
0104015	011-05	31.3	238	2259	2090	3.2	1535	2.4
0104015	011-06	30.1*	235	1489	2067	3.2	1517	2.3
0104015	013-02	32.6	185	3512	1634	2.5	1275	2.0
0104015	013-03	32.0	166	2095	1462	2.2	1141	1.8
0104015	015-01	34.4	325	4108	2867	4.4	2238	3.4
0104015	015-02	33.8	314	2978	2755	4.2	2023	3.1
0104015	017-02	33.8	649	4108	5701	8.8	4186	6.4

*Take-off not successful

Table 4: Waterjet-Booster 3 tests input and output (full scale)

Test No.	Run No.	Water volume	Water pressure	Nozzle diam.	Boost thrust	Release time	Average speed before boost	Average speed after boost	Distance to take-off
		[L]	[bar]	[mm]	[kN]	[sec]	[knots]	[knots]	[m]
0104017	002-01	5000.0	43.2	60.0		34.8	26.6	33.1	534
0104017	006-01	5000.0	72.2	60.0	*20*	25.3	26.6	37.5	443
0104017	012-01	5000.0	43.2	80.0		19.0	25.8	32.1	283
0104017	016-01	5000.0	70.1	80.0	*36*	14.2	25.6	34.4	269
0104017	018-01	5000.0	73.7	40.0	*7*	56.9	26.2	33.2	870
0104017	020-01	5000.0	87.2	40.0		50.6	26.2	33.8	780
0104017	024-01	5000.0	87.3	60.0		25.3	26.0	36.7	408
0104017	026-01	5000.0	92.4	60.0		23.7	25.8	35.4	373

Table 5: Air turbine jet-Booster 4 tests input and output (full scale)

Test No.	Run No.	Turbine RPM*	Force at turbine	Release time	Average speed before boost	Average speed after boost	Distance to take-off
		[RPM]	[kN]	[sec]	[knots]	[knots]	[m]
0104018	010-04	-	19	25	27.4	35.2	458
0104018	012-02	-	29	16	27.4	36.3	311
0104018	012-04	-	29	3	27.8	29.1**	102

*Extrapolation not applied because depending on the specific turbine installed
** Take-off not successful

SPLIT-FLAPS FOR INCREASED HEEL STABILITY OF T-FOIL CONFIGURATIONS

N.S.S. Prabahar[1], nimal.sudhan.saravana.prabahar@gu.se
L. Larsson[2], lars.larsson@chalmers.se
A. Persson[3], adam.persson@sspa.se

Abstract. Horizontal T-foils allow for maximum lift generation within a given span. However, for T-foils, the lift force acts in the symmetry plane of the boat, thereby producing no righting moment. It results in a lack of transverse stability during foil-borne sailing. In this project, we propose a system, where the height-regulating flap on the trailing edge of the foil is split into a port and a starboard flap, whose angle is adjusted proportionally to the heel. Such a system generates and scales the righting moment based on the heel angle, making the boat easier to balance. The effectiveness of this system is studied for a custom-made double-handed skiff using a dynamic velocity prediction program (DVPP), especially in the take-off phase. It is shown that the split flap system generates a larger righting moment for the foiling boat than for the non-foiling one at boat speeds above 3 m/s. . The improved stability comes at a cost of additional induced resistance and this effect is computed by CFD. It is shown, using the DVPP, that the increased resistance, slows down the boat at the lowest wind speeds, but above 2.5 m/s true wind, the split flap boat is as fast as the boat with a single flap around the racecourse. Due to the higher stability, less depowering of the sails is required, and that compensates for the increased resistance.

1. NOMENCLATURE

α	Angular Acceleration (rad.s^{-2})
α_i	Angle of attack (deg)
υ	Kinematic viscosity (N s m^{-2})
ρ	Density of water (kg.m^{-3})
φ	Heel angle (deg)
ω	Angular Roll velocity (rad.s^{-1})
a	Linear acceleration (m.s^{-2})
b	Beam (m)
C_{AR}	Correction Factor on Aspect Ratio
C_D	Coefficient of Drag
$C_{D,i}$	Coefficient of Induced Drag
C_L	Coefficient of Lift
$C_{L,2D}$	2-Dimensional Lift Coefficient
$C_{L,3D}$	3-Dimensional Lift Coefficient
f	Flap angle (deg)
f_o	Optimal Flap angle (deg)
f_p	Portside Flap angle (deg)
f_s	Starboard Flap angle (deg)
g	Acceleration of gravity (m.s^{-2})
h	Height (m)
m	Mass of the boat (kg)
p	Roll rate (rad.s^{-1})
q	Pitch rate (rad.s^{-1})
r	Yaw rate (rad.s^{-1})
v	Linear boat velocity (m.s^{-1})
AR	Aspect Ratio
AR_C	Corrected Aspect Ratio
AWA	Apparent Wind Angle (deg)
AWS	Apparent Wind Speed (m.s^{-1})
CAD	Computer-Aided Design
CB	Centre of Buoyancy
CFD	Computational Fluid Dynamics
CG	Centre of Gravity
DOF	Degree of Freedom
DVPP	Dynamic *VPP*
GC	Gain Coefficient
GM	Metacentric Height (m)
RANS	Reynolds Averaged Navier-Stokes
RM	Righting Moment (N.m)
TCE	Transverse Centre of Effort of the foil
TCG	Transverse Centre of Gravity of the sailors
TE	Trailing Edge
TWA	True Wind Angle (deg)
TWS	True Wind Speed (m.s^{-1})
VMG	Velocity Made Good (m.s^{-1})
VPP	Velocity Prediction Program

2. INTRODUCTION

One of the most common foil configurations for lighter dinghies is the T-foil. It is favoured in foiling dinghies with high lift requirements and has better structural characteristics compared to cantilever foils, see *Monsonnec and Waller* [1]. Horizontal T-foil configurations usually consist of a single main horizontal foil and a vertical centreboard connecting the horizontal foil to the hull such that the lift of the foil acts in the centreplane of the boat. Such configurations also have a smaller horizontal foil attached to the rudder board to stabilize the boat in pitch. However, when foil-borne, the boat is inherently unstable in roll.

In T-foil systems, height regulation is generally achieved by using Trailing Edge (*TE*) flaps on the mail horizontal foil coupled to a height response system. Considering the lack of transverse stability with the T-foils, the idea of splitting the *TE* flap into a port and starboard flaps, to increase the dynamic righting moment is introduced in this paper. If the leeward flap is deflected more than the windward flap, the lift of the horizontal foil shifts towards the leeward side: this, in turn, produces a righting moment contributing to stabilizing the boat.

[1] Ph.D. Student, Department of Marine Sciences, University of Gothenburg, Gothenburg, Sweden
[2] Professor, Department of Mechanics and Maritime Sciences, Chalmers University of Technology, Gothenburg, Sweden
[3] Ph.D. Student, Chalmers University of Technology, Gothenburg, Sweden & Project Manager, SSPA Sweden, Gothenburg, Sweden

The increased stability, however, comes at a cost. There is a very complex flow between the flaps, generating a trailing vortex, and the strength of the tip vortices may also be increased. So, there will be an increase in induced resistance.

In this paper, the forces acting on a T-foil dinghy with a single *TE* flap and with split *TE* flaps are reported. The split flap system is applied to a custom-made skiff "*Linnea*" and the stability and dynamics are studied using static and dynamic Velocity Prediction Programs (*VPPs*) as a demonstration case. Since the increase in induced resistance is of concern it is investigated in detail using CFD computations.

3. STABILITY CHARACTERISTICS OF T-FOIL CONFIGURATIONS

3.1. Foiling with a single Trailing Edge (TE) flap

During the take-off phase, the hydrodynamic lift generated from the foils will lift the hull out of the water due to the force imbalance created in the z-direction. If this lift is not regulated and reduced as the hull lifts out, the upward acceleration would not stop and would result in the boat flying out of water. Lift reduction may be achieved in many ways. Such as angled surface-piercing foils which change wetted area depending on the height or manually pitching the boat to reduce the angle of attack of the foils with height, see *Monsonnec and Waller* [1]. In the case of fully submerged foils, the lift force is automatically reduced when approaching the free surface, so there is a natural control mechanism, but it is very week. A very common solution in such a case is to use trailing edge (*TE*) flaps. These flaps are deflected up or down depending on the heave displacement: up to reduce lift when the hull tends to rise out of the water. The flaps are often deflected by an automated mechanical or electronic control system to maintain a stable heave displacement.

Assuming the windward side to be the port side, the heeling moment from the sail and appendage side forces (S_{sail} & S_{app}) will be clockwise, seen from behind. See Figure 3.1. At rest, when the hull heels to starboard, the Centre of Buoyancy (*CB*) also shifts to starboard, producing a righting moment that stabilizes the system (figure 3.1). When the boat is in motion, the lift force (*L*) generated by the foils would act normal to the surface of the foils. Thus, while heeling, the horizontal component of the lift (L_h) needs to be compensated by the vertical appendages by increasing the leeway angle. This results in an additional side force (S_{lift}) occurring only during heeling. In upright conditions, this additional side force is non-existent. The horizontal component of this side force ($S_{lift,h}$) equals L_h and the sum of the vertical component of S_{lift} ($S_{lift,v}$) and the vertical component of lift (L_v) equals the gravity force. With a rectangular planform of the centreboard, S_{lift} acts at the mid-span of the submerged centreboard span.

$$S_{lift,h} = L_h = L \sin \varphi \qquad (1)$$

$$S_{lift,v} = m * g - L \cos \varphi \qquad (2)$$

$$S_{lift} = \sqrt{S_{lift,v}^2 + S_{lift,h}^2} \qquad (3)$$

φ is the heel angle, *m* is the mass of the boat and *g* is the acceleration of gravity.

When heeling to leeward, the additional side force S_{lift} will produce a heeling moment, whilst windward heel will produce an additional righting moment. The lever of S_{lift} from the Centre of Gravity (*CG*) of the boat is dependent on the centreboard span and the height of the boat above the water surface (hull clearance). While a shorter span will reduce the moment lever, and thus reduce the additional heeling moment, a very low span will limit the hull clearance and the foil submergence; low foil submergence results in rapid lift loss due to free surface interaction effects as shown in *Faltinsen* [2]. Considering this, finding a centreboard span that offers a reasonable compromise between these contradicting requirements, is critical in achieving optimum performance of the boat.

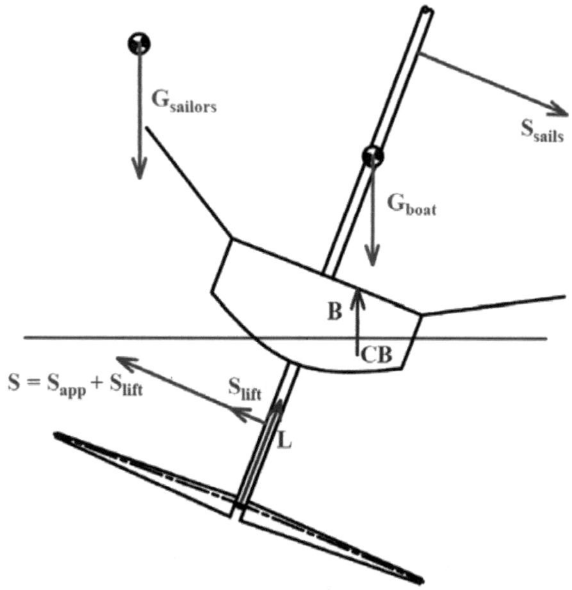

Figure 3.1 Forces on the heeled hull-borne boat with a single *TE* flap

If the hull is fully foiling (figure 3.2), the righting moment from the hull buoyancy is lost. The heeling moment from the sail and appendage side force couple now has to be balanced by the crew weight only. This calls for very skilled sailors, particularly for a double-handed dinghy.

3.2. Foiling with split TE Flaps

The concept proposed is to split the trailing edge flaps span-wise in the centreline into starboard and port flaps (as shown in figure 4.3) and adjust each as a separate element proportional to both the heave displacement and the heel angle of the boat. Each side receives two inputs in deciding the flap deflection angle: heave displacement and heel. When heeling starboard, the starboard flap is deflected downward, and the port flap is deflected proportionally upwards. The increased lift on the starboard side shifts the net *CE* of the lift force from the foils starboard (as seen in figure 3.3). This in turn creates a dynamic righting moment which helps to stabilize the boat and aims to increase the stability in foiling mode, achieving equal or greater stability in comparison to displacement mode. This righting moment is proportional to the lift produced by the foils, which in turn is proportional to the square of the speed. Hence, the righting moment is also proportional to the square of the boat speed (v).

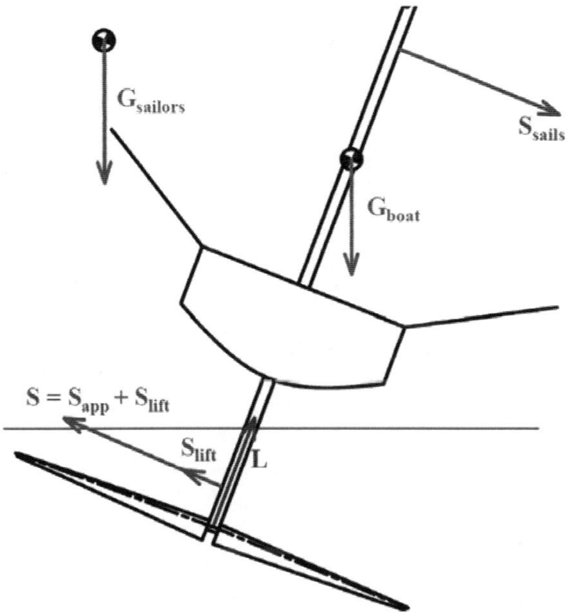

Figure 3.2 Forces on the foil-borne boat with a single *TE* flaps

The drawback of the split flap system is the increase in induced drag, as pointed out in the introduction.

It would be particularly interesting to study the stability of such a system when the boat transitions from displacement mode to foil-supported mode (take-off), as there are several dynamic effects. With the increase in boat speed the hydrodynamic lift increases, which reduces the amount of hydrostatic lift required, eventually lifting the hull out of the water. The hydrodynamic resistance of the hull is gradually reduced to zero, resulting in a rapid acceleration in surge. The increase in boat speed further increases the hydrodynamic lift, forming a positive feedback loop, which can cause instabilities in heave. The surge acceleration increases the Apparent Wind Speed (*AWS*) and reduces the Apparent Wind Angle (*AWA*), which in turn increases the sail forces and the sail heeling moment, making transverse stability critical in the take-off phase. The combined instability in heave and heel make the take-off phase of particular interest for the study of the stability characteristics of foiling boats.

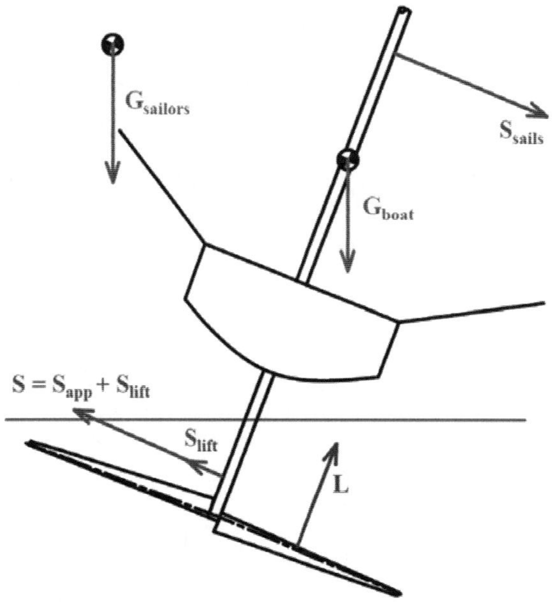

Figure 3.3 Forces on the foiling boat with the split-flaps on horizontal foils

4. TEST CASE

The test case for this study is a 4.6 m long skiff designed and built by the students from the Naval Architecture and Ocean Engineering Master's programme at Chalmers University of Technology to compete for the *1001Vela Cup* in *2018*. This double-handed skiff "*Linnea*" has a displacement of 149 kg excluding the sailors when fully rigged. The design of *Linnea* was inspired by the *Olympic 49er class* but adapted to suit the low wind average speeds at the race venue, with a taller mast and larger upwind sailplan, see *Acerbi et al* [3]. These large sails and rigging, in turn, shift the Centre of Gravity (*CG*) upwards making the boat highly unstable, even without foils. The original design was a non-foiling skiff, but recent changes in regulations to allow high-performance materials supports the idea of foiling. The project for the year 2020 was to add foils to the existing hull and T-foil configurations were considered to maximize the lift generated. Being a double-handed skiff with negative GM *Linnea* is very difficult to balance even without foils, and in foil-borne conditions, it would hardly be possible using a standard T-foil configuration. Hence the present development of the split-flap system.

4.1. Design of the Foils

One of the biggest challenges is the low prevailing wind speeds at the venue and the skiff is designed in compliance with the R3 class rules of *1001Vela Cup* [4], which for instance states that the overall breadth including appendages is limited to 2.1 m[4]. Thus, any foil configuration should make use of the entire span to generate a reasonable lift. Configurations with separate port and starboard foils would half the lift generated if

[4] The class rules have been modified now to remove the span restriction, however, this happened quite late in the project when the design was already finalized

only one side is lowered and would be hard to handle with just a crew of two.

Owing to the aforementioned challenges, a horizontal foil layout was chosen to utilize the full 2.1 m span for lift-generation. The foil will be submerged at all times, supported by the hull via the centreboard. The airfoil section is chosen as Wortmann FX-60100 (figure 4.1) based on a series of investigations done by *Andersson et al* [5]. With the span limited, and the Taper Ratio (*TR*) set at 0.45 for minimum induced drag, only the chord length of the foil was free for optimization. A systematic parametric study was performed using the static *VPP* to estimate the mean chord length which will result in the lowest possible take-off True Wind Speed. Based on this study, a mean chord of 0.25 m was selected.

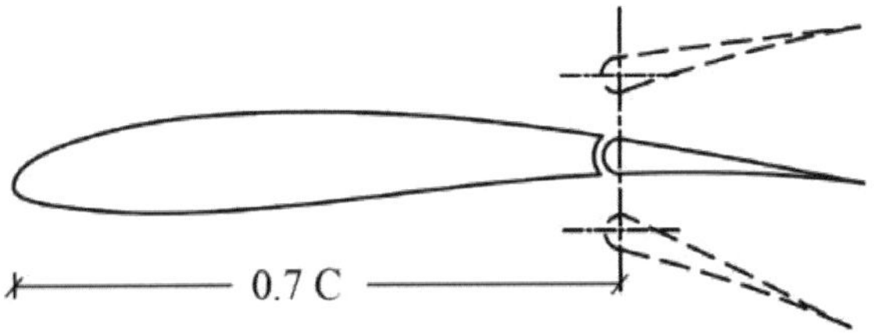

Figure 4.1 Cross-section of FX60100 showing the trailing edge flaps

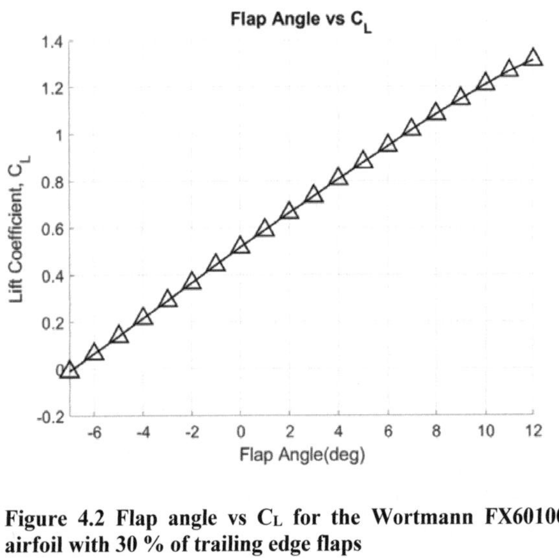

Figure 4.2 Flap angle vs C_L for the Wortmann FX60100 airfoil with 30 % of trailing edge flaps

4.1(a) Trailing Edge Flaps

The trailing edge flaps were designed to be at 30 % of the local chord length to ensure optimal control over the lift. The lift coefficient (C_L) of the airfoil changes almost linearly with the flap angle for this airfoil section as seen in figure 4.2. The flap would ideally deflect between +12 degrees and -7 degrees (positive downward). The limits are set based on the reasoning that a flap deflection of -6.8 degrees corresponds to zero lift generation for this airfoil and flap length, and any flap angle greater than +10 degrees will increase drag and flow separation could occur. The optimal initial flap angle (f_o) when the hull is fully submerged is also critical in determining the take-off speed of the boat. A high initial angle would generate a high lift at the cost of increased profile and induced drag. The optimal value for the initial flap angle was found to be +2 degrees using the static VPP. . The flaps would span the foil to have maximum control. Usually, the entire length of the flap is deflected as a single element proportional to the heave displacement of the boat. Figure 4.3 shows the top view and isometric view of the foil with the trailing edge flaps split and deflected at different angles.

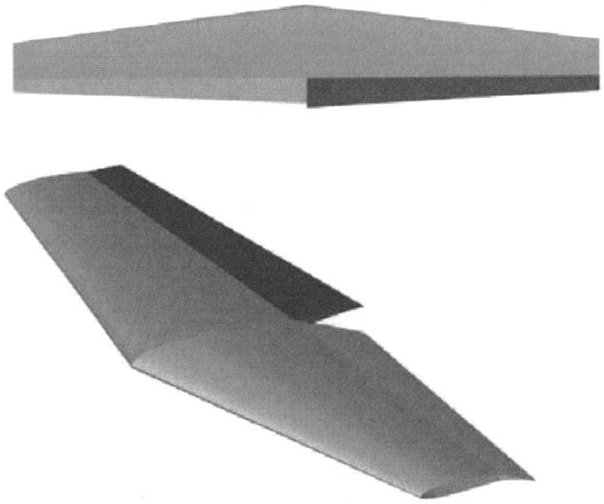

Figure 4.3 Horizontal foil showing the separate port and starboard flaps.

The performance of the boat with the split flaps is studied in detail using the static and dynamic *VPP* in section 5.

5. VELOCITY PREDICTION PROGRAM

5.1. Static Force Models

To carry out the investigations on the static and dynamic stability characteristics, the system is modelled using a combination of CFD results, semi-empirical and theoretical methods. Sail coefficients and models are taken from the Offshore Racing Congress's *International Measuring System 2019* model [6]. Resistance data is derived from CFD computations on the bare hull carried out by *Acerbi, T. et al* [3]. The change of frictional hull resistance with the draft of the boat is assumed to be proportional to the wetted surface area, while the change in wave-making resistance is assumed to be proportional to the change in displacement. A simple windage drag model is implemented based on the frontal area of the sailors, mast, rigging, and hull. The appendage forces and other force components are calculated with the same procedures followed by *Andersson et al* [5]. Simulations from *X-Foil* [7] provided the 2-dimensional force coefficients for the appendage sections as a function of the angle of attack, the flap angle, and the Reynolds number.

The 3-dimensional lift coefficient and induced drag are calculated based on lifting-line theory (equations 4 and 5 respectively).

$$C_{L,3D} = C_{L,2D}/(1 + \frac{2}{AR}) \quad (4)$$

$$C_{D,i} = C_{L,3D}^2/(2\pi AR) \quad (5)$$

The Aspect Ratio (*AR*) plays a significant part in equations 4 and 5. In the case of the split *TE* flaps at different deflection angles, the loading on the starboard and port sides are different; it is intuitive to consider the port and starboard parts of the foil as separate wings with half the span and two free tips. Such an approximation reduces the *AR* by a factor of 2, thereby increasing $C_{D,i}$, and reducing $C_{L,3D}$. Considering each side as a wing with two free tips would, however, be too conservative an approach, since the flow from the pressure side to the suction side is only geometrically possible in the space between the split flaps, where the overflow is somewhat restricted. Another approach is to consider each side as a wing with half the span of the foil, but with one free tip and a symmetry plane on the centreline of the foil (figure 5.1). Then there will be no overflow in the space between the flaps. The symmetry plane allows mirroring of the span of the individual wings whilst estimating the *AR* – thereby not affecting the *AR*. This approach is non-conservative, so we have an upper and a lower limit for the effective aspect ratio. To obtain a more exact effective aspect ratio, CFD computations were carried out, as seen below. The result is a correction between the two limits.

In the static *VPP*, the net forces and moments are estimated for each *DOF* and the variables such as speed, leeway angle, sailor position, etc. are varied until equilibrium is achieved in each *DOF*. The resulting boat speed is plotted in a polar plot against the True Wind Speed (*TWS*) and Angle (*TWA*).

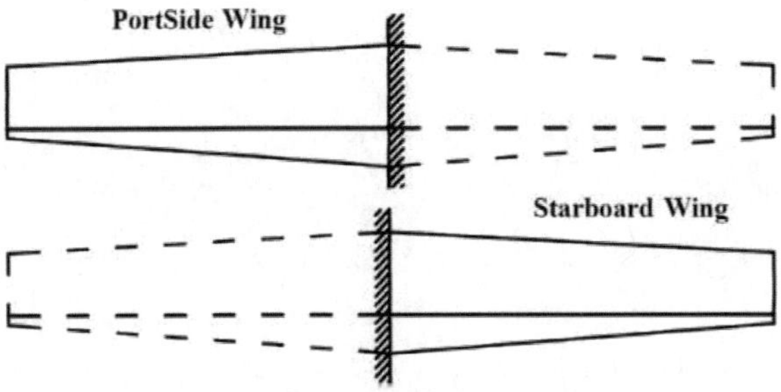

Figure 5.1 Port and starboard foil considered as separate wings with a symmetry plane on the centerline

5.2. Dynamics

Solving the static force balance equations does not consider any dynamic effects, and the take-off procedure cannot be studied effectively. Thus, the static *VPP* was developed into a dynamic *VPP* (*DVPP*). Figure 5.2 shows the coordinate systems used; the body-fixed coordinate system is fixed at the net *CG* of the crew and the boat, and the inertial frame is fixed at *IF*. The net force (F_{net}) and moment residuals (M_{net}) on each *DOF* are estimated and the acceleration in the same is obtained using the second law of motion (equations 6 and 7), *Horel and Durand* [8]. The accelerations are integrated in time to obtain the velocities and displacements in each *DOF*.

$$F_{net} = m^* a + m\omega \times a \quad (6)$$

$$M_{net} = I^* \alpha + \omega \times I^* \alpha \quad (7)$$

Where, $m^* = m + A$, and m is the total mass of the boat, and A is the added mass matrix in the respective *DOFs*. $I^* = I + I_A$, and I is the mass moment of inertia along the axis of rotations of the *DOFs* and I_A is the added inertia. F_{net} and M_{net} are the net force and moment residuals respectively, a and α are the linear and angular accelerations, respectively. Added mass and inertia for the hull are neglected in heel, as the change in the section shape with heel is minor for this hull. Added mass and inertias for unit chord length for the appendages are calculated using strip-theory approximation for a flat plate (equation 8), *Eggert et al* [9].

$$m_{22} = \pi\rho t^2 \qquad m_{66} = 1/8\pi\rho t^4 \quad (8)$$

where $2t$ is the strip width of the appendage, m_{22} and m_{66} are the added mass in vertical and roll accelerations, respectively. *Eggert et al* [9] suggest accounting for damping as the quasi-static change in the angle of attack (α_i) of the flow to the appendages and sails. When the boat moves upwards, the foil experiences a downward flow, thereby reducing the angle of attack and the lift produced. The opposite happens when the boat moves down. For roll-damping, when the boat rolls to port, the port side moves down and faces an upward flow, increasing the angle of attack and the lift force on the port side. The appendages are split into chord-wise strips and the instantaneous flow angle and velocity at the strips are estimated. Using the instantaneous α_i, the forces on each strip are calculated, which are added to other strips span-wise to obtain the total forces on the appendages.

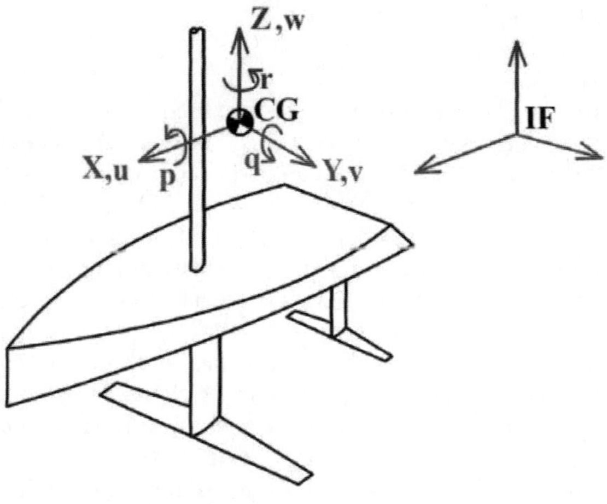

Figure 5.2 Coordinate systems used in the DVPP

To run dynamic simulations, an automated sailor control is used to mimic practical sailing by adjusting the Transverse Centre of Gravity (*TCG*) of the sailors, *flat* factor, and *reef* factor of the sails [6]. The rate of change of these variables is dependent on multiple parameters such as the heel angle, angular velocity, and acceleration. The response coefficients of these parameters are optimized through an iterative process.

To allow the user to experience sailing the boat with and without the split *TE* flaps, a sailor control was also designed with an option to accept real-time user inputs. Paired with a simple display (figure 5.3), it enables the user to subjectively evaluate the stability of the boat.

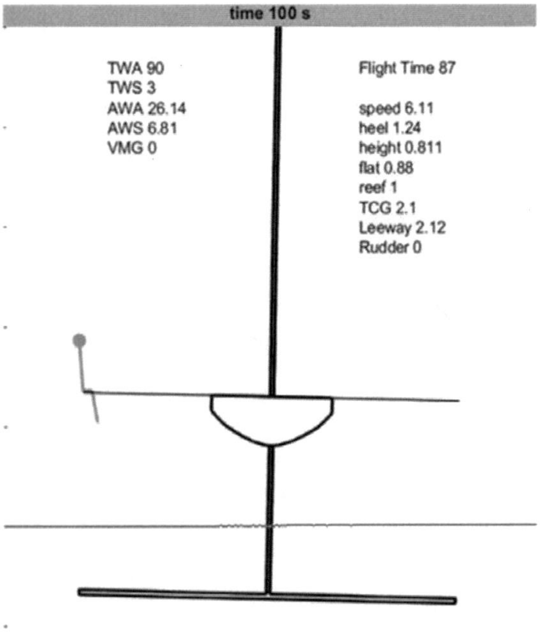

Figure 5.3 Simulator display showing the boat and various sailing parameters

6. RESULTS

6.1. Control System

6.1(a) Ride height Response

Ride height is defined as the displacement of the net *CG* of the boat in the body-fixed z-axis. There have been both electronic and mechanical control systems used to adjust the flaps. However, due to R3 class compliance, which prohibits the use of stored power, the system was designed to be purely mechanical. One of the most common ways is to use a surface skimming wand that follows the water surface. The wand is connected to the flaps using mechanical links. Since the wand skims the surface, when the boat rises, the wand is lowered, which in turn reduces the flap angle and consequently the lift force. The opposite happens when the boat falls. This system has been successfully applied in foiling moths for a long time, *Beaver and Zseleczky* [10].

The response of the flap angle to the change in height, i.e., the slope of the ride height vs flap angle curve: termed as gain coefficient (GC) is quite critical to ensure a stable ride. A low slope system would be less sensitive to disturbances, but it will be reducing the flaps at a lower rate with the ride height. This would result in a higher ride height. Whereas a high slope system would reduce the flap angle rapidly with ride height and result in a lower hull clearance, with high sensitivity to disturbances. Figure 6.1 shows ride height vs flap angle response curves with linear and non-linear profiles. A linear variation of the flap angle also ensures a linear variation of the lift coefficient in this case (figure 4.2). Zero ride height signifies a fully submerged hull, and a positive value represents a displacement of the hull upwards. Since the draft of the boat is 0.182 m, a ride height greater than 0.182 at upright conditions signifies a foil-supported hull. The flap angle starts at an optimal +2 degrees at full displacement and is reduced with increase in ride height until the lift reaches zero. The dotted line represents the angle at which the lift coefficient is zero for this airfoil. In the present work, we have adopted a linear relation between the flap angle (*f*) and the ride height (*h*), where f_0 is the optimum angle at full displacement.

$$f = -GC * h + f_o \qquad (9)$$

It is also common to use non-linear response curves to make the system more or less sensitive at different phases (figure 6.1). In this case, the gain coefficient is also varied with height. Such non-linear responses could be designed to produce more lift before take-off, good sensitivity to balance height disturbances while taking-off, and a quick reduction in lift once a critical ride height is reached. As in figure 6.1, it could practically take advantage of both a high and low slope linear system.

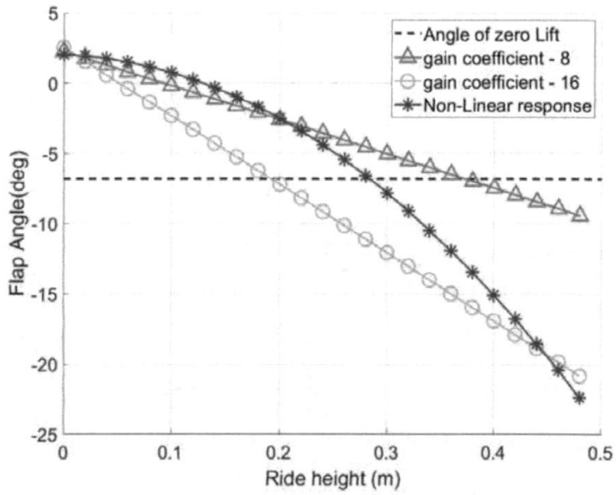

Figure 6.1 Ride height vs flap angle for linear and non-linear responses

Ideally one would prefer to have a high enough ride height such that the hull is clear of the waves. Also, it should not be too high such that the foils come close to the water surface as it would reduce the lift coefficient, see *Faltinsen* [2]. Hence, setting the optimal gain coefficient is quite critical.

6.1(b)) Combined Ride height - Heel Response

When there is no heel, both flaps are set at the same angle depending on the ride height. As the boat heels, the change of the lift coefficient on the port and starboard sides is determined by two wands: one on the port extreme and one on the starboard extreme instead of a single wand. Each wand would be connected only to the respective side of the split *TE* flap, such that each wand behaves as an individual ride height response system. When there is no heel and there is a change in the ride height, both wands are displaced in sync to adjust both flaps in proportion only to the ride height. When the hull heels, one wand is displaced upwards and the other downwards, and this deflects the respective flaps upwards and downwards (figure 6.2).

Let h be the ride height at a given instant without heeling. When the boat heels φ degrees to starboard, the heights (in the body-fixed system) at starboard and port would be $h - b * 0.5 * tan(\varphi)$ and $h + b * 0.5 * tan(\varphi)$ respectively, where b is the full beam of the boat. The linear response curve in equation 8 is applied for the individual starboard and port wand-flap system as,

$$f_s = -GC * (h - b * 0.5 * \tan(\varphi)) + f_o \quad (10)$$

$$f_p = -GC * (h + b * 0.5 * \tan(\varphi)) + f_o \quad (11)$$

It could be inferred that the higher the difference between the port and starboard flaps, the higher the righting moment produced. The difference, according to equations 10 and 11 is only proportional to the gain coefficient for a given heel angle. Though the flap angle in the two-wand system is dependent on the ride height and the heel angle, the proportionality constant (gain coefficient) is the same for both variables. It implies that if the gain coefficient is changed to facilitate a better response for one variable, it might adversely affect the response of the other.

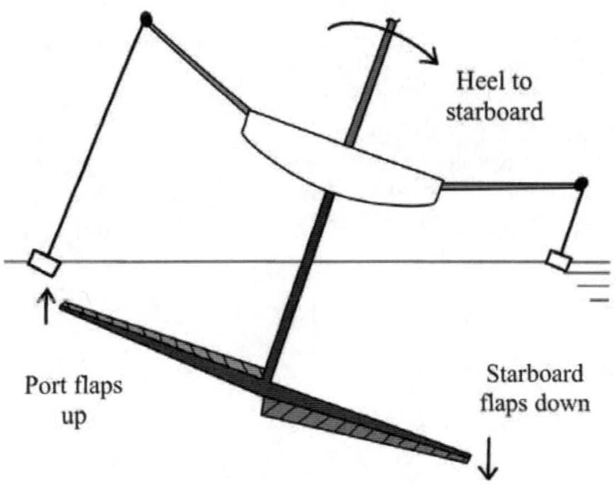

Figure 6.2 Split-flap control system with two wands

Figure 6.3 shows the flap angles as a function of the heel angle at a constant ride height. It simulates heeling with a linear or non-linear height response curves (same response curves as in figure 6.1).

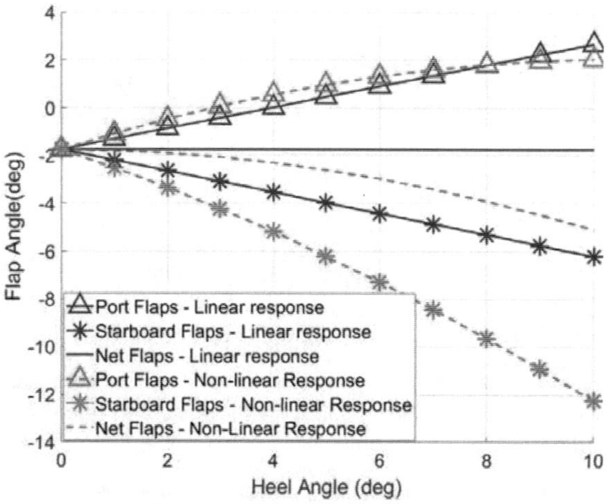

Figure 6.3 Flap angle vs heel angle for linear and non-linear responses at a constant ride height

In figure 6.3, the flap angle at that constant ride height and zero heel is shown as the horizontal line. The net flap angle for the heeled linear system remains the same as without any heel angle, thereby not affecting the total lift generated. The non-linear response provides an increasing slope on the flap difference with higher heels, making the system more stable in the heel *DOF*. However, the net flap angle deviates from the flap angle without heel (in this case lower). It would result in a lower lift force than without heel, reducing the ride height.

Figure 6.4 shows the aforementioned phenomenon with plots of the heel and ride height for the same linear and non-linear systems as a function of time, produced by the dynamic *VPP* (*DVPP*). The boat starts from a state of equilibrium, foiling at 2.75 m/s of true wind at 90 degrees. After the first 30 seconds, the true wind speed is rapidly increased to 3.5 m/s and kept constant to simulate the boat to heel leeward. The plots presented are for the boat without split flaps (W/O SF), with split flaps and a linear gain (W/ SF – Linear Gain), and with split flaps and a non-linear gain (W/ SF – Non-Linear Gain). Although the ride height increases initially due to the higher wind speed, it decreases as the boat heels. This is attributed to the loss of the apparent wind speed for the heeled boat. Then the boat attains a stable ride height for the increased wind speed. The linear response curve gives the largest raise in ride height, followed by the non-linear response and the single flap configuration. The reason for the increase in ride height is an increase in speed. When the speed increases the lift will be too large for a given flap angle, i.e., for a given ride height, and the boat will raise. The reason why it does not raise as much for the non-linear response is that the net lift coefficient is reduced, as shown in Figure 6.3. For the single flap, the speed increase is smaller, hence the smaller raise.

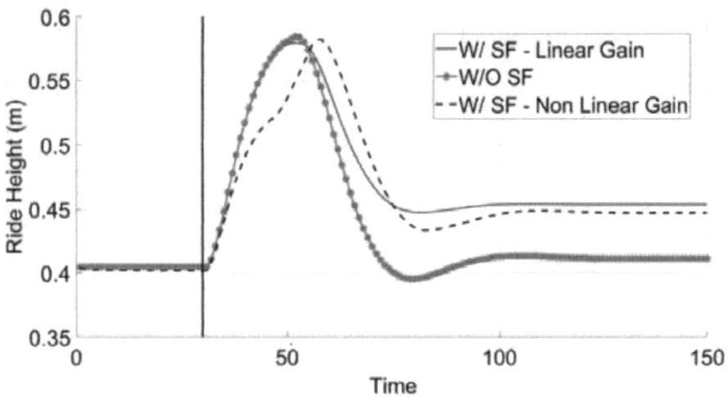

Figure 6.4 Heel angle and ride height as a function of time

6.2. Parameter Study of the Stability

Earlier, two critical parameters were identified to affect the stability in heel: centreboard span and the gain coefficient.

6.2(a) Effects of Reduction of Centreboard span
As mentioned in section 3.1, reducing the centreboard span (total draft) will reduce the lever of the heeling moment from the additional side force (*S*). Figure 6.5 shows the variation of C_L due to the free surface effects comparing it to the 3D lift coefficient at infinite depth for the Wortmann FX60100 airfoil as a function of the foil submergence calculated based on the experimental investigations from, *Faltinsen* [2]. It could be inferred that the lift coefficient decreases as the foil moves closer to the surface. Also, *Hoerner* [11] demonstrated the increase in induced drag when the foils get closer to the free surface, as the free surface acts as a mirroring plane at low Froude numbers. For evaluating the optimal centreboard span, the gain coefficient was fixed as 8 degrees/meter.

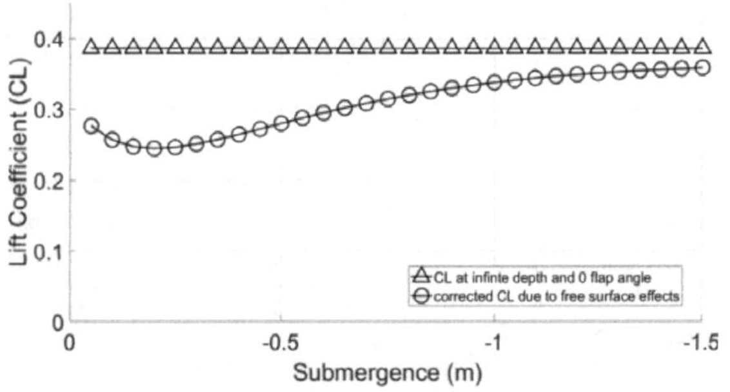

Figure 6.5 Variation of the 3D lift coefficient with the foil submergence for the Wortmann FX60100 [2]

Figure 6.6 shows the stability curves for the different centreboard spans of 0.5, 1.0 and 1.5 m. These plots were recorded at a constant heel angle of 10 degrees and with increasing boat speed for Linnea. The lightship without rigging or foils had a positive righting moment (*Linnea-Lightship*), which became negative with rigging (*Linnea-Rigged*). As the speed increases, the foils will generate lift and the hull goes out of the water, which implies that the righting moment from the hull is lost. The dynamic righting moment plots represent the net righting moment of the combined hull and split-flaps system as the boat transitions from displacement to foil-borne conditions.

It can be seen in figure 6.6 that, although there is an initial decrease in *RM,* it starts to increase as the boat becomes supported by foils. At a centreboard span of 1.5 m, the boat becomes more stable than sailing without foils at a speed of 4 m/s and higher. Decreasing the centreboard span to 1 m and 0.5 m, while keeping the same gain coefficient results in improved righting moment curves as seen in figure 6.6. For a span of 1 m, the boat has a positive *RM* for a speed of 4 m/s and above, also the maximum negative *RM* is reduced. With a centreboard span of 0.5 m, the minimum *RM* is not significantly different from the displacement boat, the foil-supported boat is almost always more stable than the displacement boat.

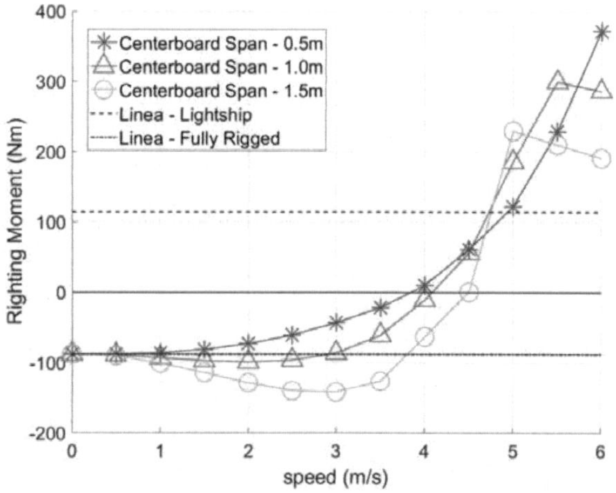

Figure 6.6 *RM* of Linnea for different centreboard spans of 0.5, 1.0 and 1.5 m with a gain coefficient of 8 and a heel of 10 degrees

A reduction of *RM* at higher speeds can be observed from the results. This is attributed to the reduction of the foil submergence (i.e., the foil comes close to the surface), due to the increase in the lift generated and ride height. Figure 6.7 shows the reduction of foil submergence with speed for the different centreboard spans.

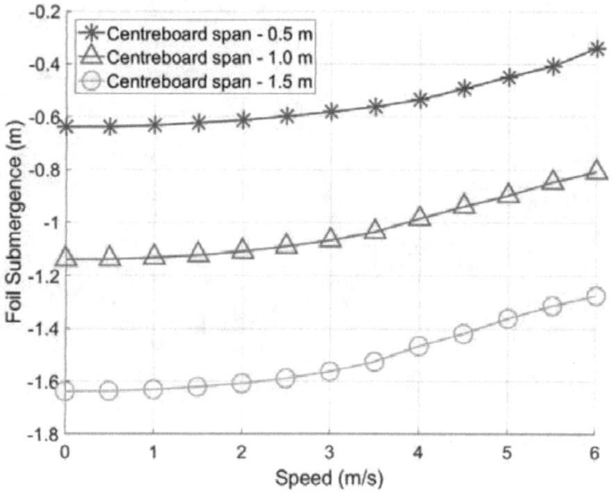

Figure 6.7 Foil submergence vs speed for different centreboard spans of 0.5, 1.0 and 1.5 m with a gain coefficient of 8

The lower centreboard span infers that the foil would be close to the free surface. As per observations from *Faltinsen* [2], it is not advised for the foil to be closer than one chord length from the surface. With a chord of 0.25 m (minimum foil submergence of 0.25 m) and a hull draft of 0.171 m, a centreboard span of 0.5 m would mean the hull clearance would be low. Thus, even though it provided high stability, the 0.5 m centreboard span was disregarded in favour of the 1 m centreboard span. In the latter, there would be more clearance for the foiling hull from waves.

6.2(b) Effects of changing the gain coefficient

As seen in equation (11), a high gain coefficient implies a large response to a given change in height, which would also result in a larger difference in flap angles for a given heel angle. To investigate this effect, the gain coefficient was increased by keeping the centreboard span constant at 1 m. Figure 6.8 shows the *RM* plots for gain coefficients of 8, 12 and 16, respectively. Compared with the gain coefficient of 8, a gain coefficient of 12 shows much-improved stability plots. Over a large part of the speed range, the foil-supported boat has a higher *RM* than the displacement boat when a gain coefficient of 12 is used. With a gain coefficient of 16, the *RM* is higher than the displacement boat for the entire range of speeds.

The take-off speeds for the gain coefficients 8, 12 and 16 are 3.6, 4.2 and 4.1 m/s respectively. Systems with high gains will reduce the flap angle quickly relative to the height thereby reducing overall lift rapidly and consequently increasing the take-off speed. However, the GC of 16 has a lower take-off speed than the GC of 12. This could be due to the reduction of foil induced and profile resistance in the high gain systems due to lower flap angles. Lower resistance results in increased boat speed and lift force from the foil thereby reducing the take-off speed.

The reduction of *RM* at higher speeds here is attributed to the negative limit on the flap angle. The negative limit of -7 degrees corresponds to a slightly negative lift coefficient. At high ride heights (speed) or high heels, one or both of the *TE* flaps can reach the minimum limit. Figure 6.9 shows the flap angles at each speed from the *DVPP* for the different gain coefficients at a heel of 10 degrees to leeward. With a high gain coefficient, the net flap angle is lower than with a lower gain coefficient at the same ride height. Heeling with an already low net flap angle reduces the windward side's flap angle even further and it reaches the negative limit of -7 degrees rapidly. Heeling further will increase the righting moment at a lower rate as the windward flap is locked at -7 degrees and only the leeward flap increases. If the speed increases in such a case, the net ride height increases, thereby decreasing the net flap angle. As the windward side is already set at -7 degrees, only the leeward side would be decreased, thereby decreasing the difference between the two flap angles. This explains the reduction of *RM* at high speeds and high gain coefficients.

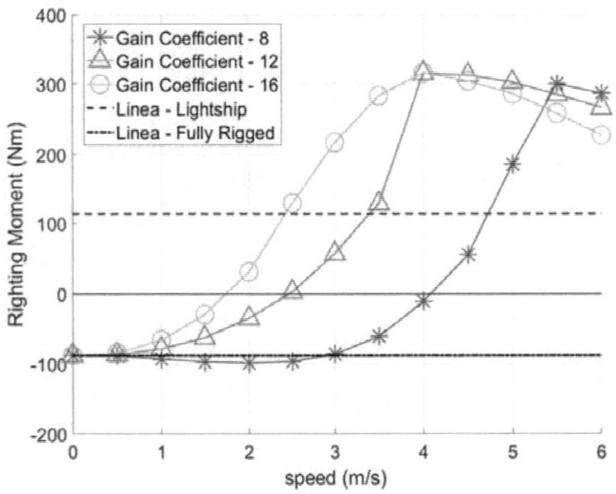

Figure 6.8 RM of Linnea for a draft of 1.0 m and gain coefficients of 8, 12, and 16 and a heel of 10 degrees

The above issue could be resolved by removing the negative limit on the flap angle. This solution would result in two complications. One would be that the windward side will be producing negative lift, which is a loss of efficiency as the lift-induced drag would remain despite the negative lift coefficient. Secondly, a too high righting moment would put an immense load on the rigging. Making the rigging stronger could require increasing the weight of the rigs, which would further decrease the stability. Hence, the negative limit of -7 degrees was retained in the design.

The issue with having higher gain coefficients apart from the increase in induced drag would be the impact on the ride height. The control system uses the same gain coefficient for the heel and height responses as seen in equations 10 and 11. A higher value corresponds to a large difference in flaps angles and higher *RM*, but it would also be highly responsive to the changes in the net ride height. It implies that the hull would have lower ride heights while foiling. From the *DVPP*, the ride height at a *TWS* of 2.75 m/s and *TWA* of 90 degrees is around 0.382 m with a gain coefficient of 8. When the coefficient is doubled to 16, the ride height drops to around 0.222 m (hull clearance of 0.04 m). Such low hull clearance would be practically impossible to sail at, and extremely sensitive to waves. Hence the gain coefficient of 8 should be used to maintain a reasonable ride height response. However, the *RM* would be almost the same irrespective of the gain coefficient at higher speeds for greater gain coefficients as seen in figure 6.8.

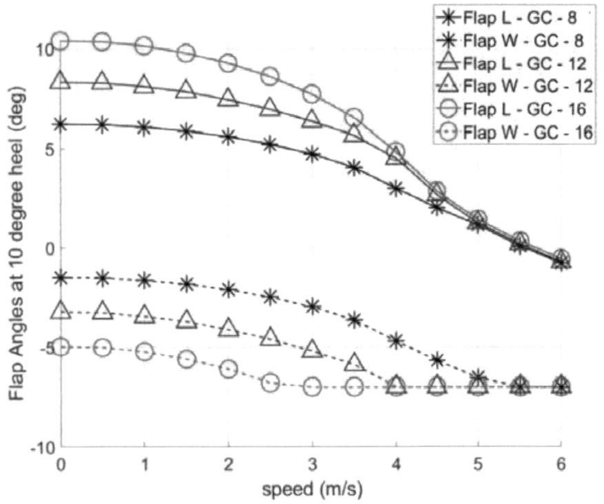

Figure 6.9 Flap angles at each speed with the different gain coefficients of 8, 12 and 16 at a heel angle of 10 degrees

6.3. Decoupled System

The response limitation is only applicable to the mechanical two-wand control system proposed here. However, if a system could separate the variables of heel and height and apply individual gain coefficients to them, a more efficient system could be developed. It could have a gain coefficient of 16 on the heel response to achieve a stability plot better than non-foiling at all speeds (figure 6.8). While the height response could still retain the lower gain coefficient of 8, to have an optimal ride height and hull clearance. An example is given by equations 12 and 13, where *GC1* and *GC2* are gain coefficients for height and heel, respectively. In such a case, the net flap angle (thereby C_L) would remain the same as without having a heel angle. The height response could also be set non-linear with the decoupled system. The mechanical implementation of such a system has been made, but it is outside of the scope of this paper.

$$f_p = GC1 * h + f_o + GC2 * \varphi \qquad (12)$$

$$f_s = GC1 * h + f_s - GC2 * \varphi \qquad (13)$$

6.4. Performance

Figures 6.10 and 6.11 show the polar plots for the displacement boat and foil-supported boat with split flaps, respectively. It can be observed that the boat can foil at wind speeds as low as 2.5 m/s. Table 6.1 contains the maximum velocity made good (*VMG*) in upwind and downwind sailing for the displacement and foil supported boats. Represented in the table is also the *VMG* for a foiling boat without the split-flap system (single trailing edge flaps).

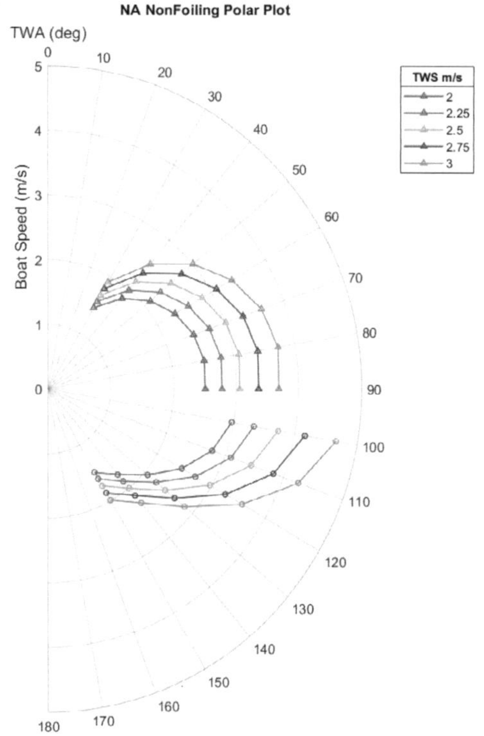

Figure 6.10 Polar plot for the displacement boat

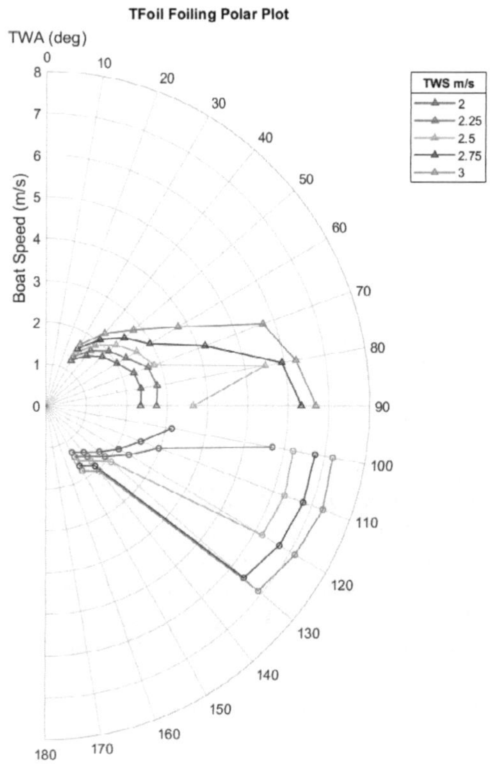

Figure 6.11 Polar plot for the final foiling boat

Table 6.1 Maximum VMG upwind and downwind for the displacement and foil supported boats.

TWS	VMG UD	Heel UD	VMG UF	Heel UF	VMG USF	Heel USF
2.0	1.40	0	1.24	0	1.20	0
2.25	1.53	0	1.39	0	1.32	10
2.5	1.66	0	1.56	0	1.45	10
2.75	1.79	0	1.78	0	1.62	10
3.0	1.93	0	1.98	0	1.96	10
TWS	**VMG DD**	**Heel DD**	**VMG DF**	**Heel DF**	**VMG DSF**	**Heel DSF**
2.0	-1.33	0	-1.15	0	-1.12	0
2.25	-1.44	0	-1.91	0	-1.23	10
2.5	-1.56	0	-3.00	0	-3.10	10
2.75	-1.69	0	-4.00	0	-4.12	10
3.0	-1.82	0	-4.37	0	-4.43	10
TWS	**VMG Avg**	**-**	**VMG Avg**	**-**	**VMG DSF**	**-**
2.0	1.37	-	1.20	-	1.16	-
2.25	1.49	-	1.65	-	1.27	-
2.5	1.61	-	2.28	-	2.28	-
2.75	1.74	-	2.89	-	2.87	-
3.0	1.88	-	3.18	-	3.19	-

UD – Upwind Displacement
UF – Upwind Foiling with single flaps
USF – Upwind Foiling with split flaps
DD – Downwind Displacement
DF – Downwind Foiling with single flaps
DSF – Downwind Foiling with split flaps
Avg – Average of upwind and downwind VMG

Upwind, the split flap boat is slower than the single flap boat in the speed range tested. Downwind, the split flap is faster above 2.5 m/s true wind. The average speed is about the same for both cases above this wind speed. So, although there is an increase in induced drag due to the flap angle difference in the case of the split-flap system, there is no disadvantage in speed, except in very light wind. The maximum transverse distance the crew can move out to produce righting moment is when they stand on the racks. If the corresponding righting moment is not large enough, they have to depower the sails. With the split-flaps, due to the additional righting moment from the foils, the sailors do not need to reduce the sail forces as much as they would for the case without split-flaps. Thus, the sail drive force is higher, and this compensates for the increased resistance.

The comparisons of Table 6.1 are between a single flap boat at zero degrees of heel and a split flap boat at 10 degrees of heel. It would be interesting to compare the boats at windward heel, which is normal in the Moth class. If the split flap system is provided with an offset control, the stability increase could be achieved also at windward heel, which could result in a higher speed. Such a system was proposed by *Boman* [12], but it has not yet been tested.

It can also be inferred from Table 6.1 that the foiling alternatives have a larger average VMG over the course for wind speeds from 2.5 m/s upwards (from 2.25 m/s for the single flap boat).

6.5. Effects of the Split Flaps on the Induced Drag – a CFD Study

As explained above, the split flaps generate more induced resistance. Looking at this in more detail, while heeling leeward, the two flaps are set at different angles as shown in figure 4.3 with a higher flap angle on the leeward side. This implies that the C_L will be higher on the leeward side of the foil than the windward side. At such conditions, due to the space between the two flaps in the middle, additional vortices (referred here as flap vortices) will develop, leading to increased drag. Due to the higher loading on the leeward side, the leeward side vortices will be much stronger than the windward side. The flow in the space between the flaps is complicated as one strong flap vortex (leeward) and a weaker flap vortex (windward) tend to form in proximity. Also, the geometrical presence of the windward flap could hinder the development of the leeward flap vortex and vice versa. Empirical methods are generally valid for an elliptical lift distribution over the span of the wing. With different C_L on each side of the wing, the lift distribution will not be symmetric and would be far from elliptical. Hence, the lift force can deviate from that predicted by empirical methods and a more exact numerical investigation is required.

The above effects were studied with numerical *RANS* simulations. Friendship Systems *CAESES* [13] was used to make the *CAD* model of the foils and *Star CCM +* [14] was used to carry out the CFD simulations. The *k-ω* turbulence model was used with the foils in a double body setup. The simulations were carried out at a constant speed of 5 m/s with a difference in flap angles of 0, 4, 8, 12, and 16 degrees (table 6.2). The results from the numerical simulations are compared with empirical estimations in table 6.2 and Figure 6.12. As the shear drag is not of concern in this study, only the drag due to the pressure effects are considered. The total pressure drag in the empirical method is estimated as the sum of the profile pressure drag and lift induced drag. The black dashed curve represents the CFD results for symmetrically loaded flaps, i.e., where the leeward and windward flaps are displaced equally much, but in different directions. The red curve represents the lower limit in the model, assuming a symmetry plane between the flaps. The yellow curve includes a correction to the minimum curve, such that the empirical formula fits the CFD results as well as possible. This is the method used in the VPP to make the polar plots in figures 6.10 and 6.11. The correction factor (C_{AR}) on the aspect ratio to account for the flow between the flaps (yellow curve in figure 6.12) is given by the equation 14 and 15. Note that this correction is derived from a symmetric displacement of the flaps. To cover also asymmetric cases more CFD results are required, and this is a matter for further study.

$$C_{AR} = 1.9338 * |(f_p - f_s)| - 2.0332 \qquad (14)$$

$$AR_C = AR * (1 - \frac{C_{AR}}{100}) \qquad (15)$$

Table 6.2 Comparison of the empirical and numerical pressure drag for flap differentials of 0, 4, 8, 12, and 16 degrees.

Flap Difference (deg)	f_P	f_S	Empirical Pressure Drag (N)	CFD Pressure Drag(N)
0	0	0	63	64
4	-2	2	70	70
8	-4	4	90	90
12	-6	6	121	122
16	-8	8	166	169

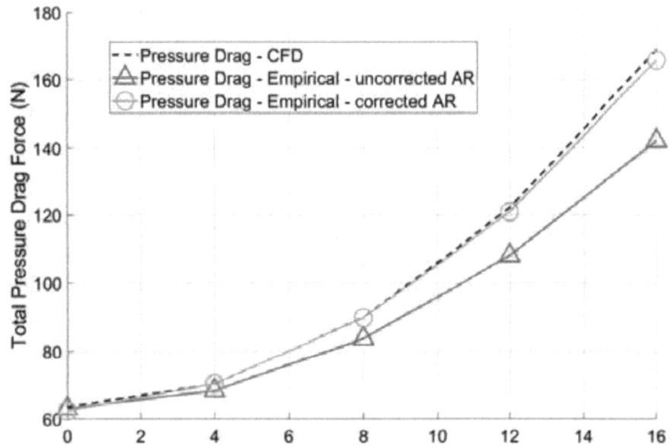

Figure 6.12 Foil drag force vs flap angle difference for empirical and CFD estimations

Figure 6.13 shows streamlines in the space between the two flaps with a flap difference of 16 degrees. From the figure, it can be seen that there is an additional vortex developed at the space between the flaps, which contributes to the discrepancy between the CFD and empirical methods.

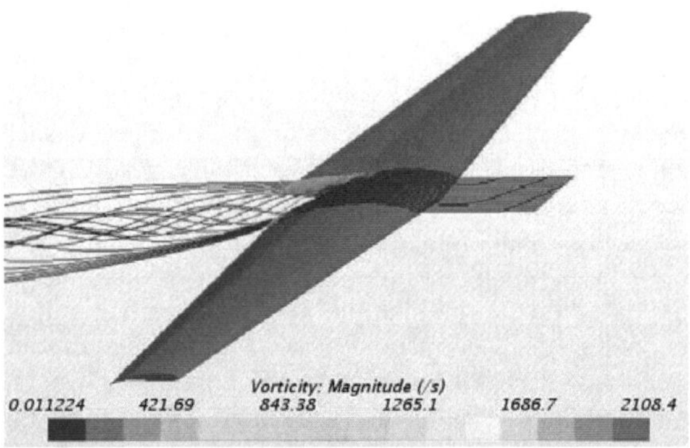

Figure 6.13 streamlines the interface between the split flaps

To study the interaction effects in detail, two additional simulations were carried out with the starboard flap set at 0 degrees and the port flaps set at -3 degrees and -10.7 degrees. The reasoning behind this being that the flap angle of -3 degrees and -10.7 degrees has a C_L of 0.2866 and -0.2866 respectively. Theoretically, the induced drag is proportional to C_L^2 and hence the empirical $C_{D,i}$ should be the same for both the cases. The results from the numerical and empirical simulations are given in table 9.2. The difference in the pressure distribution between the starboard and port sides due to the flap angle differential can be seen in figure 6.14.

Table 6.3 Comparison of the empirical and numerical pressure drag for a starboard flap of 0 degrees and a port flap at -3 and -10.7 degrees.

Case	1	2
Starboard Flap (deg)	0	0
Port Flap (deg)	-3	-10.7
Empirical Pressure Drag (N)	44	54
Numerical Pressure Drag (N)	46	61
Empirical Lift (N)	2444	657
Numerical Lift (N)	2404	565

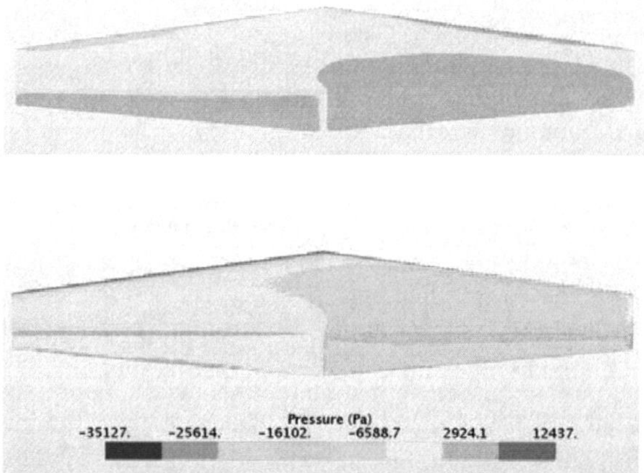

Figure 6.14 Pressure contours on the pressure side (top figure) and suction side (bottom figure) with the starboard flaps set at +8 degrees and port flaps set at -8 degrees

It is seen that there is a strong interference effect between the two sides when the port and starboard flaps are set at 0 and -10.7 degrees, respectively. Figures 6.15 and 6.16 show the velocity vectors on the transverse plane at the trailing edge of the foil viewed from behind for case 1 and case 2, respectively. The plots are zoomed in closer to the space between the two flaps.

In case 1, there was no strong vortex formation in the space between the flaps (figure 6.15). There is a lack of space for vortex formation as each flap hinders the overflow from the other flap to some extent. Also, both the flaps are loaded positively with pressure on the bottom and suction on the top. Thus, on the starboard side, the tip vortex tends to roll clockwise in the space between the flaps. However, the port tip vortex tends to roll in the counterclockwise direction. These two vortices seem to oppose each other and could hinder the formation of a stronger vortex.

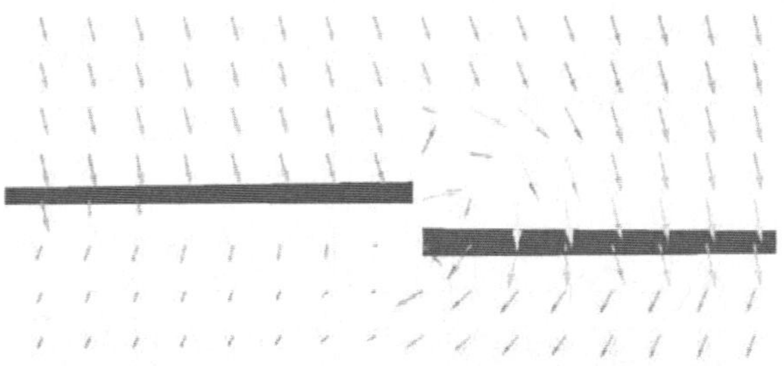

Figure 6.15 Velocity vectors at the Trailing Edge of the foil with starboard flaps at 0 degrees and port flaps at -3 deg (case 1)

In case 2, the vortex in the space between the flaps is much stronger than case 1 (figure 6.16). Despite the presence of a larger space to aid vortex formation, there is also a negative loading on the port side i.e., high pressure on the top and low pressure on the bottom due to the negative lift coefficient. Thus, tip vortices from the port side will try to roll up clockwise in the space between the flaps. Since the starboard vortex also rolls clockwise, these two vortices seem to complement each other, contributing to a stronger vortex in the space between the flaps. Though in both cases the starboard flap is at the same angle, the pressure below the flap has been significantly reduced in case 2 due to the flap vortex. This contributes to a reduction in the lift coefficient as well. The correlation error between the numerical and empirical estimations for case 2 is significantly larger than case 1 and all other simulations represented in table 9.1. The errors in the pressure drag and lift estimations are 12.9 % and -14.0 % respectively.

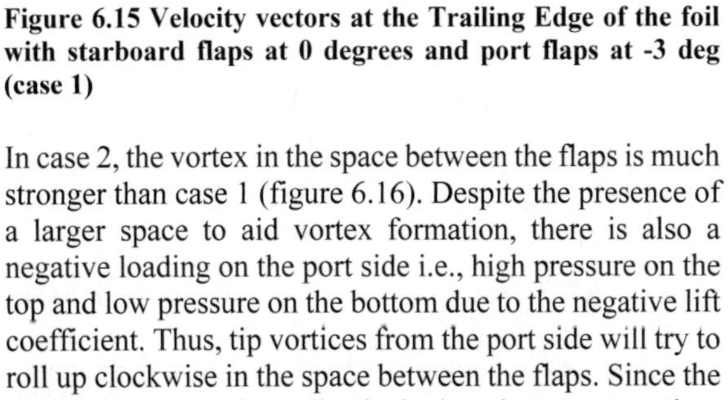

Figure 6.16 Velocity vectors at the Trailing Edge of the foil with starboard flaps at 0 degrees and port flaps at -10.7 deg (case 2)

The case with +8 starboard and -8 port flaps has an error of 1.7 % on the pressure drag. Though the flap angle difference here is larger than case 2, the correlation is better. This could be attributed to the fact that a port flap of -8 degrees corresponds to an approximately zero (slightly negative) lift coefficient. The starboard flap at 8 degrees has a positive lift coefficient and a vortex that rolls up clockwise in the space between the flaps. There would be no significant vortex formed from the port side and thus no complimentary effect to the clockwise port flap vortex. Therefore, it could be inferred that it is not only the difference between the flap angles but also the loading patterns of the two flaps, that is critical in increasing the pressure drag. As observed from the studies, a negative loading on one of the flaps increases the vortex in the space between the flaps and a non-negative loading on both the flaps tends to decrease the vortex.

7. CONCLUSIONS

Dinghies supported by T-foils are inherently unstable, but by splitting a trailing edge flap into starboard and port flaps, the transverse stability is greatly improved. The split-flap control system adjusts the flap angle difference in proportion to the heel angle, thereby increasing the transverse stability of the boat.

When the boat heels to leeward, there is an additional side force to compensate for the horizontal component of the lift force from the foil. The centreboard span affects the lever of the heeling moment from this additional side force. Therefore, it should thus be as short as possible without bringing the foils too close to the free surface.

The extreme skiff (*Linnea*) was investigated as a demonstration case for the split flaps in this study. It has high negative stability due to the heavy rigging. Evaluations show that, with a proper design, the split-flaps can make the hull even more stable than the hull without foils. A two-wand mechanical control system is used to measure and adjust flap angles based on height and heel by responding to the height at the maximum beam on both sides. The flap control system is modelled in the computations using response functions to transfer the changes in ride height to the flaps.

The gain coefficient, which is the slope of the height-flap response curve is a critical parameter for transverse stability. A higher gain coefficient would give a larger difference between port and starboard flaps while heeling, thereby larger *RM*. Since it is coupled with the height also, a large gain coefficient results in a much lower ride height. An optimal value of 8 degrees/meter was set as the gain coefficient for this coupled system. If the height and heel responses could be decoupled, then a high gain could be placed on the heel, resulting in better stability than the non-foiling hull at all speeds. At the same time, the gain on the height could be set lower to have an optimal ride height. Though the mechanical design of such a system is complicated, the problem has been solved with a system that can generate individual responses for height and heel, but the description of this system is beyond the scope of this paper.

Since there is an increase in induced resistance with the split flap system, a systematic CFD investigation of the flow around the main foil was carried out. It showed that an empirical estimation, assuming a symmetry plane between the two flaps underestimates the induced drag, so a correction was derived from the CFD results. There is a very complex flow in the gap between the flaps and it depends on the flap loading. A reduced additional drag is observed when both flaps have a non-negative lift coefficient. A negative lift coefficient on one of the flaps increases the strength of the additional vortex and thereby the additional drag.

Increased stability means less depowering of the sails and thus more thrust. Therefore, the increase in induced resistance is compensated fully for true wind speeds above 2.5 m/s, where the split foil boat is as fast as the boat with a single flap. It should be mentioned, however, that the comparison depends on the modelling of the induced resistance. More CFD results are required to develop a model that covers all possible flap angle configurations.

Acknowledgments

The authors would like to acknowledge the help, expertise, and support provided by,

1. Chalmers Formula Sailing team of 2018 for designing and building *Linnea.*
2. Studies carried out by the Chalmers Formula Sailing team of 2019-20.
3. Chalmers University of Technology for providing software licenses and resources.

The CFD computations were performed on resources at the Chalmers Centre for Computational Science and Engineering (C3SE), provided by the Swedish National Infrastructure for Computing (SNIC).

References

1. Monsonnec, F. and Waller, N. (2015) *The Foil Alphabet*. Available at: http://proafile.com/multihull-boats/article/the-foil-alphabet.
2. M. Faltinsen, O. (2005) 'Steady free-surface effects on a foil', in *Hydrodynamics of High-Speed Marine Vehicles*. Cambridge University Press, pp. 199–202.
3. Acerbi, T. *et al.* (2017) *Chalmers Formula Sailing.* Chalmers University of Technology.
4. 1001Vela Cup. Class Rules "R3" 2020-2022. url: http://www.1001velacup.eu/regolamento/regolamento-di-classe.html
5. Andersson, A. *et al.* (2018) 'Design of A Foiling Optimist', *Journal of Sailboat Technology*, pp. 1–24.
6. 'International Measurement System' (2019) in. Offshore Racing Congress, Ltd. Available at: www.orc.org
7. Drela, M. and Youngren, H. (2013) 'xfoil'. Available at: http://web.mit.edu/drela/Public/web/xfoil/
8. Horel, B., & Durand, M. (2019). Application of System-ased Modelling and Simplified-FSI to a Foiling Open 60 Monohull. In *the 23rd Chesapeake Sailing Yacht Symposium* (pp. 90–107). Annapolis, Maryland.
9. Eggert, F., Henrichs, J., Hansen, H., & Hochkirch, K. (2020). Flight Dynamics and Stability Assessment for an International Moth. *INNOVSAIL 2020.*
10. Beaver, B. and Zseleczky, J. (2009) 'Full Scale Measurements on a Hydrofoil International Moth', *Chesapeake Sailing Yacht Symposium.* Annapolis, Maryland.
11. Hoerner, S. F. (1957) *Fluid Dynamic Drag.* Albuquerque, USA.
12. Boman, A. (2020) 'Design of a flight-control system for foiling dinghies with a T-wing and individual flaps'. BSc Thesis, Chalmers University of Technology.
13. *CAESES* (2019). url: https://www.caeses.com/
14. *Star CCM* + (2018). url:

    https://mdx.plm.automation.siemens.com/starccm-plus.

ANALYTICAL STUDY OF A FOILING CATAMARAN IN WAVES

Ignacio Castañeda Sabadell[1], icastanedasabadell@sailgp.com, **Mike Drummond**[2], mdrummond@sailgp.com
Ellie Copeland[3], ellie.copeland@outlook.com, **Tomer Simhony**[4], tsimhony@sailgp.com

Abstract. A study was performed to understand why the foiling catamarans used in the SailGP circuit are harder to sail under certain headings in waves. The study was done using a Lifting Line method combined with regular wave theory. The Lifting Line method used discretizes the span of the geometries (foil and rudders) using a set of horseshoe vortices placed on the quarter chord. This particular code was coupled with an orbital velocities calculation of regular waves. The combination of the boat speed with the orbital velocities allowed the calculation of the local effect on the force distribution due to the change in the local flow velocity and angle of attack. Those local velocities are a function of boat speed, geometry orientation, depth and position on the wave. This code can be used to analyse any foil configuration from foiling Moth to *ultime* trimarans, however, results in this study are presented for the SailGP F50 catamaran. The first part of the study was carried out with one appendage with different wave amplitudes and periods to understand the effect in the force distribution of the appendage for the different wave conditions. Then the same methodology was applied to a set of three appendages (foil and two rudders with elevators) to capture the effect on the boat appendage position relative to different parts of the waves. A set of three different headings were used (upwind, reaching and downwind) to analyse the effects of the geometry facing different wave directions. It showed that for different headings, the phase relationship between the vertical force and the free surface wave elevation is a function of the boat heading to the waves.

1. INTRODUCTION

SailGP is a sailing based sports league that is conducted at venues all over the world. Its F50 catamarans are high-tech boats that have broken the 50 knot mark and are powered by nature and the SailGP athletes. These high speeds are achieved, in part, due to the F50's highly efficient wing sail, the hydrodynamic drag reduction possible through foiling, and by increasing the boat's righting moment through differential in rudder rakes.

Figure 1: The F50 hydrofoiling catamarans sailing in Sydney harbour 2019

The F50 athletes must work together to accurately position the various control surfaces on the boat to optimally contend with the environmental factors of wind, current and sea state as well as the tactical factors of racing other boats. Getting this wrong may lead to falling off the foils and losing significant time on your opposition or crashing the boat with the potential for damage and injury.

On the F50 catamarans there is often a trade-off the athletes have at their disposal between speed and stability, where the more talented and precise athletes have an advantage. One particular area that proves to be quite challenging is sailing in a large sea state. Although foiling upwind seems to be only marginally affected, sailing downwind has been shown to be especially difficult.

This study explores the effect on forces generated by a foiling catamaran sailing in waves with respect to wave orbital motion, wave amplitude, and wave length as it applies to the F50 boat. A modified lifting-line code is developed to quantify vertical and side forces for any foiling configuration which enables the code to be used to analyse modern foiling sailing boats such as Moths, A classes and Nacra 17s or open water "*Ultime*" trimarans like Gitana or Sodebo 3. Results are shown for the SailGP F50 appendages alone and in unison. Similar studies had been carried out with other types of vessels or devices such as a conventional hydrofoil ship in [2] or an oscillating 2D section in [1].

2. METHODOLOGY

2.1. Wave modelling

The motion of a wave induces local flow speed variations, called orbital velocities [3]. In deep waters, the velocity of a water particle at a wave crest points forward in the

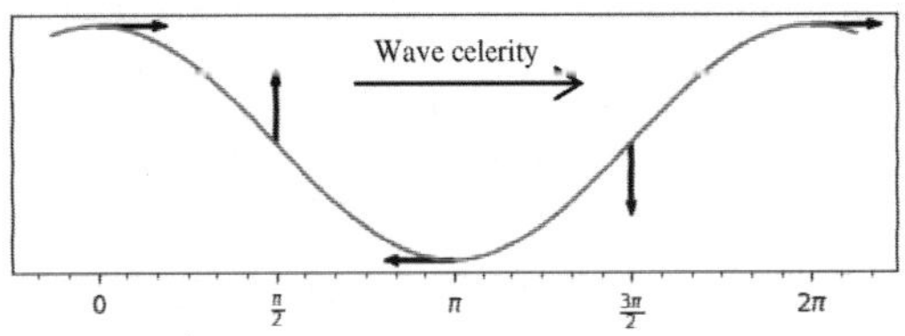

Figure 2: Wave orbital velocity (shown in black) imposed over a sinusoid wave approximation

---

[1] Designer, SailGP
[2] Design Director, SailGP
[3] Former Design Engineer in SailGP
[4] Design Engineer for SailGP

direction of the wave propagation, after which it processes in a an approximately circular path (clockwise in the example of Figure 2) as the wave progresses. Orbital velocities impact the flow velocity and the angle of attack seen by the F50 foil in wavy conditions.

2.2. Lifting Line

The lifting line theory was independently developed by Prandtl [5] and Lanchester [4] in the early 1900s. It allows calculation of the lift distribution over large aspect ratio wings at a small angle of attack. The formulation used in this particular paper is presented by Tom Speer in [7] but adapted to include the effect on the local flow velocity and angle of the orbital velocities of the waves.

In this method, each lifting surface is discretized by a set of horseshoe vortices located in the quarter chord along the span. Each horseshoe vortex is formed by three vortices with the same strength: a bound vortex and two trail vortices rotating in opposite directions and aligned with the flow direction.

Each trailing vortex shed into the wake affects the local flow velocity by changing the local angle of attack and shifting the local lift vector. These induced velocities can be analysed at the Trefftz plane which is located at infinity downstream of the lifting surface and is perpendicular to the flow direction.

The bound vortices have no effect in the Trefftz plane (as it is located at infinity) and only the trailing vortices are considered (as the wake is shed to infinity). Each pair of trailing vortices intersects the Trefftz plane in two locations forming a segment. A control point is located in each segment where the normal induced velocities generated by each trailing vortex of the wake are solved allowing determination of the vortex strengths and the lift distribution.

To model the free surface, a linearized boundary condition is used considering infinite Froude number (high speed) assuming a constant pressure in the surface. This is achieved by using a mirror of the vortex system with the same rotation direction.

The wave orbital velocities affect the local velocity of the flow by changing its angle and norm. This is modelled within the lifting line formulation as a corrective ratio K_{ext} based on the orbital velocity and the free stream velocity, and a change in local angle of attack, W_{ext}, modelled as a local twist, shown in Equation 1.

$$K_{ext} = 1 - \frac{V_{orbital}}{V_\infty} \quad (1)$$

The final equation analysis formulation derived in [7] will be transformed to Equation 2:

$$\Gamma_j = \left(I + \frac{a_i \cdot c_i \cdot K_{ext\_i}}{2} \cdot M\right)^{-1} \cdot \left[\frac{a_i \cdot (\alpha - \propto_{0i} + \tau_i + W_{ext\_i}) \cdot c_i}{K_{ext\_i}}\right] \quad (2)$$

Where Γ_j is the vortex strength, I the identity matrix, a_i is the section lift slope, c_i is the local chord, α is the local angle of attack, $\propto_{0i}$ is the angle of attack for $C_l = 0$ of the local section, τ_i is the local section twist and M is the analysis problem matrix (please refer to Lifting Line Method section in [3]).

2.3. Orbital effect on hydrodynamic forces

Both the rudders and foils of the F50 generate a vertical and side force in the boat reference frame. Wave orbital velocities affect both of these forces through an angle of attack effect and a flow speed effect, and affect the boat differently depending on whether the boat velocity is travelling towards the wave (upwind) or away from the wave (downwind).

The angle of attack effect occurs when the orbital velocity has a component perpendicular to the free stream flow direction and is greatest at the ¼ and ¾ points of a wave as seen in Figures 3 and 4.

The flow speed effect occurs when the orbital velocity has a component in the free stream direction, increasing or decreasing the resultant flow. Although lift is dependent on the square of the velocity, this effect tends to be smaller than the angle of attack effect for the wave celerity to boat speed ratios of foiling boats.

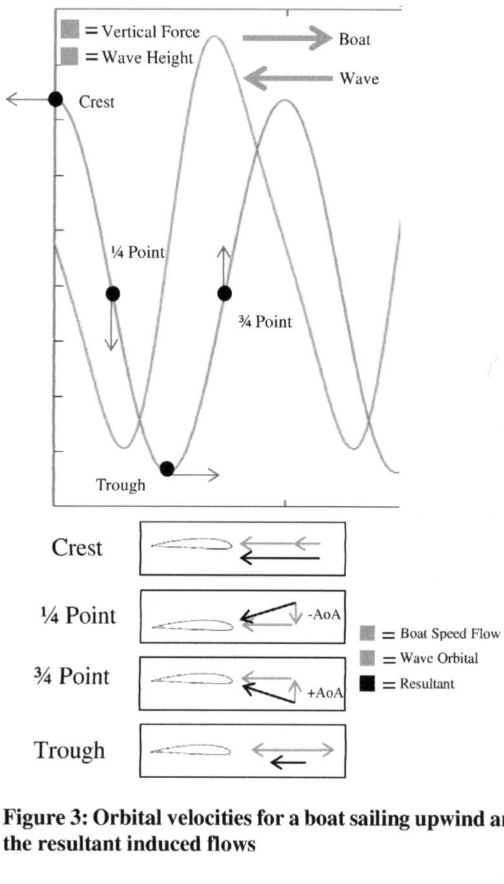

Figure 3: Orbital velocities for a boat sailing upwind and the resultant induced flows

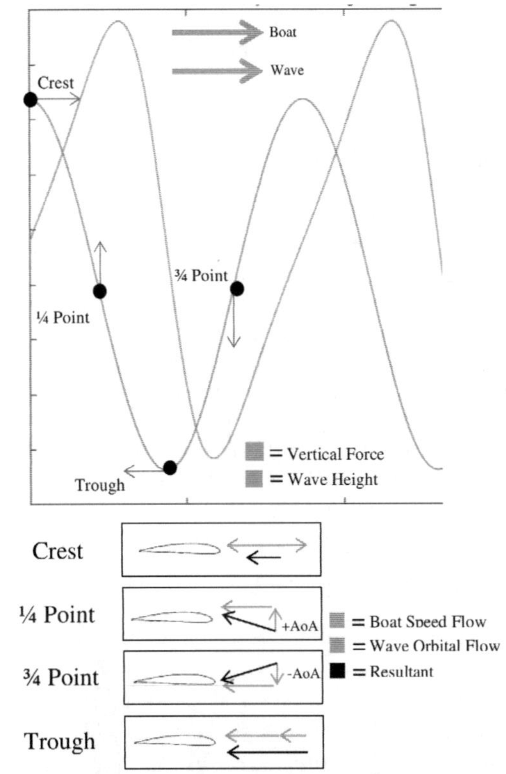

Figure 4: Orbital velocities for a boat sailing downwind and the resultant induced flows

2.3(a) Orbital Effect upon vertical lift force

The vertical lift force is affected by both the angle of attack effect and the flow velocity effect.

Figure 3 shows the vertical force of a horizontal lifting surface as it encounters a sinusoidal wave approaching in the opposite direction. The figure shows that there is a small phase shift between the vertical lift force and the wave height, however, for a boat sailing upwind the amount of lift produced decreases as the foil moves towards a trough, and increases as it moves towards a crest.

Figure 4 is similar, but shows the vertical lift for a boat sailing downwind. Here the phase shifting is greater which means that lift increases when travelling from crest to trough as the foil is already moving closer to the free surface The risk of the elevator coming out of the water is made worse by the fact that the foil produces more lift as it moves towards the trough, due to the wave orbital motion inducing a higher angle of attack on the foil.

2.3(b) Orbital effect upon side force

Side force is impacted by the horizontal orbital velocity at the wave crest and trough. The orbital velocities have an effect upon the angle of attack and the flow speed over the foil. See Figure 5

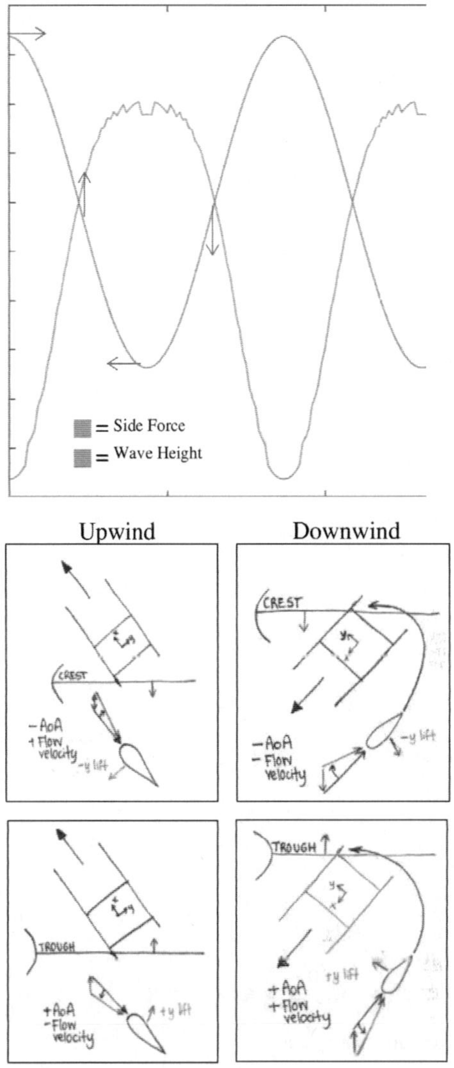

Figure 5: Orbital velocities effect on side force for a boat sailing upwind (left), and downwind (right).

Side force decrease at a crest is greater than the increase at a trough, because the lateral foil area decreases. The angle of attack pattern is the same upwind and downwind: at a wave crest the angle of attack reduces, causing a drop in side force, which will really be manifested as an increase in leeway. At a trough the angle of attack increases, causing the opposite effect. Meanwhile the resulting flow speed effect is opposite upwind and downwind: upwind the flow speed increases at a crest and downwind it decreases at a crest. These changes are overwhelmed by the more significant angle of attack effect.

2.4. Testing

Tests have been run to analyse the magnitude of the wave orbital velocity effects and the combined effects of the forces from each appendage (foil and both rudders) upon the boat overall.

All tests were conducted on proprietary F50 geometry and section shapes. A repeating sinusoidal wave set was simulated for all tests with the following amplitude and wave length parameters and at different incidence angles to the boat (Tables 1-3). The boat speed was fixed at 36 knots for all tests.

Table 1: Changing wave amplitude

Wave amplitude [m]	0, 0.15, 0.3, 0.675
Wave period [s]	2.3
Incidence angle [°]	0, 45, 90, 135, 180
Rudder rake [°]	0

Table 2: Changing wave period

Wave amplitude [m]	0.5
Wave period [s]	2.5, 5, 7.5
Incidence angle [°]	0, 45, 90, 135, 180
Rudder rake [°]	0

Table 3: Overall boat effects

Wave amplitude [m]	0.3
Wave period [s]	3
Incidence angle [°]	45, 90, 150
Board rake (port) [°]	0
Rudder rake (port) [°]	3.22
Rudder rake (stbd) [°]	-0.78

3. SINGLE APPENDAGE RESULTS

3.1. Wave amplitude tests

Tests were run at wave amplitudes of 0.675m, 0.3m, 0.15m and a no-wave case. All tests were at a constant wave period of 2.3s (8.3m wavelength).

3.1(a) Rudder elevator vertical force

The maximum vertical force increases as wave amplitude increases with a constant wave period. The greatest divergence (increase or decrease) in vertical force when compared to a no-wave case occurs at the ¼ wave and ¾ wave points, where the orbital velocities act vertically.

Figure 6 shows that doubling the wave amplitude from 0.15m to 0.3m amplitude causes the maximum vertical force to more than double. The largest wave case (0.675m amplitude) result showed that for a wave crest at 45 degrees True Wind angle (TWA) the maximum vertical force increase at the ¾ wave point is around 745 kg (or 4.7 times) more vertical force than with no wave condition.

Thus, the results suggest that as the waves become steeper (greater amplitude with the same period), it becomes more difficult to foil given the greater fluctuations in vertical force. This is particularly the case downwind as discussed earlier, given that the vertical force starts to increase as the boat moves down from a wave crest to a trough, driving the horizontal lifting surface of the foil to rise even closer to the water surface.

3.1(b) Vertical force - flow speed effect vs. angle of attack effect

A change in the angle of attack of the flow or the flow speed will alter the amount of lift produced by the foil. The angle of attack effect upon vertical force that occurs at the mid wave points has a greater impact upon vertical force than the speed effect that occurs at a peak or a trough.
The difference between the added vertical force due to wave orbital velocities at a wave midpoint, compared to a peak or a trough, increases as the wave size increases. For a 0.675m wave at 45 degrees TWA, the vertical force due to the angle of attack effect at the mid wave point is an additional 4.7 times that at a peak or trough due to the speed effect.

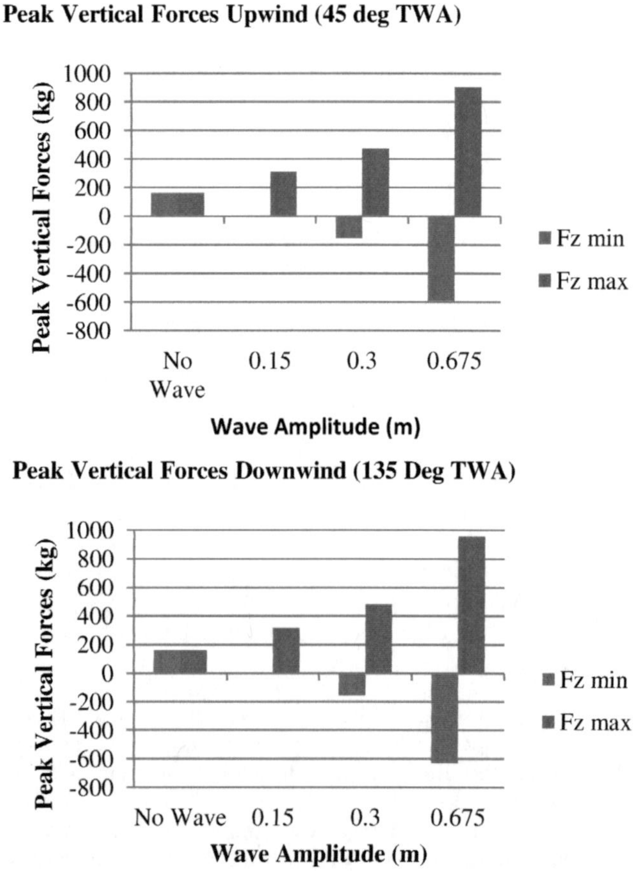

Figure 6: Peak vertical forces upwind and downwind for the changing wave amplitude tests

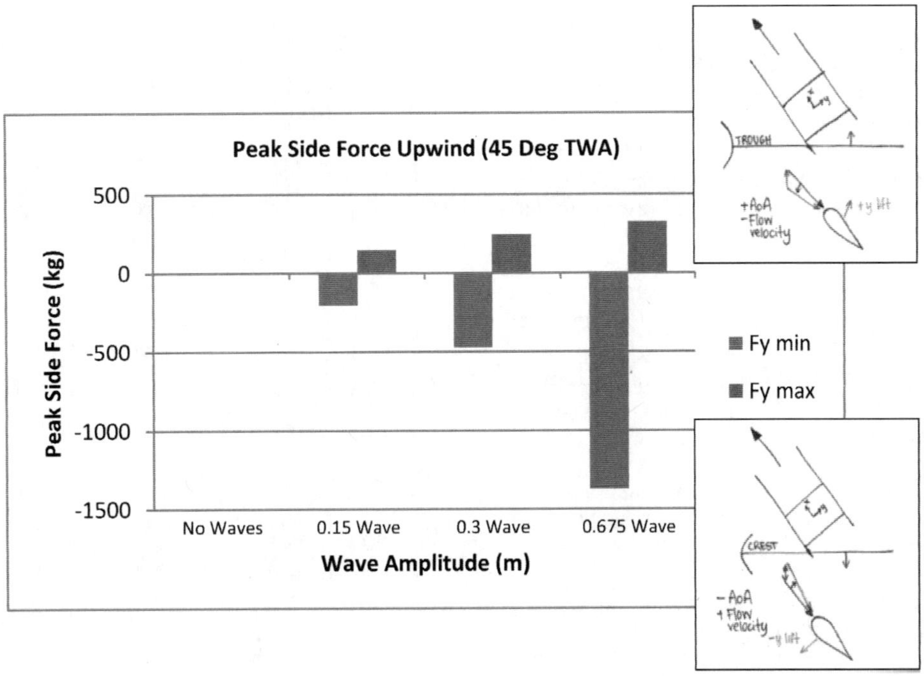

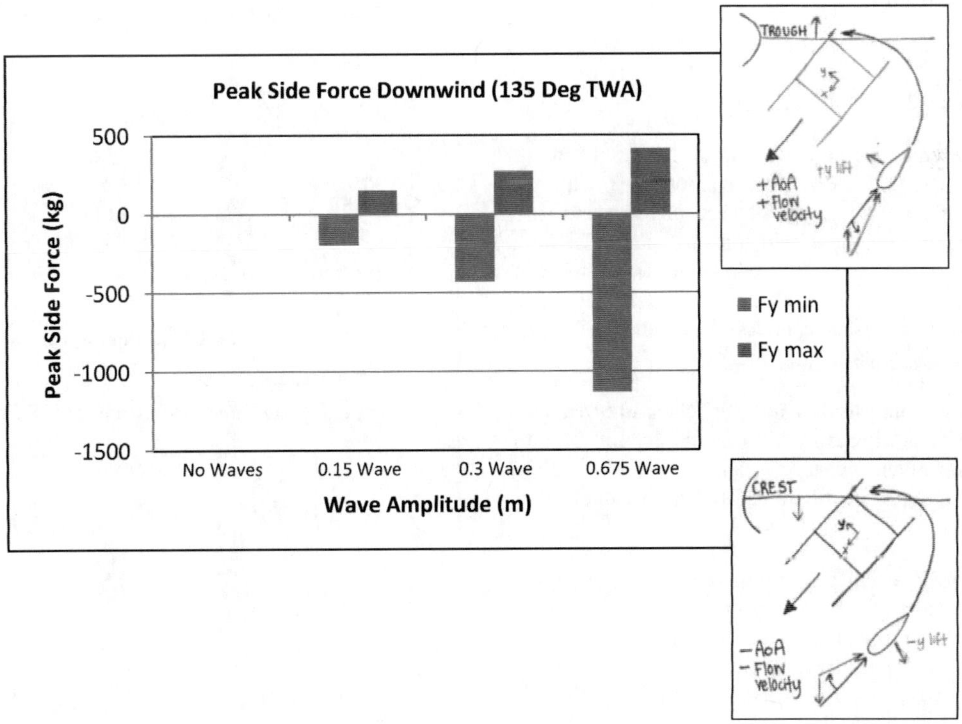

Figure 7: Side force variations increase with increasing wave amplitude. The side force decreases from nominal more than it increases.

3.1(c) Side Force

Side force increases as wave amplitude increases with a constant wave period. This analysis did not take into account leeway, so the side force in the no wave condition is zero. The largest wave case (0.675m amplitude) result showed that for a wave crest at 45 degrees TWA the maximum side force decrease at a wave crest is around 1300 kg.

The fact that the side force decreases at the crest of a wave on both tacks, both upwind and downwind, means that the boat is more likely to want to slip sideways at a wave crest. This is consistent with many sailors' experience of sailing in waves, even for non-foiling boats. This effect increases as wave size increases.

The side force decrease at a crest is greater than the increase at a trough, due to the greater amount of foil surface area in the water at a crest.

The side force decrease at a wave crest is more significant upwind than it is downwind, because the boat is moving against the direction of the wave orbital at that point, increasing the flow speed across the foil. Conversely, the side force increase at a wave trough is more significant downwind than upwind, for the same reason (Figure 7).

3.2. Wave period tests

Tests were run at 2.5s, 5s and 7.5s wave period, corresponding to the below wavelengths. Tests were run with constant wave amplitude of 0.5m.

Results, presented in Figure 8, show that as the wave period decreases with constant wave amplitude (creating a steeper wave), the magnitude of vertical force and side force due to wave orbital velocities will increase, both upwind and downwind. This makes sense as we can imagine that in a steeper wave the water will need to move the same vertical distance in less time, increasing the orbital speed and resulting lift forces.

However, this analysis has not yet taken into account 'unsteady effects'. In reality it takes time for lift force to fully develop given a change in angle of attack or flow speed creating a phase shift in the lift response due to that delay.

Marchaj in [6] presents a simple model where 90% of the lift is developed after a change in angle of attack when the lifting surface has travelled 6 chord lengths. In the case of the F50 at 36 knots, travelling 6 chord lengths takes approximately one tenth of a second.

It is expected that the force magnitude will decrease and the phase will change as the wave period reduces (increasing the frequency of the local flow changes) due to the lag in lift generation after an angle of attack or speed change. This could be investigated in future work.

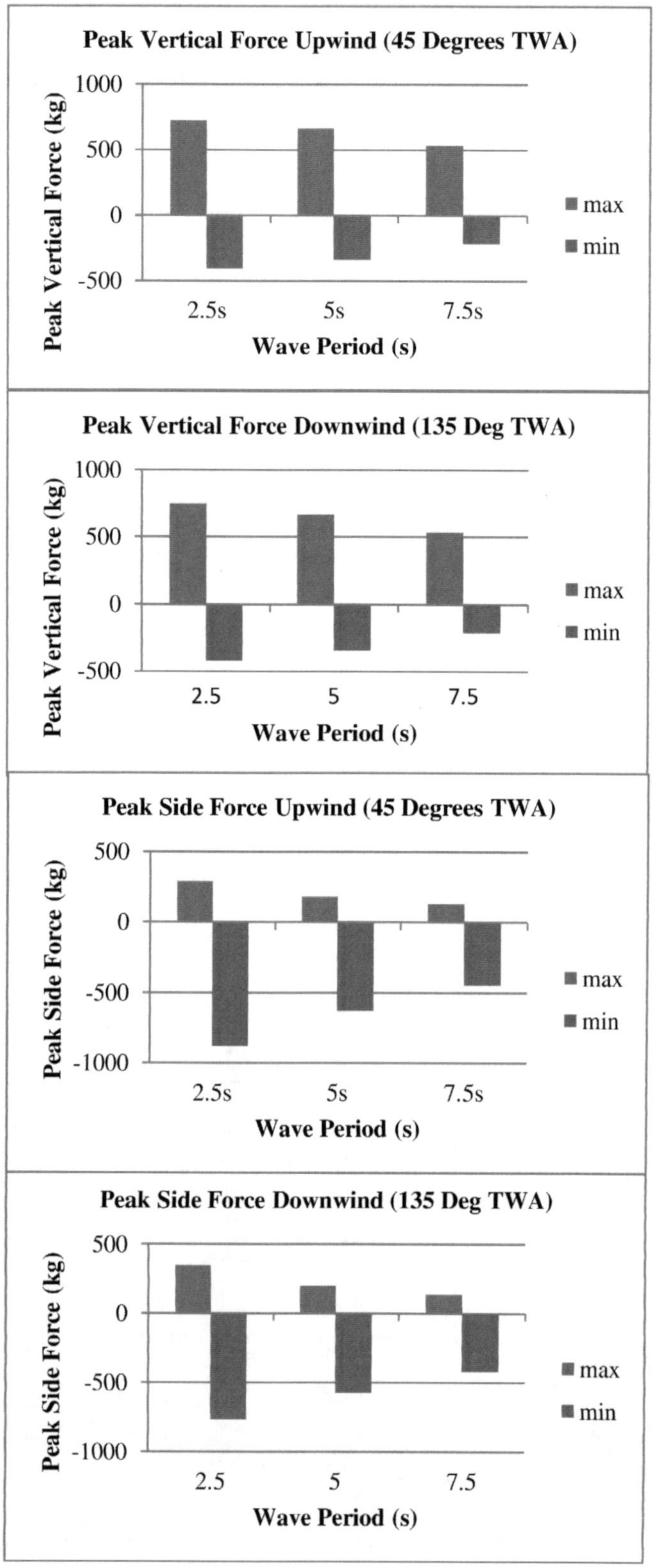

Figure 8: Vertical and side force results, upwind and downwind for different wave periods.

4. OVERALL BOAT FORCES RESULTS

Tests were run to analyse the combined effects of the wave orbital induced forces upon the two rudders and one board upon vertical and side forces in one wave condition. Cases were run with a boat on starboard tack at 45, 90 and 150 degrees TWA with consistent wave conditions of 0.3m amplitude and 3s period (14m wavelength). This means that the port foil and port rudder elevator were raked to generate lift, while the starboard

Rudder elevator was set at a negative angle in order to increase the boat's righting moment.

It is important to point out that the flight controller on the boat only actively changes the rake of the board, not the rudder elevators.

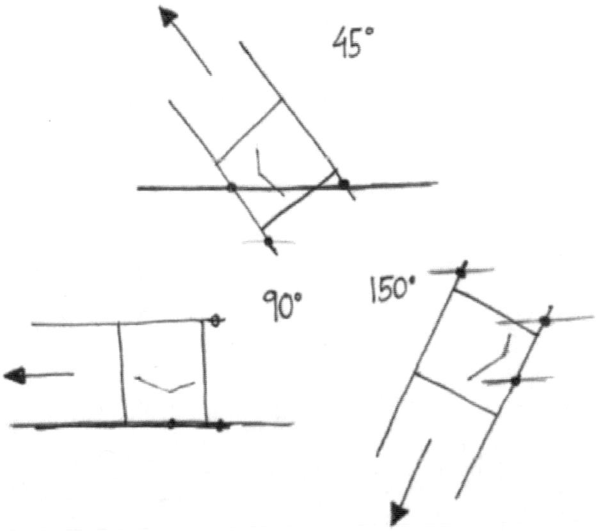

Figure 9: alignment of the boat appendages in relation to a horizontal wave set

4.1. 45 Degrees TWA

With this wave condition, the 45 degree angle causes the port foil and starboard rudder to be exactly in phase (see Figure 9). The port rudder is significantly phase shifted. The fluctuation in the sum of vertical forces is shown in Figure 10 with the greatest fluctuation being approximately 2500kg. Figure 10 also shows that the starboard rudder elevator, set to have down force, actually fluctuates between negative and positive vertical lift force. What this means in practice is that as the lift of the foil increases with the rudder moving towards a wave crest it can start to reduce the boat's righting moment, requiring the helmsperson or wing trimmer to suitably manage the side force through boat heading or wing twist. This effect is universal across the tests carried out in this study.

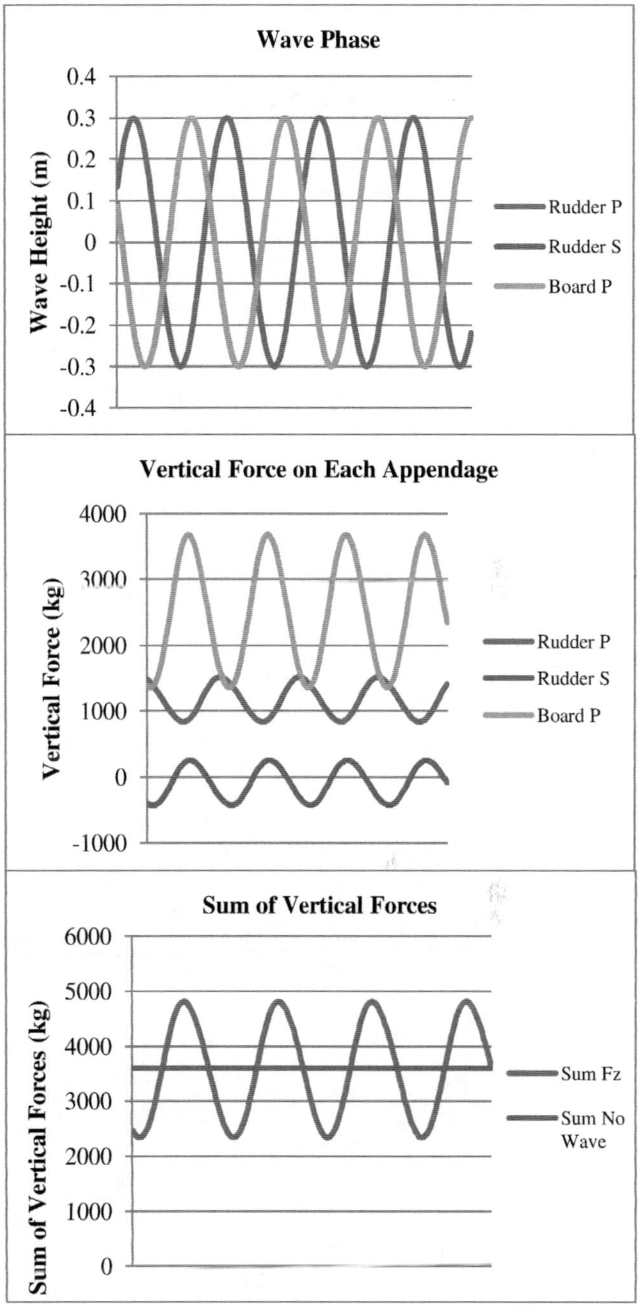

Figure 10: Overall results, for a boat sailing upwind (45 degree TWA)

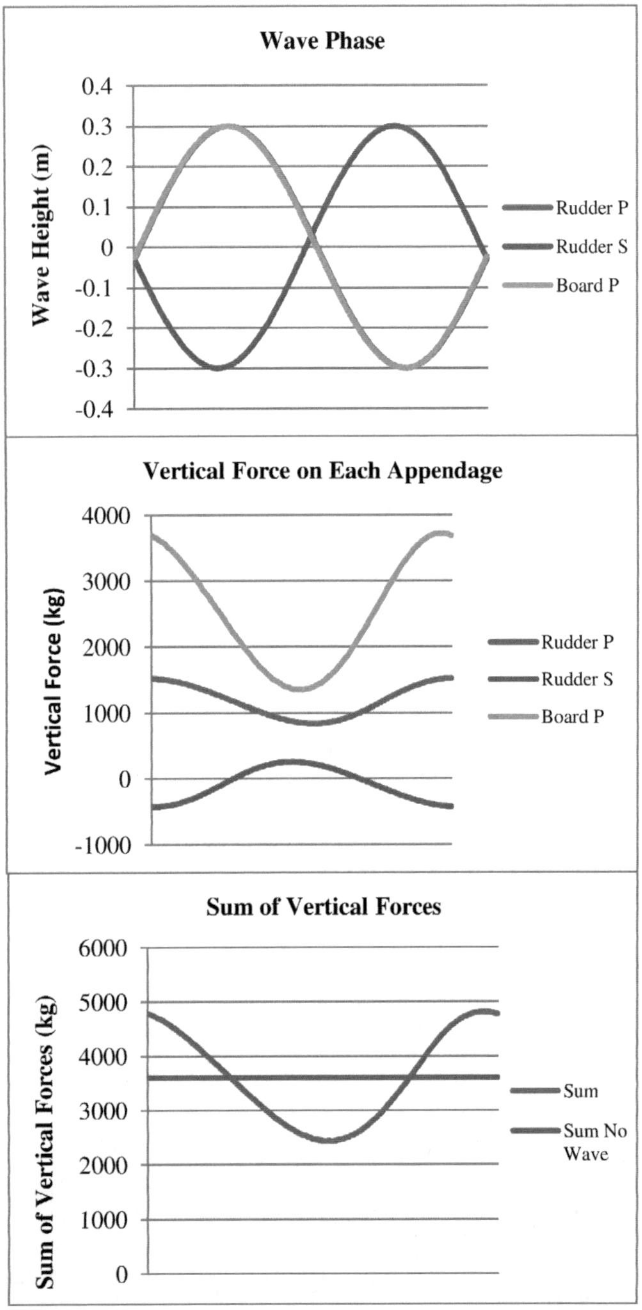

Figure 11: Overall results, for a boat sailing reaching (90 degree TWA)

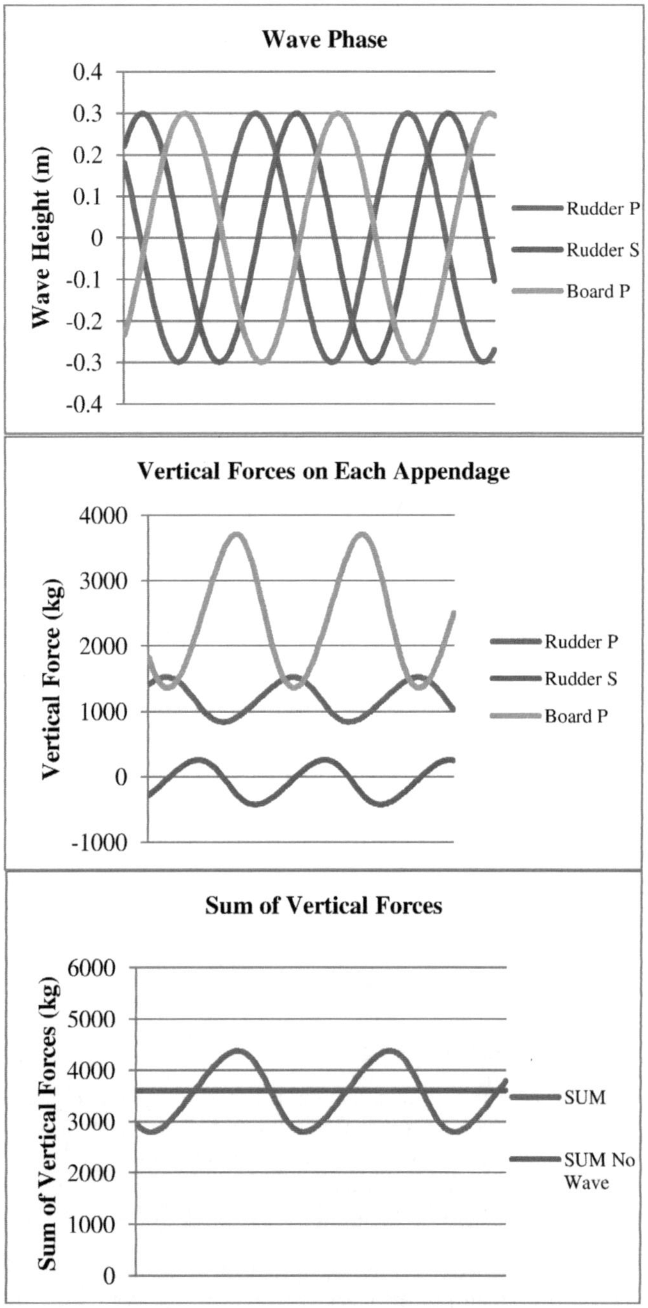

Figure 12: Overall results, for a boat sailing downwind (150 degree TWA)

4.2. 90 Degrees TWA

The port board and port rudder are in phase at 90 degrees TWA (Figure 9) and, in this wave condition, the starboard rudder is almost exactly out of phase with the port appendages (Figure 11). The maximum fluctuation in the sum of vertical force is 2380 kg. In this wave condition with the port and starboard appendage forces completely opposing each other, the boat will want to roll as the waves move past it.

4.3. 150 Degrees TWA

As can be seen in Figure 8 and 11, all three appendages are out of phase on the wave. The asynchronicity makes the port rudder partially dampen the response of the port board (Figure 12), causing the maximum fluctuation in the sum of vertical forces to be 1580kg which is less than on the upwind or reaching points of sail. However, the out of phase response of the three foils means that the athletes on board the F50 must control the appendage lift and righting moment without a visual cue of the physical wave.

5. CONCLUSIONS

Results suggest that a significant factor contributing to the greater difficulty of downwind foiling compared with upwind is likely to be the fact that, due to the upwards wave orbital velocity at the ¼ wave point, the vertical lift on the foil will be increasing as the boat moves down from a wave crest towards a trough, increasing the chance of the elevator breaking the water surface. When sailing upwind the foil lift decreases as the boat moves from crest to trough, so there is not the same problem.

Increasing the wave amplitude or decreasing the wave period both essentially make the wave steeper and produce greater vertical and side force changes due to wave orbital velocities. These results support the idea that steep waves are particularly difficult to foil in.

Although results show that the vertical and side forces will continue to increase with a decreasing wave period, in reality this would not be the case. For small wave periods the angle of attack and speed changes would be too rapid for the foil lift to develop fully. These unsteady effects must be considered in order to analyse the actual wave period that will maximise the forces produced due to the wave orbital velocities.

The vertical lift and side force produced by the board is greater than that of the rudders, so the board lift tends to drive the overall pattern of the sum of all appendage forces. This means that considering where the foil is on each wave and interpreting the effects at that point is probably a reasonable estimation.

Changing the boat heading does not change the magnitude of the angle of attack effect (from the vertical orbital velocities) upon vertical lift, but it does alter the duration of the effect. Since the boat is moving in the same direction as the wave when going downwind, the relative speed between the boat and the wave is lower. Therefore, at a downwind angle the foil will be in the vertical orbital part of the wave for longer than in the upwind case. This may contribute towards the downwind foiling problem, as the foil has longer to develop its maximum lift when moving from wave crest to trough. If this is the case then it would follow that the sailors could point lower on the 'downhill' part of the wave to minimise the amount of time spent affected by the angle of attack change, but this is speculative and should be investigated further.

Changing the boat heading does change the magnitude of the side force. The side force is maximised at 90 degrees TWA, and reduces as the heading is closer to 0 or 180 degrees TWA. The wave orbital at a crest causes the side force to reduce, making the boat want to slip sideways. This effect will therefore be most prevalent at angles closer to 90 degrees TWA. This could be considered in developing an understanding of how to best 'mode' the boat over waves in rough sea states or swells.

Future work

This analysis does not take into account dynamic effects. In reality it takes time for lift to fully develop, and as discussed with regard to the changing wave period tests, this will have an impact upon the results.

In order to run this analysis, the flow over each appendage had to be approximated as being at one point in the wave. In reality the board and rudders have span and each section of that span would be at a different point in the wave, experiencing slightly different orbital flows.

Therefore this study would be improved with the following topics:

- Test a greater range of wave conditions for the 3 appendages to find wavelength, amplitude, boat angle and wave speed to boat speed ratios provide the worst-case scenarios.
- Investigate the resulting moments acting on the boat due to the force imbalances caused by wave orbital velocities on each appendage (from the 3 appendage tests).
- Investigate the impact of unsteady effects upon the forces resulting from wave orbital velocities, and integrate this into the code
- Develop strategies for sailors that may help with sailing in wavy conditions
  - Downwind sailing techniques
  - Suggested average ride height for given wave conditions
- Consider wave effects in future design of flight control systems

Acknowledgements

The authors would like to thank Tom Speer for his help in the implementation of the lifting line code and wave method derivation.

References

[1] Grue, J., Mo, A. and Palm, E., Propulsion of a foil moving in water waves, 1988, Journal of fluid mechanics, 186, pp.393-417.

[2] Hirsch, I. A., Prediction of the Seakeeping Characterstics of Hydrofoil Ships, 1968, Journal of Hydronautics 2.1: 8-13.

[3] Holthuijsen, L. H., Waves in oceanic and coastal waters, 2010, Cambridge University Press.

[4] Lanchester, F. W., Aerodynamics, 1907, Constable (ed.).

[5] Prandtl, L., Tragflügeltheorie, 1919 Göttinger Nachrichten, mathematischphysikalische Klasse, 451–477.

[6] Marchaj, C.A., Aero-Hydrodynamics of Sailing,1988, 2nd Ed., Adlard Coles Nautical, London, UK.

[7] Speer, T., Minimum Induced Drag of Sail Rigs and Hydrofoils, 2011, http://www.tspeer.com/Planforms/Planar.htm

F50 WING 2.0: DEVELOPING A NEXT-GENERATION WING SAIL CONTROL SYSTEM

Edward Hawthorne[1], ehawthorne@sailgp.com
Jonathan Nicols[2], nicolsjt@gmail.com

Abstract. Wing sails have been in use for a number of decades, and the control systems used to trim them have come to be a point of differentiation. This paper explores the development of a novel control system used in the F50 foiling catamaran's new modular wing sail, introduced through Season 2 of SailGP. The new wing sail design sees a shift from legacy systems that were largely controlled by mechanical and rigging arrangements to a new system of hydraulic actuators on independent control arms with feedback-based control schemes. This discussion introduces the control system design and comments on the methods of position control used, the power distribution scheme, human-machine interface used and a method of defining twist profile. Additionally, a plant model is presented and the merits of utilising simulation tools in control system development are explored.

1. INTRODUCTION

Wing sails of various forms have found use in high performance sailing for a number of decades, popularised early on by developments within the C-Class of catamarans [1][2], and later by several America's Cup designs - USA-1, USA-17, AC72 and AC50. When it comes to the design and development of such wing sails, the more traditional aspects of rig design focusing on aerodynamics and structures are now met with the additional challenge of control systems. Whilst there exists a limited body of literature covering the aerodynamic aspects of wing sail design [3][4][5], there is almost no information in the public domain pertaining to the control systems used to operate these wing sails.

The wing sail in question is characterised by a two-element 'slotted flap' arrangement. The forward-most element, referred to as the *main element*, is rigid throughout its entire span, whilst the aft-most element typically consists of a series of semi-rigid *flaps* hinged to the main element by *control arms* (abbreviated to CA when referring to numerical position, and identified by subscripts starting with 1 at the bottom). Each control arm rotates about an axis near the trailing edge of the main element, allowing for camber to be added to the aerofoil section, as illustrated in Figure 1. The trailing edge of the flap panels are vertically pinned to adjacent flaps such that the entire flap element remains continuous when twisted. Choosing different camber angles at the top and bottom of a flap forces the structure to twist along its axis with its aerofoil sections remaining unchanged. If the wing is made up of multiple flap elements, then it is possible for the control system to achieve some degree of variation over the shape of the overall twist or leech profile.

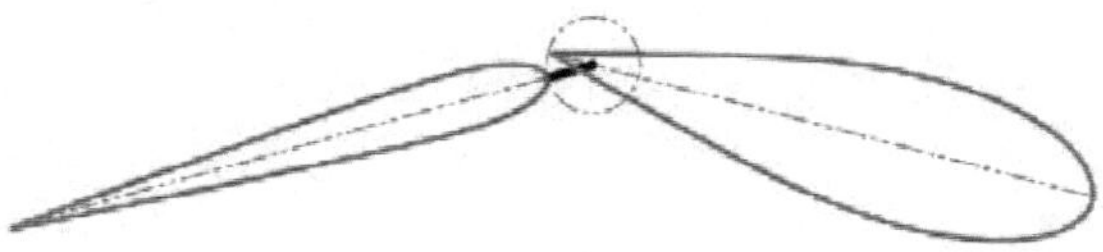

Figure 1. Plan view of a two-element wing sail, starboard tack with positive camber. The hinged flap is located to the left and main element to the right.

Figure 2. The fleet of F50 yachts racing on Sydney Harbour in February, 2020.

A multi-element wing sail requires a control system that ties together the manipulation of each control arm in order to generate a three-dimensional aerofoil shape with both camber and twist. These two qualities – camber and twist – along with wing rotation controlled by a sheet arrangement, are used to trim the wing for optimum performance. Over time, these systems have come to be a point of differentiation and have also led to different styles of trimming. Earlier designs relied heavily on a combination of cables and mechanical systems in order to provide twist and camber control. Some level of complexity arises because of the need to harvest forces in the lower part of the wing, to force twist into the upper flaps. It is possible to have the upper flaps inverted in order to produce aerodynamic force in the opposite direction relative to the lower part of the wing. This is typical in strong winds for optimum performance. For larger sized wings with increased loads, most of these controls are manipulated by hydraulic actuators.

In 2020, SailGP introduced a new modular wing sail design for its fleet of one-design F50 catamarans; which we will refer to as Wing 2.0. Although there are several new features of the wing having been redesigned entirely from the ground up, perhaps the most significant change is the departure from legacy control systems. Wing 2.0 is based on a system of hydraulic actuators coupled directly

[1] Systems Engineer, SailGP
[2] Systems Engineer

to independent control arms with feedback-based control schemes and a fly-by-wire interface.

Many of the innovations found in this design have only been made possible by the approach taken by SailGP in diverging from the AC50 class rule that the F50 was originally based on. Permitting stored power and feedback in control systems has allowed SailGP to adopt a number of technologies that ultimately enable the yachts to be sailed faster and more safely.

Furthermore, by replacing rigging with electronic wiring harnesses and hydraulic hoses, the wing has been able to be designed in a modular way that can be assembled rapidly and reliably for a range of wind conditions. Three wing configurations are possible, assembled from five separate 'modules', as illustrated in Figure 3. With a structural limitation of 2.5° of twist per metre of span, total twist achievable is approximately 45°, 60° and 70° for the Heavy Air Wing (HAW), All Purpose Wing (APW) and Light Air Wing (LAW) configurations respectively.

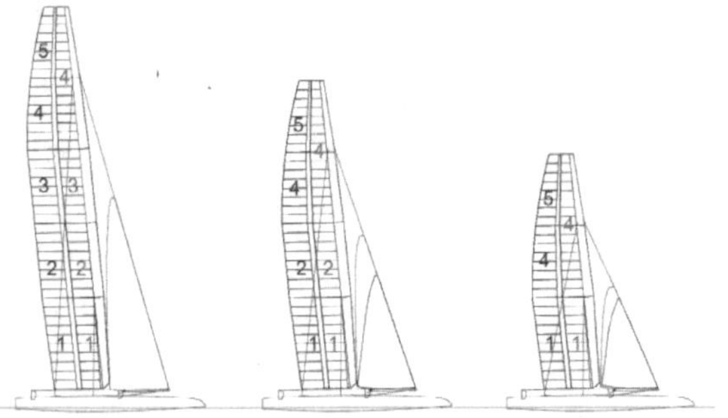

Figure 3. Modular wing configurations - 29m LAW, 24m APW, 18m HAW from left-to-right respectively. Flap panels are numbered in blue over the flaps; module sections are numbered in red over the main element.

Table 1 presents the mass and vertical centre of gravity (VCG) for each of these configurations, as well as the AC50 class rule and F50 legacy wing. Despite the increase in hydraulic system complexity, including increasing the number of hydraulic actuators, the mass of the APW configuration has remained similar to the comparably sized F50 W1.0, and VCG has improved marginally.

Table 1. Mass properties

Wing Configuration	Mass (kg)	VCG (m)
AC50 Class Rule (min.)	445	9.15
F50 W1.0	462	9.30
F50 W2.0 APW	461	8.87
F50 W2.0 LAW	557	11.46
F50 W2.0 HAW	363	6.33

2. CONTROL SYSTEM DESIGN EVALUATION

When evaluating wing control systems, a variety of factors come into consideration. Of these, one of the most important is controllability – how easy is it to achieve desired wing shape in terms of camber and twist?

With regard to camber control, wing systems can be characterized as being *driven*, *floating* or a *hybrid* of the two. A driven camber system uses an actuator, or system of actuators, to induce camber into the bottom, and then hold position. In contrast, a floating camber system relies on external forces such as sheet and aerodynamic load to induce camber while a separate system is used only to limit and vary the maximum camber. A hybrid concept may be employed where one or more additional actuators are used temporarily to induce camber while inverting the wing through a tack or gybe, but are generally not relied upon for holding position. Without the ability to drive and hold camber, wings can be subject to decambering in light winds when aerodynamic load is insufficient to overcome the 'wind-up' effect created by rotation of the main element against the standing rigging.

The mechanisms employed for the control of twist seem to vary the most between different wing systems, however almost all rely on the same approach to effect camber at each of the upper control arms. Each control arm is equipped with a quadrant to effect rotation, with cables running down to a mechanical system coupled to the bottom control arm by some means. The principal role of this mechanical system is to transfer the bottom camber angle to the upper control arms via some adjustable ratio, thereby creating twist in the flap elements. Factors differentiating twist control systems include how twist is induced into the wing. Some designs utilise sheet load in order to induce twist, independent of bottom camber, whilst others create twist purely as a function of bottom camber. Furthermore, certain designs may allow for control arms at the top of the wing to twist past zero degrees to the point where they are inverted relative to the bottom camber (known as an *inverted head*), thereby reducing heeling moment, or conversely may suffer from not being able to achieve zero twist.

Other important factors include:

- *Dynamics* – how quickly and smoothly can trim adjustments be made?
- *Tunability* – how easily can the twist profile or other aspects be tuned?
- *Energy efficiency* – how much energy is required to control the wing?
- *Weight and centre of gravity*
- *Serviceability and reliability*
- *Logistics* – how does the wing system affect the launch and retrieval of the wing as well as setup and pack-up?
- *Cost*

The Wing 2.0 control system aims to address each of the above factors in a way that is specifically tailored to the format of SailGP's global race league. Controllability was considered of foremost importance – and as a result, the system was designed with the aim of allowing any combination of camber and twist within the design limits to be achieved quickly and accurately.

A fly-by-wire style control interface gives rise to infinite possibilities for tuning various features of the wing setup, especially the twist profile. Such adjustments can even be made in real-time whilst sailing, allowing for repeatable sailing mode changes.

3. HYDRAULIC SYSTEM DESIGN

Each control arm has its own hydraulic actuator and valve manifold, controlled independently of the others. All hydraulic actuators are of double-acting, through-rod design, such that the cylinder areas on each side are balanced and therefore the same amount of fluid and force is used to move the piston a certain distance in either direction. In-cylinder linear displacement sensors with dual output are used, offering accurate and reliable position measurement with redundancy. Pressure sensors are used liberally throughout the system, providing measurements for all actuator ports, manifold inlet ports, and accumulators. Each of the proportional position control valves are driven by a current control device. Actuators pivot about a fixed point on the main element and an articulated point on the control arm, as shown in Figure 4.

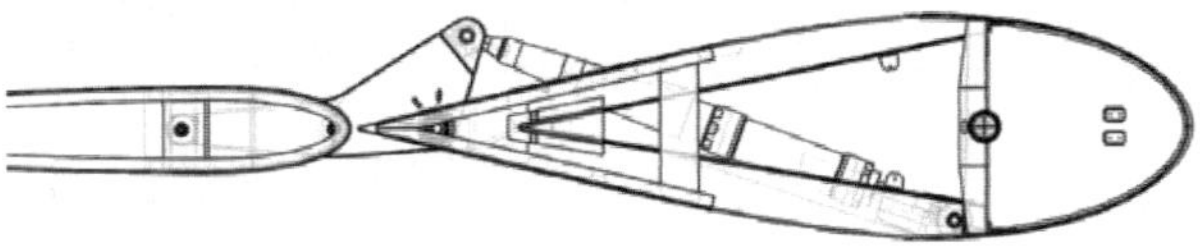

Figure 4. CAD illustration of the bottom control arm and actuator geometry.

For the bottom control arm, which has a particularly high flow requirement of up to 50 lpm due to its large size, the hydraulic circuit employs an H-Bridge valve arrangement with two directional poppet valves on the pressure side and two proportional poppet valves on the return side, illustrated in Figure 5.

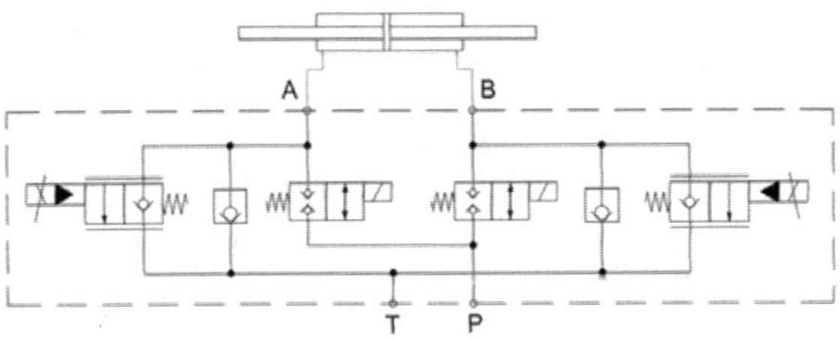

Figure 5. CA1 manifold schematic.

For the remaining control arms, the hydraulic circuit, illustrated in Figure 6, consists of a 4-way directional spool valve for flow control, a normally open poppet valve for isolating the manifold from system pressure and a normally closed poppet valve for shorting the two actuator ports, thereby facilitating a 'free-float' function. Additionally, pilot-operated check valves offer a passive locking functionality to ensure that the control arms stay in position after power down to facilitate launching, retrieval and storage.

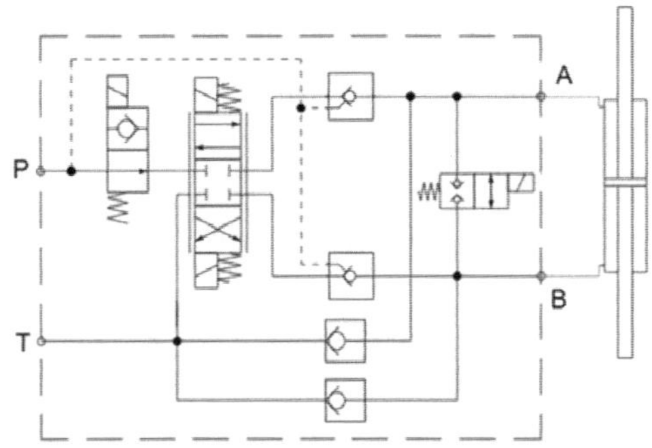

Figure 6. CA2-6 manifold schematic.

Both styles of control manifolds feature check-valves in an 'anti-cavitation' configuration, permitting flow from the return side to both of the actuator ports. This addition facilitates the operation of the manifold in an 'unpowered' mode, where fluid is able to flow within the manifold from one side of the actuator to the other, via the flow control valve(s).

4. POSITION CONTROL

Accurate position control of each control arm actuator is maintained by an architecture of independent closed-loop controllers, driven by a higher level of target generation and arbitration algorithms, illustrated in Figure 7.

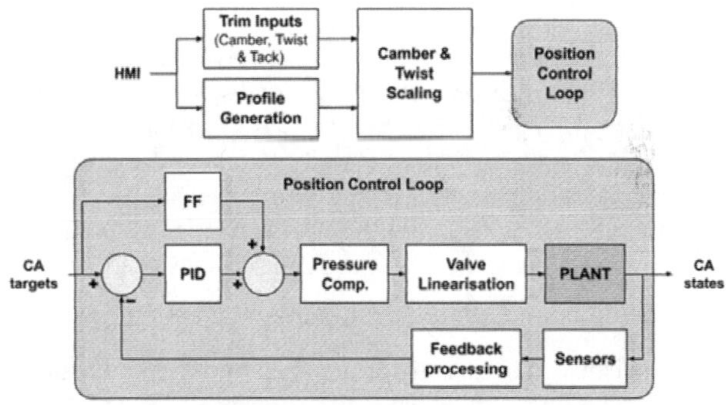

Figure 7. Block diagram of the control arm position control scheme.

In order to make effective use of well-understood linear controllers, the position control loop should not consist of any non-linear components. Angular targets are first converted to linear actuator displacements in order to avoid introducing the non-linearity of the control arm geometry. Proportional control valves often exhibit certain non-linearities, including hysteresis due to spool stiction and non-linear flow characteristics due to internal valve geometry. The aim is to linearise the flow output by applying a linearisation function to the controller output before being sent to the valve driver. Furthermore, the output is compensated for pressure drop to provide consistent performance across a range of operating pressures – especially when switching between the powered and unpowered operating modes.

A combination of a Proportional-Integral-Derivative (PID) controller and Feed Forward (FF) is used to achieve position control for each control arm actuator. The FF output is made up of components representing set point velocity, acceleration and jerk. As a result of the linearisation attempts described above, the feed forward components alone are effective in providing a rudimentary velocity control when the first derivative of the set point is continuous.

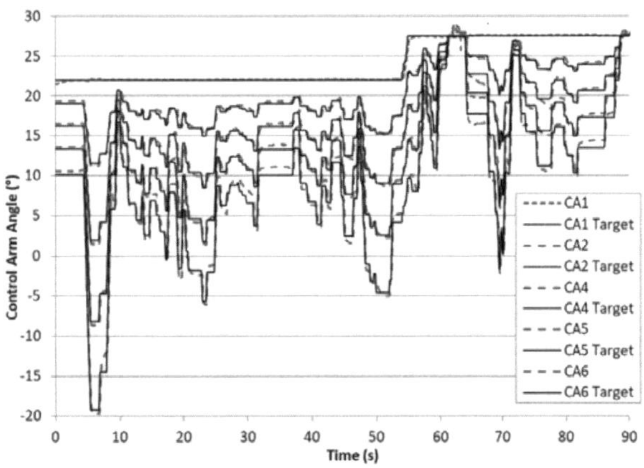

Figure 8. Target camber angles against measured camber angles. Due to the monotonic nature of the twist profile, the control arms are always shown in order from CA1 at the top of the chart, to CA6 at the bottom of the chart. The spread between the camber angles is representative of the magnitude of twist. Note that for the APW, no CA3 exists.

Repeatability of angular position has only been verified by the system's measurement of actuator displacement, rather than any flying shape analysis tool, as is often the case for soft sails. However, due to the rigid nature of the wing sail structure, we posit that flying shape can be effectively estimated from camber arm angle measurements alone. Figure 8 demonstrates the speed and accuracy of trimming that can be obtained.

The displacement sensors are claimed by the manufacturer to have a linearity deviation of $\leq \pm 0.04$ % F.S. Analogue voltage output is measured by an Analogue to Digital Converter (ADC) device with a 12-bit resolution over a range of 5V, equating to 1.25mV. As the full scale output voltage over the range of movement (approximately 74°) is 0.5V to 4.5V (a characteristic built in to the sensor by the manufacturer), we effectively have an angular resolution of 0.023 degrees. Additionally, based on mechanical design analysis, we are confident that the mechanical linkages involved in the control arm and flap assembly result in less than 0.2° of backlash.

5. POWER DISTRIBUTION/EFFICIENCY

Although the hydraulic systems are no longer powered by human grinders, instead being serviced by battery-powered electric pumps, hydraulic power management and efficiency is still of significant importance to our application. A single 2.1 kW 24V brushless DC motor is dedicated to the wing system. After efficiency losses are taken into account, the maximum flow that the pump can deliver is approximately 5.6 lpm. Given the sizing of the hydraulic actuators, the flows demanded when inverting the wing at high speed are well in excess of that achievable by the pump alone, which is therefore supplemented by flow from several hydraulic accumulators. Managing the tacking flows in an efficient manner is a significant challenge to the development of a successful system. For example, the flow required to move the bottom camber arm at 40°/s is approximately 45 lpm, adding up to a total of almost 75 lpm for all camber arms in the largest of the three wing configurations. The accumulators are plumbed in such a way that the oil demands of the significantly larger volumed bottom control arm actuator can be isolated from the rest of the system. Figure 9 presents a sample of how the hydraulic system looks installed in the wing.

Figure 9. Photo of the bottom wing module, showing a) pump, b) accumulators, c) distribution manifold and d) CA1 control manifold.

The system automatically evaluates the hydraulic power demands of each commanded actuator movement, and determines the most efficient method of fulfilling those demands in real-time. Each control manifold has a number of discrete operational modes that are selected based on estimated actuator load, accumulator pressure and controller error:

- *Locked* – selected when the position error falls within an acceptable range.
- *Powered* – selected when movement is commanded and estimated actuator load is working against the desired direction of travel.
- *Unpowered* – selected when movement is commanded and estimated actuator load is sufficient to move in the desired direction of travel.
- *Regeneration* – selected when requirements for unpowered mode are met, and additionally actuator pressure is greater than accumulator pressure (bottom control manifold only).
- *Floating* – used only for maintenance purposes.

The direction of load in the bottom control arm actuator typically acts in the opposite direction to the cylinders above it due to the force applied from the sheet at the clew. The initial phase of inversion during a tack or gybe requires the bottom control arm actuator to move against the sheet load, until the point where the combination of the aerodynamic flap moment and sheet load transfers to the opposite side. For the final phase of inversion, the bottom control arm movement is being aided by the sheet load, and instead of sourcing oil from the accumulators, we force the actuator to operate in the unpowered mode. Furthermore, if the load applied to the actuator during the second phase of inversion is sufficient such that the pressure produced is greater than that currently in the accumulators, this excess pressure can be directed back into the pressure circuit to recharge the accumulators. This is known as the regenerative mode.

Having separate accumulator circuits allows for different precharge pressures to be selected in order to cater for different operating ranges, and allows for increased efficiency in the charging of accumulators. Pressures during a large portion of an inversion have been found typically to be significantly lower than those required for straight-line trimming. Reducing the precharge and minimum operating pressures for a hydraulic accumulator increases its effective volume. Since a full inversion requires a much greater volume of oil than the typical high-frequency trimming adjustments, we can optimise the precharge pressure of one accumulator circuit to better suit the high-flow, low pressure demands of inversion, while having another accumulator circuit optimised for the low-flow, high pressure demands of straight-line trimming.

During any 'powered' actuator movement, the pump is driven to match the flow demand, in an attempt to maintain a constant accumulator volume and pressure (using only open-loop control). The high pressure accumulator is principally used as a buffer to cover any flow inaccuracies and latencies of the on-demand pump, and is automatically opened to the system pressure line when any of the upper control manifolds are operating in the 'powered' mode. During any camber adjustment, when the bottom control manifold is operating in the 'powered' mode, the low pressure accumulator is opened to the bottom control manifold. Depending on the actuator loads, 'powered' movements can be supplied by either the high or low pressure accumulators, or in the case of loads exceeding both accumulator pressures, by the pump directly. For example, during an inversion, the system will typically utilise the low pressure accumulator to service a significant amount of the oil demand, until the point at which the wing loads increase and supply will be switched to the high pressure accumulator.

In the standard mode of operation, all control arm targets are tied to the *tack input* explained in Section 7, however an additional mode of operation aimed at improving the power efficiency during manoeuvres exists. In this mode, once the bottom camber has reached a predetermined angle (nominally, zero camber), the manifold is forced into an unpowered mode and the measured bottom camber angle is then used to drive the *tack input* for the remainder of the control arms. This mode emulates some of the legacy styles of floating wing systems where the upper control arms are driven by the bottom camber by means of cables and other mechanical linkages.

6. HUMAN MACHINE INTERFACE

A key requirement for the human-machine interface (HMI) was to facilitate an easy skill transfer for the teams, aiming to manage the learning curve for both experienced wing trimmers and new ones alike. With limited practice opportunities, a drastically different HMI concept would potentially pose a challenge for the teams and affect the quality of racing.

All of the legacy wing systems used in Season 1 of SailGP had separate control inputs for bottom camber limit, twist and powered inversion, and were all open-loop systems. As used in AC35, some of these systems offered further flexibility in their control of twist, but these features were removed in an attempt to make each of the grandfathered wing systems more similar for one design racing and simpler for teams to operate.

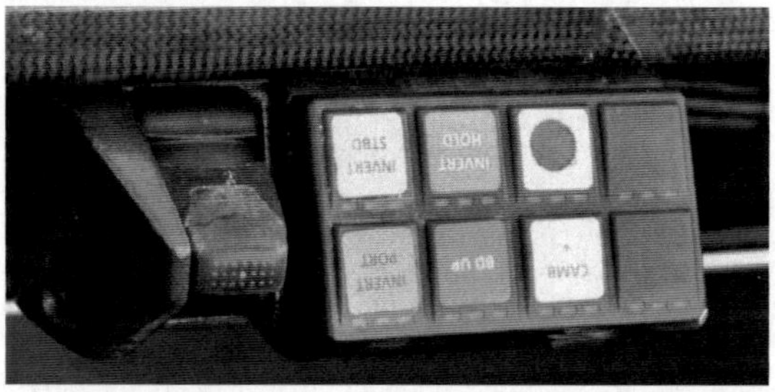

Figure 10. Photo of minimal wing trim HMI, located under the outboard gunwale of each hull where the wing trimmer is positioned.

The initial intention was to reuse the existing HMI devices installed on the F50s, emulating the same set of control inputs that the teams were already accustomed to. Original HMI featured momentary buttons linked directly to the control of the camber limiting ram and inversion rams, and a single-axis proportional joystick for proportional speed control of the twist ram, pictured in Figure 10.

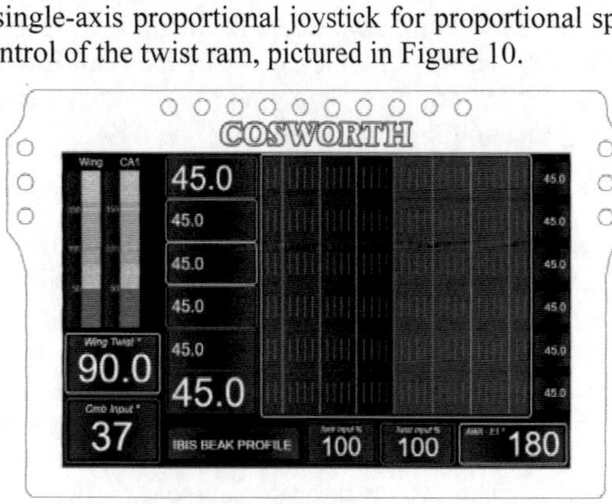

Figure 11. Screenshot of display for wing trimmer feedback, located on the inboard side of each hull in view of the wing trimmer.

Where the legacy systems had separate hydraulic actuators for controlling camber, twist and inversion, we have replaced these with a concept of virtual control inputs, collectively referred to as the *trim inputs*:

- *Camber Input* – momentary buttons adjust this value between zero and the maximum achievable bottom camber angle.
- *Twist Input* – single axis joystick adjusts the total amount of twist between 0% and 100% at a proportional rate.
- *Tack Input* – momentary buttons adjust this parameter between -100% and 100%, corresponding to maximum negative bottom camber and maximum positive bottom camber, respectively, as per the *camber input*. This control is used to tack the wing in the same way that the legacy inverter function works.

In addition to these control inputs, other parameters used for tuning the control system are available for adjustment via a touch screen interface. A colour display, presented in Figure 11, provides real-time feedback to the wing trimmer on the state of the wing system.

7. TWIST PROFILE

With the advent of having control arm position control that is both repeatable and accurate to within one degree, the possibility of dynamic twist profile manipulation becomes a reality.

Having up to six independent control arm targets is likely to require too much cognitive bandwidth for a human operator, and therefore a scheme was required to simplify the set of inputs exposed to the sailors, whilst still maintaining a degree of control over the twist profile in real-time and allowing for easy scalability to the different wing configurations.

For our purposes, we refer to the twist profile as being a curve that provides an offset from the bottom camber angle for each control arm, where the offset at the bottom control arm is always zero. To compute the camber angle at each control arm, we simply add the bottom camber angle to each of the twist values.

7.1. Twist profile as a Bézier curve

A method was chosen to define a nominal twist profile that could be expressed as a function of a small handful of intuitive parameters. We choose to define the twist profile as a curve, with the span-wise positions represented on the y-axis, and camber angle represented on the x-axis, as in Figure 12. The quadratic Bézier curve was chosen, allowing a curve between two predetermined points to be defined by just two additional parameters. These two parameters represent the coordinates of a point on the Cartesian plane, known as the *control point*. The end points of the curve are defined simply as thus: one point lies at the origin - i.e. zero camber at the bottom of the wing; and the other at the maximum twist point - i.e. maximum twist at the top of the wing. One can intuitively control the twist profile by simply 'dragging' the control point in the x- and y-axes.

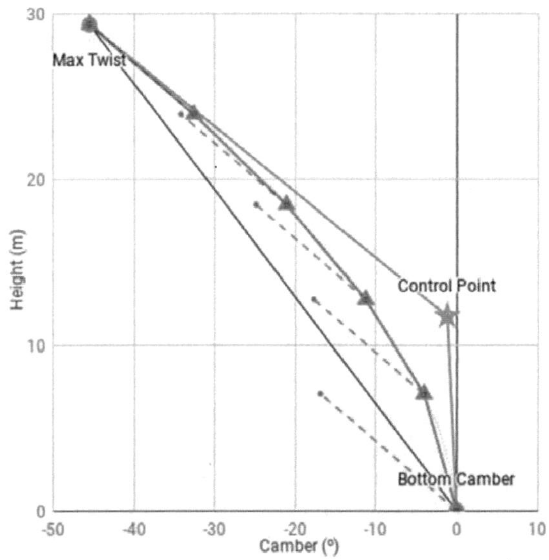

Figure 12. Bézier curve for the LAW configuration with *Max Twist* = 65% (45.5°), *Profile X* = 100%, *Profile Y* = 40%. The control point is illustrated by the green star-shaped marker, while the blue line triangle markers represent the twist profile. A linear twist profile is illustrated in black for reference, and spans between the circular marker at the origin and maximum twist point. Individual flap twist limits are represented by red dashed lines.

The equation of a quadratic Bézier curve is:

$$\mathbf{Q}(t) = (1-t)^2\mathbf{P}_0 + 2t(1-t)\mathbf{P}_1 + t^2\mathbf{P}_2, \quad 0 \le t \le 1 \qquad (1)$$

where $\mathbf{P}_0$ and $\mathbf{P}_2$ are the end points of the curve, and $\mathbf{P}_1$ is the control point.

Letting $\mathbf{P}_0 = (\theta_t, h_t)$, $\mathbf{P}_1 = (\theta_c, h_c)$ and $\mathbf{P}_2 = (0,0)$, we form two separate equations for h and θ, where h is used to refer to the vertical axis representing height, and θ is used to refer to the horizontal axis representing camber angle. Subscript t refers to the top point representing maximum twist, while subscript c refers to the control point.

$$h(t) = h_c t(-t+1) + (-t+1)\big(h_c t + h_t(-t+1)\big) \qquad (2)$$

$$\theta(t) = \theta_c t(-t+1) + (-t+1)(\theta_c t + \theta_t(-t+1)) \qquad (3)$$

Rearranging the above equations and then solving to eliminate t, we arrive at the following equation for θ as a function of h, h_c, θ_c, h_t and θ_t:

$$a = \sqrt{-2hh_c + hh_t + h_c^2}$$

$$\theta(h) = \frac{(h_c - a)\big(2\theta_c(h_c - h_t + a) + \theta_t(h_c - a)\big)}{(2h_c - h_t)^2} \qquad (4)$$

This equation is used to compute the nominal twist value at each control arm height, given the coordinates of the top point and the control point.

7.2. Twist profile constraint

Whilst the Bézier curve method described above is capable of defining a range of radical shapes, in practice we must place some constraints on the curve generation. The principal constraint for a wing system such as the one described here is the maximum flap panel twist limit imposed for structural reasons. Here we assume that the span-wise twist rate, or cumulative angle of twist per span-wise length, is uniform for all flap panels, regardless of their individual span, and therefore for the wing as a whole. If the curve is monotonic, then as long as both the bottom and top flap panels' twist constraints are observed, all intermediary flap panels will also be within their twist constraints. For the quadratic Bézier curve described, we can assure monotonicity with the following control point constraints:

$$\theta_t > \theta_c > 0$$
$$h_t > h_c > 0$$

i.e. the control point must remain inside the rectangular bounds created by the top control point and the origin.

7.3. Twist profile parameters

Rather than having the user directly input the control point coordinates, the system exposes three parameters that provide a layer of abstraction to the profile generation:

- *Max Twist %* – allows the user to specify the maximum wing twist at the top of the wing as a percentage of the maximum twist available for the given configuration. Together with the height of the wing configuration in use, this parameter determines the coordinates of $\mathbf{P}_0$.
- *Profile Y %* – allows the user to specify the height of the control point as a percentage of the wing height for the given configuration, determining the y-component of $\mathbf{P}_1$, h_c.
- *Profile X %* – this parameter is used to determine the x-component of $\mathbf{P}_1$, θ_c, and represents a deviation from laying on the linear twist profile to the left or right depending on the value's sign. A value of 0% always results in a linear twist profile, whilst -100% and 100% result in the left and right extremums of the valid range of control point coordinates, respectively. This is done to make the adjustment of the *Profile X* parameter more intuitive, since the midpoint of the valid range does not necessarily result in a linear twist profile.

7.4. Parameter constraints

Firstly, we compute the range of values that θ_c can take while observing the twist limit in all panels. As the control point moves towards the left, the bottom flap panel twist increases and becomes the greatest of all the flap panels. The value for θ_c at which the bottom flap panel twist reaches its limit therefore becomes the lower limit for θ_c. Conversely, as the control point moves to the right, the top flap panel becomes the limiting factor, and the value for θ_c at which the top flap panel twist reaches its limit becomes the upper limit for θ_c. The combination of these limits is illustrated for the APW configuration in Figure 13.

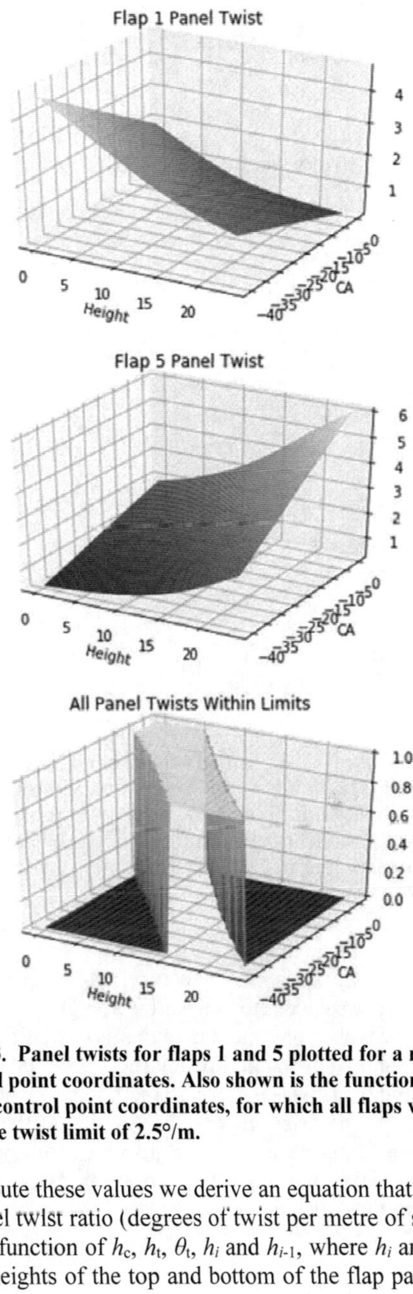

Figure 13. Panel twists for flaps 1 and 5 plotted for a range of control point coordinates. Also shown is the function space of control point coordinates, for which all flaps will be within the twist limit of 2.5°/m.

To compute these values we derive an equation that gives flap panel twist ratio (degrees of twist per metre of span), T_p', as a function of h_c, h_t, θ_t, h_i and h_{i-1}, where h_i and h_{i-1} are the heights of the top and bottom of the flap panel in question.

$$T_p' = \frac{\theta_i - \theta_{i-1}}{h_i - h_{i-1}} \quad (5)$$

Solving the flap panel twist equation for θ_c yields a function that can then applied to both the top and bottom flap panel geometries in order to obtain a valid range for θ_c.

The final value for θ_c is obtained by mapping the *Profile X* percentage to the valid range using the following set of equations:

$$l = \frac{h_c \theta_t}{h_t}$$

$$\theta_c = \begin{cases} x(\theta_{c,\max} + l) + l, & x > 0 \\ -x(\theta_{c,\min} + l) + l, & x \le 0 \end{cases} \quad (6)$$

where x is the *Profile X* parameter expressed as a fraction.

7.5. Twist profile scaling and shifting

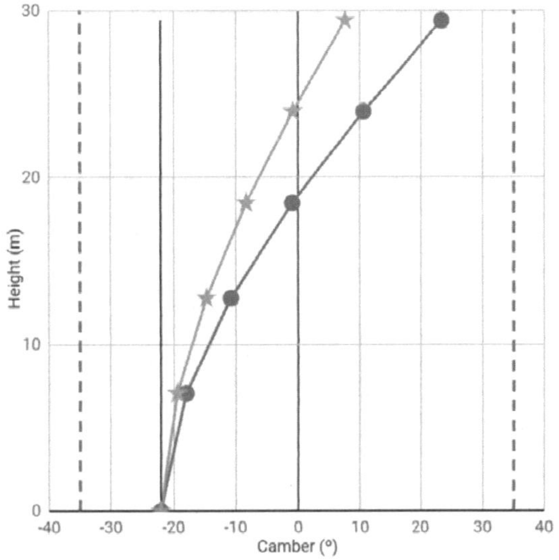

Figure 14. Scaled and shifted twist profile (22° camber, 50% twist, -100% tack). The shifted nominal twist profile is shown in blue with circular markers, whilst the scaled twist curve is shown in orange with star-shaped markers. The vertical black line at -22° represents a wing profile with zero twist.

Following the generation of the nominal twist profile, we must set out a means to control its magnitude. For simplicity, we choose to only define a single nominal twist profile, which represents the maximum twist applied. A simple approach to modify this curve for intermediate magnitudes of twist is to scale the nominal twist profile linearly by the *twist input* factor. So long as the nominal curve is monotonic and meets the constraints for maximum flap twist, then all intermediate *twist inputs* will also meet the constraints.

Finally, we require a means for applying the combination of camber and twist in a way that can be inverted symmetrically, providing a set of trajectories to move from one tack to the other. For this purpose, we introduce the *tack input* parameter, which is used to interpolate both the bottom camber and twist from one tack to the other. The following two equations are used to compute the camber angles for the bottom control arm and upper control arms, respectively. T_i represents the nominal twist curve value evaluated at the height of control arm *i*, while C_i represents the camber angle for control arm *i*. For brevity, we drop the '*input*' from the names of the three *trim input* variables – i.e. *camber input* becomes *camber*.

$$C_1 = camber \times tack \quad (7)$$

$$C_i = C_1 + T_i \times twist \times tack \quad (8)$$

This implementation of the tack scaling results in all control arms passing through zero simultaneously in an inversion. An example of the scaled and shifted curves is shown in Figure 14.

7.6. Maximum camber constraint

The additional constraint of the mechanical limits of the camber arms requires the *twist input* to be limited as a function of the *camber input*. For example, if the twist profile has been set up to achieve a maximum of 50° at the top control arm, and the maximum achievable camber angle for any control arm is 35°, then the *twist input* will need to be reduced for any *camber input* of less than 15°. The following equation is used to compute a limited *twist input* percentage:

$$twist_{\text{lim}} = \min\left(\frac{|C_1| + C_{\text{lim}}}{T_{max} \times tack}, twist\right) \quad (9)$$

8. PLANT MODELLING AND SIMULATION

As on-water testing is often constrained by cost and logistics, the use of simulation tools has become a popular method of developing and validating design concepts. Using the Mathworks Simscape environment, we have developed a high-fidelity 'plant model' of the Wing 2.0 control system hardware, encompassing all of the hydraulic and electronic components found in the wing. The system is modelled down to the component level, where every valve has been characterised from datasheet values and validated against real-world test data, an example of which is shown in Figure 16. A Graphical User Interface (GUI), shown in Figure 15, was also built in order to visualise the output of the simulation in real-time as well as emulating the system's HMI. The plant model is connected directly to the control system software, which is also developed using Mathworks' Simulink environment, thus creating a Software-in-the-Loop (SIL) simulation tool. The SIL simulation has proved to be an invaluable tool in developing and debugging control software. Mathworks provides a comprehensive suite of debugging features, accessible whilst running the SIL simulation, which would not otherwise be available on target hardware. Furthermore, software changes that may be considered risky to test in a shed environment or at the dock can be thoroughly tested in the safety of the simulation environment.

The SIL plant model can also be coupled with a dynamic VPP simulation, in order to provide a more realistic dynamic behaviour for the yacht's movable control surfaces and wing. As well as HMI inputs, the plant model also takes force inputs applied to each of the actuators, which when coupled with the VPP, are computed directly from the force models. The addition of force feedback within the simulation assists in identifying and addressing scenarios in which high loads might affect actuator dynamics. In particular this has been useful in developing features of the wing control software that are active during manoeuvres.

Further development is being undertaken to extend the work to run in a Hardware-in-the-Loop (HIL) configuration, where various hardware components of the system can be integrated into the simulation loop.

9. CONCLUSIONS

This paper has presented an overview of the system design for controlling the modular wing sail employed by SailGP's F50 yachts. The system has been tested and demonstrated to function well in races during the first SailGP event of the 2020 calendar year.

A plant model of the hydraulic system was developed in order to facilitate SIL and HIL testing processes. The model has been tuned and validated against real-world data, showing strong enough correlation to effectively test closed-loop control schemes. The availability of this tool will become a vital component of future development and testing activities, and will furthermore add a more realistic element to the VPP.

Acknowledgements

The authors would like to acknowledge SailGP for the opportunity to work on this exciting project, as well as several individuals who contributed significantly to the development of various aspects of the system: Alex Davis, Mike Drummond, Nathan Libby, Chris Mitchell, Alex Reid and Brad Webb.

References

1. Scherer, J. O. (1974). Aerodynamics of high-performance wing sails. *Marine Technology and SNAME News*, 11(03), 270-276.

2. Magherini, M., Turnock, S. R., & Campbell, I. M. C. (2014). Parameters affecting the performance of the c-class wingsail. *International Journal of Small Craft Technology*, 156(B1), 21-34.

3. Blakeley, A. W., Flay, R. G. J., & Richards, P. J. (2012, December). Design and optimisation of multi-element wing sails for multihull yachts. *18th Australasian Fluid Mechanics Conference*, Launcestion, Australia, 3-7.

4. Chapin, V., Gourdain, N., Verdin, N., Fiumara, A., Senter, J. (2015), "Aerodynamic Study of a Two-Elements Wingsail for High Performance Multihull Yachts", 5th High Performance Yacht Design Conference, Auckland, 10-12 March, 2015.

5. Viola, I. M., Biancolini, M. E., Sacher, M., & Cella, U. (2015, March). A CFD-based wing sail optimisation method coupled to a VPP. *5th High Performance Yacht Design Conference* (pp. 1-7).

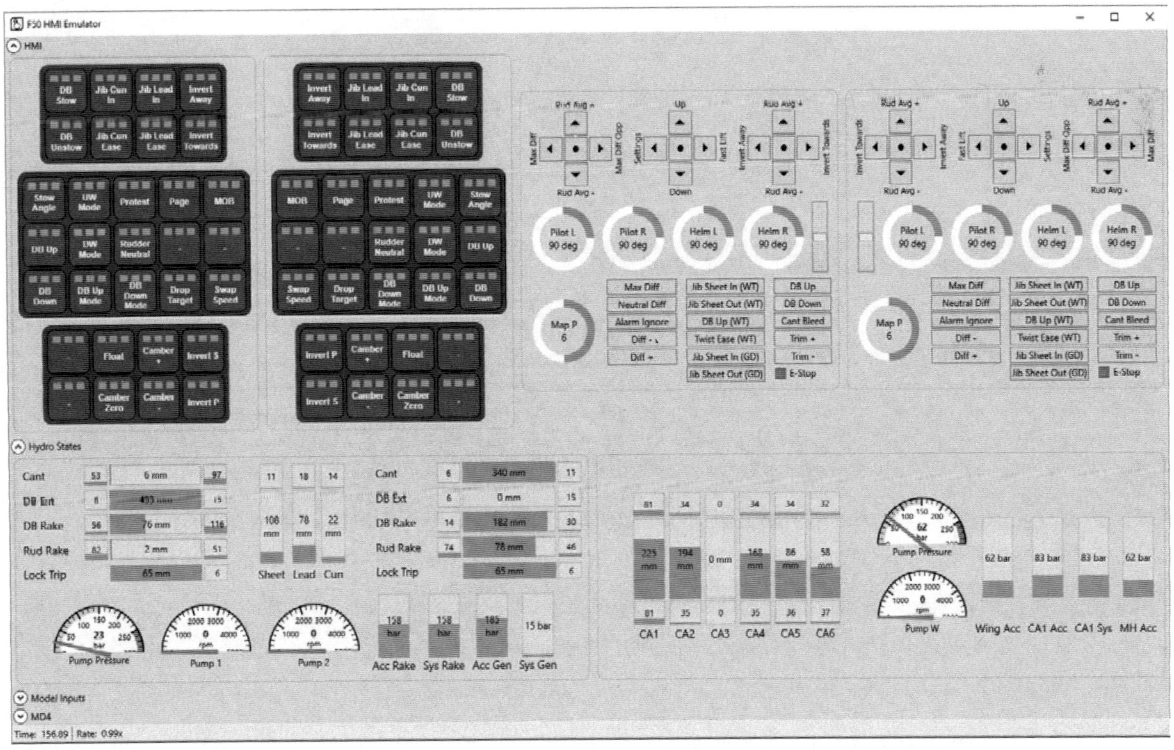

Figure 15. Screen capture of the SIL simulation GUI. HMI devices are emulated in the top half of the window, with hydraulic states shown at the bottom. The GUI, alongside the SIL simulation, can be used as a training tool for crews to familiarise themselves with the on-board HMI.

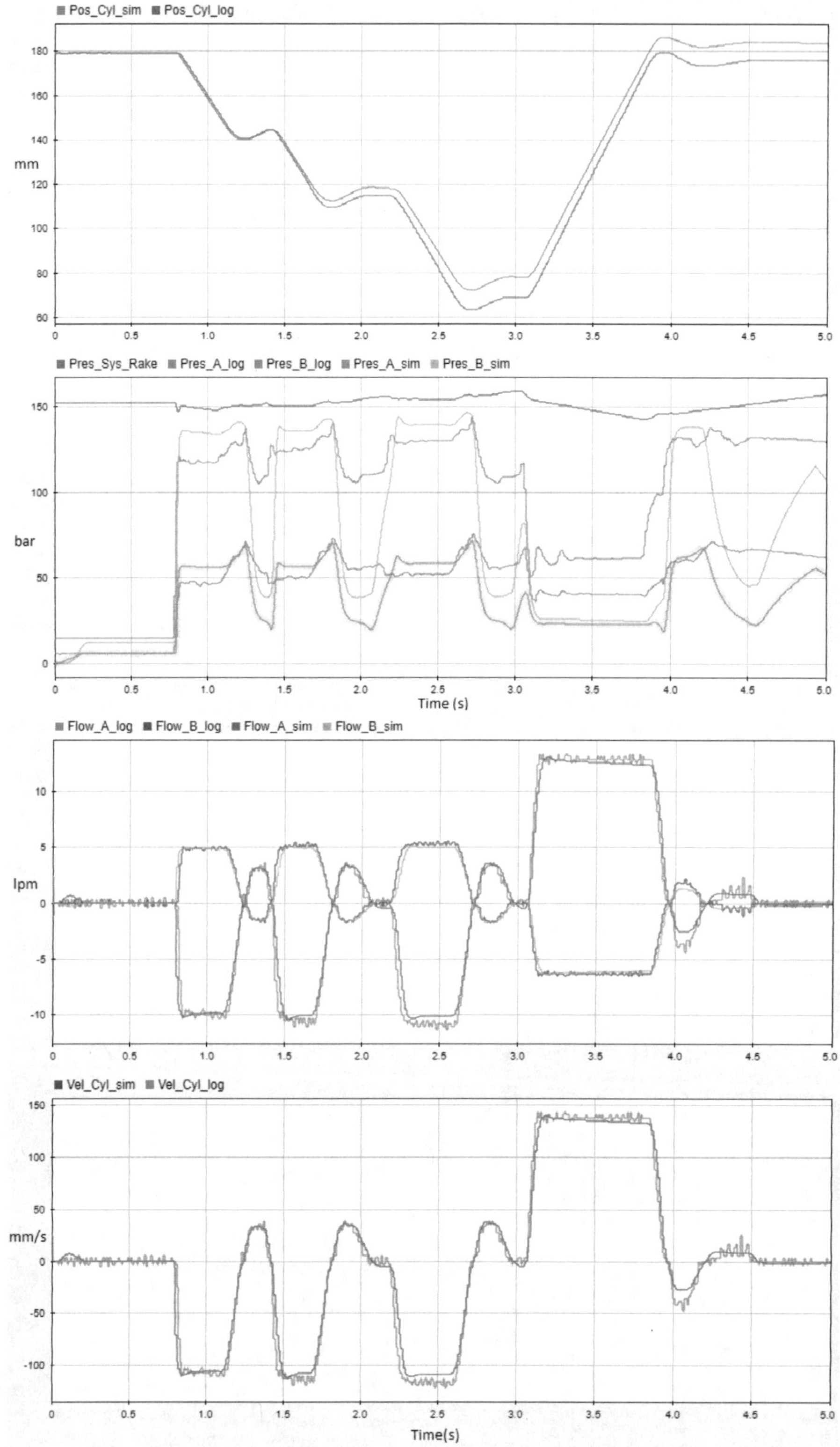

Figure 16. Measured and simulated hydraulic states from a SIL simulation for a series of logged valve commands.

PERFORMANCE PREDICTIONS OF LONG-DISTANCE SAILING VESSELS

Sofia Werner[1], sofia.werner@sspa.se
Jakob Kuttenkeuler[2], jakob@kth.se
Axel Hörteborn[3], axel.horteborn@sspa.se
Olov Lundbäck[3], olov.lundback@sspa.se
Mikael Razola[4], mikael.razola@walleniusmarine.com
Ulysse Dhomé[5], dhome@kth.se

Abstract. Designing sailing vessels for world-wide cargo transportation requires that the trade route and weather conditions are taken into consideration in the design process. This work studies various strategies for representing the weather distribution related to the expected operational profile. The balance between accuracy and design work efficiency for the different methods is discussed and demonstrated for a wind powered car carrier concept.

1. NOMENCLATURE

$\bar{V}$	Route average velocity
CFD	Computational Fluid Dynamics
EEDI	Energy Efficiency Design Index
EFD	Experimental Fluid Dynamics
IMO	International Maritime Organisation
TWS	True wind speed
TWA	True wind angle
VMG	Velocity Made Good
VPP	Velocity Prediction Program

2. INTRODUCTION

Using the wind for ship propulsion has for a century been reserved for yachtsmen. The climate change and tough requirements from IMO to reduce green-house gas emissions from shipping [1] has raised the interest in wind propulsion for commercial cargo vessels. The shipping industry has, however, developed since the time of the wind jammers. Designing a modern sailing cargo vessel means for many aspects starting from scratch again. Luckily, during the century long hibernation of sailing cargo vessels, the yacht design technology has continued to develop. In this study we use design practice from yacht design and apply them in the concept development of a sailing cargo vessel.

Sailing vessel design is all about balancing conflicting objectives. Performance must be weighed against building cost, upwind performance against downwind, lift against drag, speed against risk and so on. Finding the right balance at concept stage requires computational tools for assessment of the performance. Today, a wide range of methods, from Experimental Fluid Dynamics (EFD) [2], [3] and Computational Fluid Dynamics (CFD) [4] to lifting line methods [5] are available. Whether it takes seconds or weeks to derive, the result is usually a set of polar curves that describe the ship's speed in all possible wind conditions. The question that will be discussed here is *how* these polar curves are used in the design process. Which wind condition is the most important? How do we translate the polars to a single performance indicator? The answer will of course depend on how the sailing vessel is intended to be used. A racing yacht designed for a windward/leeward course has a narrow and well-defined design condition. A cargo vessel designed for a long-distance trade with a fixed time-schedule has a much more complex design target condition.

Weather route optimisation and route analysis tools are routinely used onboard cargo ships as well as sailing yachts [6],[7], and in concept development of wind powered cargo ships [8],[9]. This work studies various strategies for taking the route and weather conditions into consideration in the design process. The sensitivity of the choice of method is illustrated using an example of a design variation for the concept Oceanbird, shown in Figure 1. The aim is to build up a suitable process for concept development of long-distance sailing cargo vessels.

Figure 1. The Oceanbird concept

[1] Manager Strategic Research Hydrodynamics, SSPA, Sweden
[2] Professor, Royal Institute of Technology, Sweden
[3] Senior Specialist, SSPA, Sweden
[4] Project Manager, Wallenius Marine, Sweden
[5] PhD Student, Royal Institute of Technology, Sweden

3. TEST CASE

The test case is the wind powered car carrier concept Oceanbird. The ship is 200 m long, displaces around 30.000 ton and will transport 7000 car units across the Atlantic Ocean in 12 days, which corresponds to an average speed of 10 knots. Five 80 m tall (from deck) rigid wing sails will drive the ship, only occasionally supported by conventional propulsion, reducing the emissions by about 90%.

A large number of underwater appendages have been designed for the concept ship and evaluated using EFD, CFD and VPP simulations [4]. This paper focuses on the strategies for representing the operational profile, and for that purpose a simple fin will be used as an example of a design variation. The fin is a non-retractable symmetric fixed foil positioned at the ship's longitudinal centreline. The design of the fin itself is not relevant in the context, it only serves as an illustration. The Oceanbird concept will most likely be equipped with other types of appendage. The polar curves of the ship with and without fin are shown in Figure 2. They are derived using a combination of model test, CFD and lifting line for the hydrodynamics and a VLM code for the aerodynamics.

The gain of the fin is, as expected, largest in the close-hauled to broad reach regions, and the gain increases with wind speed. The positive effect on the upwind performance should be weighed against increased resistance downwind, increased building and maintenance cost and a higher risk of damage and failure in shallow water. For example, a large gain in speed performance may over-weigh the risks and increased cost. In reality, the design process is of course not digital and simplified as in the current example, it involves large number of variations of appendage size and positions.)

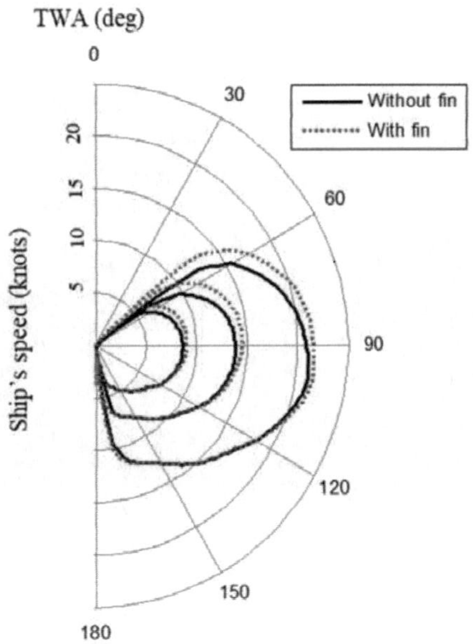

Figure 2. Performance of the ship with and without a centerline fin at 3 different wind speeds

4. METHODS

4.1. Static performance predictions

For predicting the ship's instant speed and CO2-emissions, the Oceanbird design team utilises a range of methods of varying confidence levels. Limiting the scope of the current study to steady state simulations, the tools for assessing the aerodynamic performance includes the methods listed in Table 1.

Table 1 Methods for aerodynamic performance assessment in the Oceanbird project.

Type of method	Computational time
Lifting line [10],[11]	Milliseconds
Vortex Lattice Method [12], [13]	Seconds
RANS and URANS	Hours-days
Detached Eddy Simulations [14]	Days-weeks
Wind tunnel	Weeks

For prediction of the hydrodynamic forces, the design team possesses a similar range of methods, from lifting line methods, to 3D CFD and towing tank test, covering a broad range of accuracy and cost levels.

The aerodynamic and hydrodynamic models are, as usual in sailing vessel design, combined using VPP-algorithms that solve force and moment equilibriums to compute the ship's speed. The current concept requires, however, a higher degree of complexity than a VPP for a normal sailing yacht. There are five wings, hence five sheeting angles. The wings can be retracted in steps, effectively reefing the wing-area, to restrict heel and reduce structural load, adding 5x3 more degrees of freedom. Furthermore, the auxiliary engine power can be increased from 0 to 100% power continuously.

Finding the static balance and optimum sheet angels, reef, rudder angle and engine power in the VPPs is handled in two ways, depending on the context:

a) By interpolating in a precomputed matrix. This is fast, but due to the high degree of freedom requires a matrix of many dimensions and discrete steps for each variable. The sheet angles are optimised based on purely aerodynamic performance without consideration of the hydrodynamics and engine power.
b) By optimisation within the VPP-algorithm. This requires more computational time, and it may be instable and difficult to find converged solutions. The advantage is that it avoids the interpolation between discrete steps and all variables can be optimised considering the full system.

4.2. Route evaluation methods

The design objective is to minimise the need for auxiliary engine power and hence reduce the CO2-emissions for a given crossing-time. In this simplified demonstration study, the minimum crossing time is used as design target, expressed as the ship's average velocity made good on the route $\bar{V}$, i.e. a nominal distance divided by the journey duration.

Computing $\bar{V}$ can be done in various ways, described below from the simplest to the most complex:

A. Single design wind condition
One weather condition is selected to be representative for the overall performance. Since there is only one condition to evaluate, we can afford either high-fidelity performance models such as RANS CFD, or a very large number of design variations.

B. Weather Weighted Polar
For this method we need a defined sailing route. The probabilities of meeting various wind conditions on the given route is achieved with post-processing of hind-cast data from weather providers or from open source databases. The average speed indicator $\bar{V}$ for the route is derived by multiplying each probability with the corresponding speed from the VPP:

$$\bar{V} = \sum_{j=1,i=1}^{N,M} p_{i,j} \cdot V_{i,j} \qquad (1)$$

where $p_{i,j}$ is the probability and $V_{i,j}$ the ship's velocity at true wind speed i and true wind angle j.

This approach is used in IMO's regulation for determining the Energy Efficiency Design Index (EEDI) for ships with wind propulsion technology. The "Available effective power of wind propulsion systems" is in the IMO framework derived using a "Global Wind probability matrix" that represent main global shipping routes worldwide [15].

C. Statistical route analysis
The assumed route from port to port is laid out and divided into legs. Each leg has roughly 45 combinations of wind direction and speed. The route is divided into over 40 legs, making 40^{45} possible combinations. The combinations of wind conditions are joined with estimations of the ship speed from a VPP using a Monte-Carlo simulation technique. A tacking/gybing algorithm makes the ship always sail on the largest attainable Velocity Made Good on Course. A more detailed description of this method is given in [16].

The outcome is, just as for the Weather Weighted Polar method, the most probable route average velocity $\bar{V}$. In addition, this method can also deliver the probability distribution of $\bar{V}$, which can be used to show for example the risk for delays in the trade schedule.

D. Weather optimised route analysis
Given today's availability of weather forecasts and route optimisation service providers, it is easy to assume that the ship will not sail on a fixed route as assumed in the previous method, but on a route that is constantly adapted to the current weather. This is reflected in the method with the highest level of complexity in this study. The route average velocity indicator $\bar{V}$ is here calculated using route optimisation. The algorithm is based on a forward-backwards dynamic programming scheme of a mesh-based physical space where weather data from hindcasts from the European Centre for Medium-Range Weather Forecasts ECMWF in the form of 10 m average wind. The routing motor interpolates pre-calculated performance matrices (polars) for boat performance. No manoeuvre penalties are implemented. Land and other potential no-go zones are enforced by inhibiting certain nodes. Zones with true wind speeds exceeding 20m/s is also regarded as no-go-zones based on operational practices. Hence, the algorithm is deterministic and globally convergent. However, it should be noted that no forecast prediction uncertainties are included since wind "predictions" in this implementation is always 100% accurate. $\bar{V}$ is calculated as the journey great circle distance divided by journey duration.

Table 2 lists the four mentioned route evaluation methods together with the number of times that the performance needs to be evaluated (from a VPP for example). The more realistically the operational profile is represented, the larger number of instant performance computations are required. It means that with a simple representation of the operational weather profile we can afford high fidelity methods for the performance evaluation, or a very large number of design variations, and vice versa.

Deriving the route average speed performance indicator requires a combination of one method for deriving the hydro and aerodynamic forces from Table 1, one method for the VPP and one method for representing the weather on the route from Table 2. The design team needs to choose the most suitable combination depending on the situation: the number of design variations, the maturity of the concept, the physics of the investigated object, and the available resources.

The question that this study aims to investigate is which method to use in what situation and how sensitive the result is to the choice of methods.

Table 2 Methods for representing the operation profile and the number of times the instant velocity must be derived

Type of method	Number of performance computations
A. Single wind condition	1
B. Weather Weighted Polar	170
C. Statistical route analysis	1 800
D. Weather optimised route analysis	1 000 000

5. SIMULATIONS

To investigate the sensitivity of the choice of methods for representing the operational condition, the test case described in section 3 is evaluated using the same set of polar curves (Figure 2) for all methods. The performance of the fin is assessed by comparing the route average speed for the case with the fin against the case without the fin.

5.1. Simulations using Weather optimised route analysis (Method D)

The numerical weather routing was done using a spatial discretization of 20 legs from start to finish and a mesh width of 31 nodes, 15 on each side of the great circle. Performance was evaluated on weekly departures in both directions for the duration of one year (2010), hence a total of 104 journeys where optimized and evaluated.

Figure 3 shows an example of routing results where the same ship is routed from England to USA two times in freezed weather (for clarity). The great circle between the journey end-points is marked as well as the mesh discretization, defined as the spacing around the grey dots. Red dots mark no-go areas with winds higher than 20 m/s. The two blue routes represent one optimal route without fin appendages (southerly) and the other route with fin used (northerly). It is clear from the Figure that both routes exploit regions with high wind speeds while avoiding the no-go-zones.

Also note that the optimal routes actually differ, where more headwind is chosen when the ship is equipped with a fin generating hydrodynamic lift. The difference in $\bar{V}$ was 0.4 knots in this specific example.

The resulting wind statistics over the year on the optimised routes are given in Figure 4, B.2 and B.3. The optimiser makes the ship to sail between closed hauled and beam reach. It is interesting to note that adding the fin also changes the optimised route slightly towards closed-hauled. Obviously, this pays off when the side force generation is better.

5.2. Simulations using Statistical route analysis (Method C)

The Statistical route analysis assumes that the ship sails on the great circle, the shortest distance between the ports.

For all wind directions where the true wind angle is less than closed hauled, i.e. maximum VMG, the ship is assumed to be beating upwind, using the tacking algorithm.

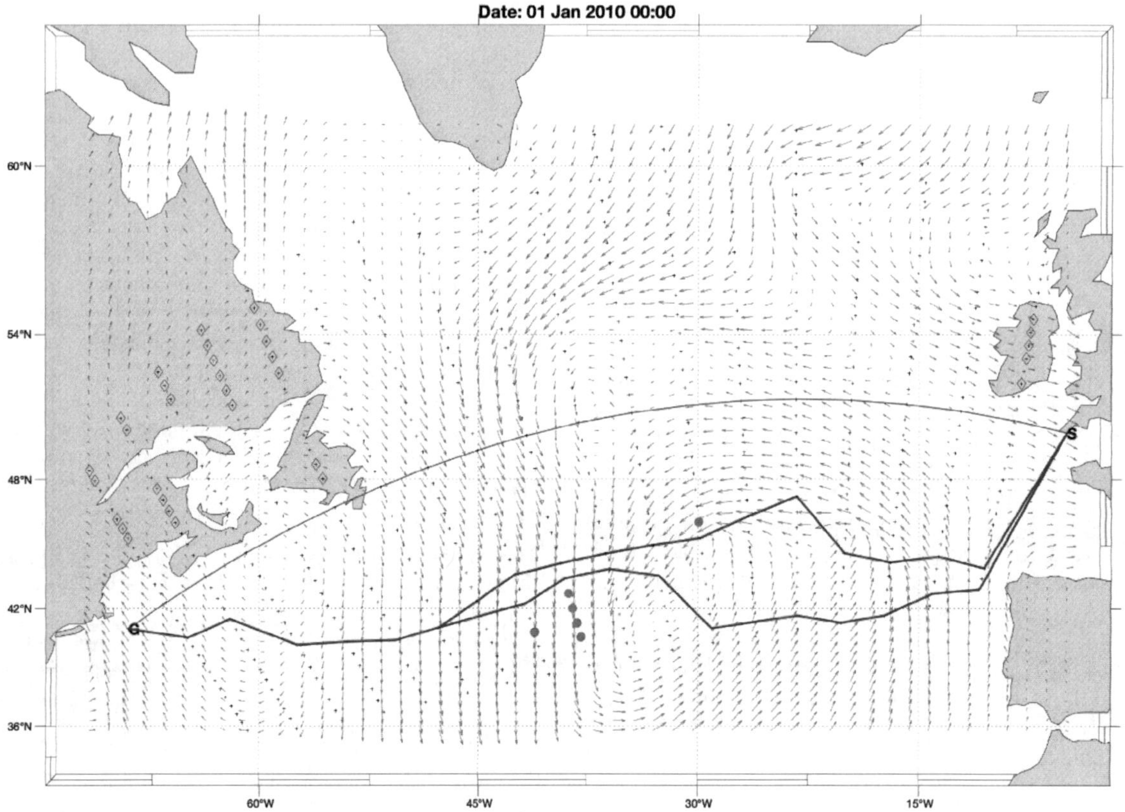

Figure 3. Example of routing results in freezed weather where the same ship is routed from England to USA with and without appendage fin deployed.

5.3. Simulations using Weather Weighted Polar (Method B)

For the Weather Weighted Polar method we need to assemble weather statistics for an assumed route, or a collection of routes. For the sake of comparison, weather statistics have here been assembled in several ways:

- B.1 great circle from England to USA – yearly average weather
- B.2 optimised routes from Method D, covering 1 year of sailing for the ship without the fin
- B.3 same as B.2 but for the ship *with* the fin
- B.4 same as B.2 but assuming that the engine will be used as soon as the ship's speed is less than 10 knots
- B.5 the EEDI Global Weather matrix

The distributions of wind speed and direction for these options are shown in Figure 4. All wind directions in the upwind sector is set equal to the closed hauled direction, i.e. where maximum VMG is obtained for this ship.

The wind direction for the Atlantic great circle route (B.1 in Figure 4) is rather evenly distributed with an overweight of head wind and following wind die to the predominant westerly wind directions. The EEDI Global Weather Matrix is rather similar to the great circle route but with slightly higher wind speeds.
Compared to the great circle route, the optimised route (B.2) gives larger true wind angles, where beam reach is the dominating direction. When the ship is instructed to use the engine whenever the speed to target is below 10 knots, the weather statistics is changed rather much (B.4). The ship will motor against the wind for true wind speeds up to 10 m/s. The dominating sail direction is still beam reach.

5.4. Simulations using Single wind condition (Method A)

The single design wind condition simulation is simply done by looking up the relevant speed in the VPP polar. The question is which wind condition would be selected to be representative.

- A.1 At the initial stage of the design process, beating upwind would be guessed to be a suitable design target condition: Headwind, 10 m/s.
- A.2 Further into the development work, a condition that represents high probability of occurrence can be selected from the route optimisation analysis: TWA=100°, TWS=6m/s

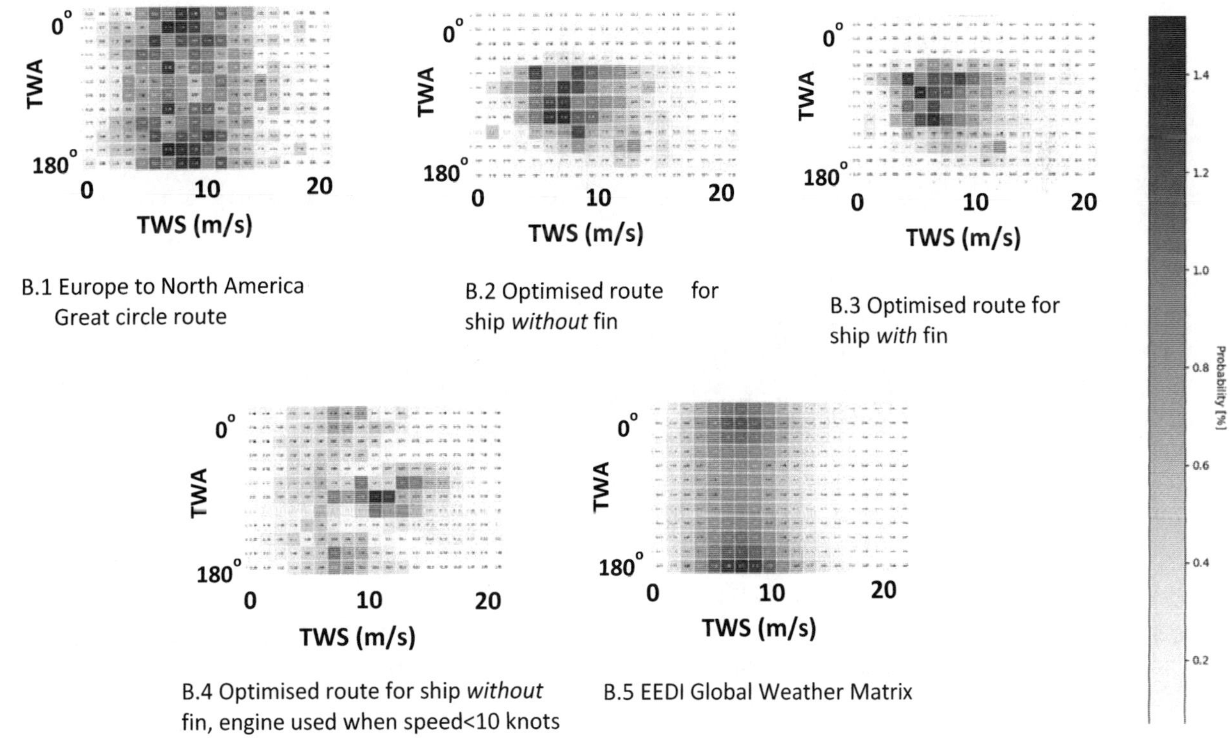

Figure 4. Yearly wind statistics, true wind *relative to ship heading*, starboard and port tack summed, used for the Weather Weighted Polar method (Method C). Darker color represents higher probability.

6. RESULTS

The predicted gain in route average speed due to the fin is shown Figure 5 for the different methods. Since the same polar curves are used for all methods, the differences between the predictions are due to the different representation of the operational weather profile only. The route optimisation, (Method D), can be regarded as the most resolved representation and serves here as the benchmark.

The very fast Weather Weighted Polar method gives similar results as the benchmark, when the wind statistics is derived from the route optimisation (B.2-B.4). This can be a very efficient way to cut the computational time in the design process.

Assuming a great circle route (B.1, C.1) instead of a weather optimised route over-predicts the gain of the fin, since upwind conditions are more frequent occurring on the great circle route than on the optimised route, where more larger wind angles are found.

The EEDI Global weather matrix is clearly not representative for the current route (B.5). The higher wind speeds makes this option to overpredict the gain of the fin compared to the benchmark method D, since the fin is more efficient in higher ship's speed when drag is less penalising

For all results mentioned so far, it has been assumed that the ship is using the wind power alone for 100% of the time. In reality, the ship's Master will choose to use auxiliary engine power in unfavourable winds to keep the schedule. The exact strategy for when to use the engine is not decided at an early design stage. Result B.4 is based on the assumption that the engine will be used to maintain a minimum instant speed of 10 knots. The result indicates that this assumption does not affect the average velocity much for this case. However, the effect of different engine strategies on the optimal design should be further investigated.

Evaluating the fin performance for a 10 m/s upwind condition only (A.1) clearly overestimate its importance, compared to the realistic route representation. The fin is the most effective when it generates side force in upwind condition, and the effectiveness increases with increasing wind speed, when drag is less important. For a large portion of the time on the optimised route the wind speed is less than 10 m/s and the wind directions more open.

The strategy to select the most frequent wind condition on the optimised route as a single point target condition (A.2) leads to an under-estimation of the speed gain, since all upwind conditions are not weighed in. Trying to represent the operational profile with one single condition is clearly not appropriate.

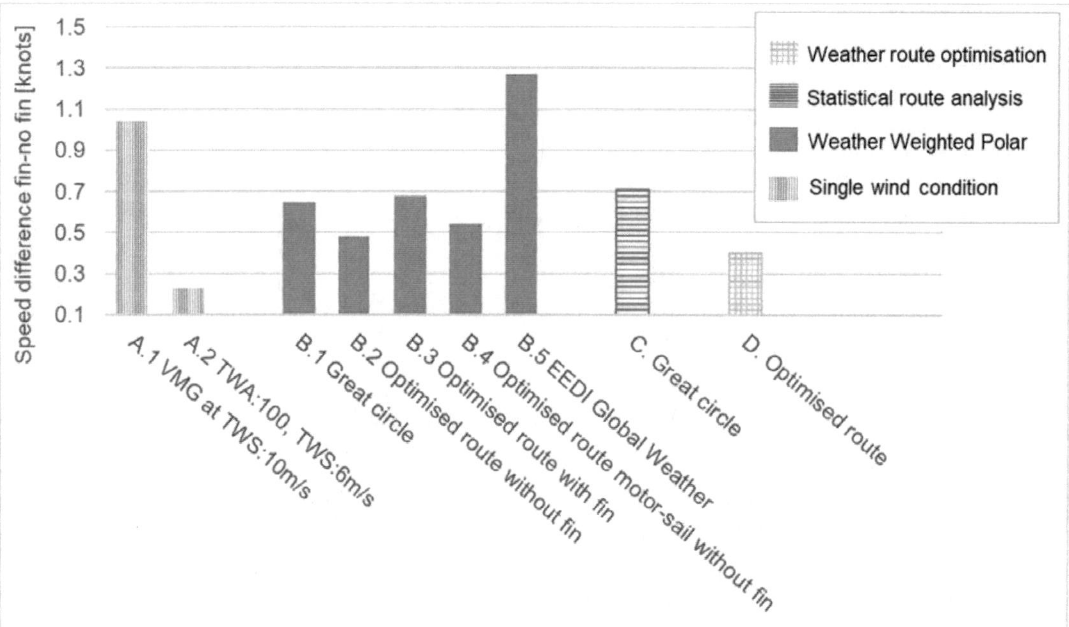

Figure 5. Performance evaluation of the fin test case. Gain of route average speed for ship with fin compared to ship without fin in knots based on different methods

7. CONCLUSIONS AND DISCUSSON

The design process for long-distance sailing vessels requires tools for predicting the speed at given wind conditions as well as methods for estimating the performance on a route considering the all-year around wind conditions. In this study, different strategies for representing the weather profile for the expected operation has been examined using a simple design variation.

For single-objective optimisation, for example parametric design variation with respect to speed, a single point representation of the operation profile may be sufficient. The target wind condition should *drive* the design in the desired direction. This is not necessarily the same as using the most *likely* weather condition. Assume for example that a route optimiser is used to define the most *likely* wind condition. The problem is that the routes found by the optimiser depend on the actual ship's design. If the parent hull does not perform well upwind, then the route optimiser will find routes with less upwind legs. When the corresponding weather statistic is used as design target, it will not favour design changes that improve the upwind performance. The design may be locked-in in a sub-optimal corner. By instead using an upwind sailing condition as the design target, the design will be driven towards better upwind performance. It will, however, over-estimate its importance in relation to other design objectives.

For a multi-object design, for example when balancing performance against cost and risks, it is more relevant to use a realistic representation of the operation profile, as a single wind condition may heavily over or under-estimate the importance of design changes.

There are potential pitfalls and simplifications associated with all the investigated methods. Using the most accurate representation of the operational weather profile requires in practice lower fidelity methods for the performance predictions and/or allows less design variations to be evaluated due to the long computational costs. This study highlights the importance of employing a network of different methods of varying fidelity levels throughout the whole design process. The following strategy for the design process is suggested:

- A weather route optimisation algorithm is used to derive realistic routes and probable weather profiles that the ship will encounter in operation. It can be used to assess design variations when an accurate representation of the overall performance on the route is requested, for example when the time on the route is weighted against other objectives.
- The weather route optimisation algorithm is also used to assess the overall performance, fuel consumption, risk for delays and many other indicators.
- Route analysis based on a *fixed* route is a time-efficient way to assess the overall performance in similar way as the route optimisation algorithm. However, it can be somewhat conservative in assessing the crossing time and it may overestimate the relative importance of upwind performance.
- Weighting the polar diagram output from a VPP with the expected occurrence on an optimised route is a very time-efficient way to accurately represent the operational weather profile. This can be used for example for multi-objective parametric design variation. One needs however, to be aware of how the selected profile drives the design.
- High-fidelity hydrodynamic and aerodynamic evaluations carried out for a few selected target wind conditions can be used to tune and improve the low-fidelity methods needed for the route evaluation. The selection of target wind conditions can be based on the route optimisation algorithm.
- A single target wind condition can be used for single-objective parametric optimisations when very time-consuming high-fidelity hydro/aerodynamic methods are required.

Finally, it can be concluded that the selection of design route may have a significant impact on the assessment and ranking of design alternative. The EEDI Global weather matrix defined by IMO regulations for wind propulsion ships represents an average of a global trade routes, but it may be very misleading to use as a design target for a specific route.

The study was performed for a sailing cargo vessel, but the conclusions regarding methodologies should be applicable to other types of sailing crafts targeting high speed on long routes, like offshore racing yachts.

Acknowledgements

The study is carried out as a part of the Oceanbird project, a research and development project financed by the Swedish Transport Agency and Wallenius Marine.

References

1. International Maritime Organization, "STRATEGIC PLAN FOR THE ORGANIZATION FOR THE SIX-YEAR PERIOD 2018 to 2023," IMO International Maritime Organization, 2017.
2. Kolk, N.J., et.al., (2019), "Part 1: Experimental validation of a RANS-CFD methodology for the hydrodynamics of wind-assisted ships operating at leeway angles", Ocean Engineering 178 (2019) 375–387

3. Eggers, R., Kisjes, A.S., (2019), "Seakeeping and manoeuvring for wind assisted ships", *Proceedings of the International Conference on Wind Propulsion*, 15th -16th October 2019, London, UK

4. Marimon Giovannetti, L. et.al. (2020), "Appendages investigation and their effects on maneuvering coefficients for applications in wind assisted ships "*Proceedings of 5th International Conference on Innovation in High Performance Sailing Yachts and Sail-Assisted Ship Propulsion,* Gothenburg, Sweden, June 2020

5. Graf, K., Hoeve, A.V., Watin, S., (2014), "Comparison of full 3D-RANS simulations with 2D-RANS/lifting line method calculations for the flow analysis of rigid wings for high performance multihulls', Ocean Engineering, No. 90, 2014, pp. 49-61

6. Philpott, A., Mason, A., (2015), "Optimising Yacht Routes under Uncertainty", *in Proceedings of the 15th Chesapeake Yacht Symposium,* 2001

7. Guillou, G., Lemarchand, L., Babau, J-P., (2019), "Sailboat routing with multiple objectives for sailing races", *in the proceedings of the 23rd Chesapeake Sailing Yacht Symposium,* 2019

8. Davies, G.D.S., Waseda, T., (2019) "Optimal routing of a wind-powered cargo vessel using ensemble weather forecast data" *in Proceedings of the International Conference on Wind Propulsion*, 15th - 16th October 2019, London, UK

9. Tillig, F., (2020) "Simulation model of a ship's energy performance and transportation costs" Doctoral thesis, Chalmers University of Technology, Sweden

10. Persson, A., et.al, "Performance predictions of wind propulsion systems using 3D CFD and route simulations," *in Proceedings of the International Conference on Wind Propulsion*, 15th -16th October 2019, London, UK

11. Malmek, K., et.al. (2020), "Comparison of two rapid numerical methods for predicting the performance of multiple rigid wing-sails", *Proceedings of 5th International Conference on Innovation in High Performance Sailing Yachts and Sail-Assisted Ship Propulsion,* Gothenburg, Sweden, June 2020

12. Helmstad, A., Larsson, T., (2013) 'An Aeroelastic Implementation for Yacht Sails and Rigs.' Master's thesis, KTH Royal Institute of Technology

13. Werner, S., (2001), Application of the vortex lattice method to yacht sails, Master's thesis, Technical University of Denmark

14. Heng, Z., (2020), "CFD Investigation of Wind-powered Ships under Extreme Condition" Master's thesis from Chalmers University of Technology

15. MEPC.1/Circ.815 - 2013 Guidance on treatment of innovative energy efficiency technologies for calculation and verification of the attained EED (MEPC.1/Circ.815 (IMO, 2013)

16. Olsson, F., et.al. (2020), "A performance depowering investigation for wind powered cargo ships along a route", *Proceedings of 5th International Conference on Innovation in High Performance Sailing Yachts and Sail-Assisted Ship Propulsion,* Gothenburg, Sweden, June 2020

VARIATION OF UNDERWATER APPENDAGES AND THEIR EFFECTS ON WIND POWERED SHIPS

Laura Marimon Giovannetti[1], Laura.Marimon@sspa.se
Kasper Ljungqvist[2], Kasper.Ljungqvist@sspa.se
Sofia Werner[3], Sofia.Werner@sspa.se

Abstract. Assessing the manoeuvring performances of a ship early in the design stage is becoming ever more important, especially now that wind-assisted propulsion is radically increasing in popularity as a measure to reduce fuel consumption. For conventionally propelled vessels most of the forces are in the longitudinal direction. Some, relatively small, drift angles and rudder angles are encountered in order to compensate for side forces due to environmental conditions. For a wind-propelled vessel on the other hand, side forces are an inevitable consequence of the propulsion choice and it is of utmost importance to balance these side forces with a matching hydrodynamic design to efficiently propel the vessel forward avoiding travelling for longer periods of time. In order to counteract those large side forces, it is necessary to add underwater appendages. Their position along the hull will determine the centre of lateral resistance (CLR) that needs to be related to the aerodynamic centre of effort (CoE) in order to achieve a balanced ship. The research herein described, compares towing tank captive tests with Virtual Captive Tests (VCT) derived from Computational Fluid Dynamics (CFD) simulations and the effects of generated side forces and moments when the inflow angles are varied. Adding a set of diagonal foils to the hull, in the investigated case reduces the leeway angle of approximately 2 degrees in close-hauled conditions.

1. INTRODUCTION

Sail-assisted vessels are recently becoming an attractive concept that is going to be able to reduce fuel consumptions to satisfy the United Nations 2050 environmental plan. The possibility of reducing the carbon emissions deriving from Pure Car and Truck Carriers (PCTC) could tackle emissions ranging between 30 and 60 tonnes of fuel per travel day [1]. New regulations on fuel emissions from the International Maritime Organization (IMO) and the European Commission (EC) starting with the 2030 Agenda and the Paris Agreement at the twenty-first session of the Conference of the Parties to the United Nations Framework Convention on Climate Change (COP 21) are beginning to have an impact on the current research and development of shipping industries [2, 3]. New interest and funding are becoming available for research on wind-assisted ships and alternative fuels. For what concerns sail-assisted vessels, estimates regarding how much fuel can be saved via the aid of the wings to the propulsive system of the ships depends on a number of factors, including the total area of the wings, the type of wings used (i.e. the shape and the arrangements along the deck) and the wind conditions and angles encountered during the routes [4, 5, 6, 7]. All those factors need to be taken into consideration when designing a sail-assisted vessel.

More in detail, for what concerns the current research, the wind conditions and angles encountered during a route are of utter importance. It is indeed necessary to assess whether a wind-assisted ship needs to provide a higher side-force than current sea-going vessels to counteract the aerodynamic forces. When adding appendages, it is extremely important to account for their effects and their interactions with the hull and between them. Wind-powered vessels indeed experience side forces from the aerodynamic characteristics of the sails/wings much larger than standard vessels. The aim of the design processes both for newly produced wind powered vessels and retrofitted ones, is ultimately to introduce at the design or retrofit stages the possibility of assessing and analysing a large number of underwater appendages, their location along and across the hull, some geometric variations (i.e. aspect ratio changes) and some performance changes (i.e. variation of angles of attack).

The effects of the added appendages when coupled with the Wind Propulsion Unit (WPU) and therefore the relationship of the location and the lead between the aerodynamic centre of effort (CoE) and the hydrodynamic centre of lateral resistance (CLR), will allow the vessels to safely manoeuvre and sail.

Due to the complexity of downwash from the upstream appendages (such as daggerboards), the flow speed and angle behind them will be affected (to change the rudder inflow). A summary of low fidelity to high fidelity models is presented (Figure 1), allowing an understanding of the drawbacks and advantages of each method. The possibility of introducing different methods and rankings for a range of appendages can affect the design strategies, allowing early assessment of a large number of load-cases and routing scenarios.

After the presentation of the possible methods, the experimental and numerical set ups and results are presented and discussed. Lastly the effect on the CLR and leeway angle of the diagonal foils configuration are shown and discussed, presenting the Velocity Prediction Program (VPP) polar for leeway.

[1] Senior Researcher, SSPA Sweden AB, SWE
[2] Project Manager, SSPA Sweden AB, SWE
[3] Manager Strategic Research Hydrodynamics, SSPA Sweden AB, SWE

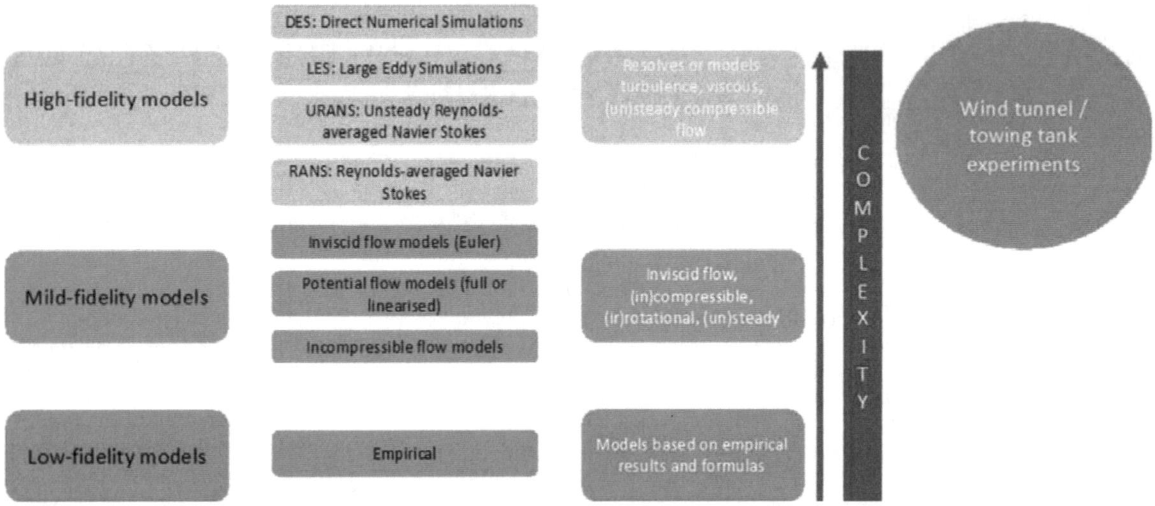

Figure 1: Aero-hydrodynamic flow models in different complexity and accuracy levels. Alongside the numerical models lie the experimental methodologies and results that, if accurate and properly performed in a controlled environment, can lead to the highest fidelity.

1.1. Low fidelity – semi empirical approach

Low-fidelity methods do not model the flow, however, they can be used to estimate the hydrodynamic performances as they rely on empirical data and they can be used for estimates of properties such as lift coefficient, drag coefficient and hydrodynamic efficiency.

For sailing yachts those models have been used extensively in the development of VPPs (for example for the Offshore Racing Congress (ORC)) both for the aerodynamic and hydrodynamic models [8, 9].

For what concerns the hydrodynamic characterization of sailing yachts, the keel to rudder interaction is deemed important to describe the downwash from the keel and how the flow speed and direction are then altered prior to reaching the rudder. In [9] the Delft Ship Hydrodynamics Department investigated the influences of the keel on the rudder, increasing the fidelity level of side-force response for the full yacht. Keel downwash will cause a change in the effective angle of attack of the rudder. The magnitude of this downwash has been determined by the experimental procedures with the hull in yawed condition, while the rudder angle has been varied. The lift generated by the rudder can then be plotted against the rudder angle δ. The difference between the leeway angle β and the rudder angle at which rudder lift becomes zero δ_0 maybe considered to be equal to the averaged downwash angle for a reference system as presented in Figure 5.
$\Phi = \beta + \delta_0$.

From the studies developed in [9] it is important to note that the lower the aspect ratio of the keel, the larger downwash influence on the rudder is measured. Furthermore, the magnitude of downwash angle related to the leeway angle on the keel decreases with increasing leeway angle. The conclusions from the empirical assessment presented in [9] as well as the presented formula are insightful for the present research as the semi-empirical formulae can be easily adapted to be included in SEAMAN, SSPA's simulation tool, for a fast semi-empirical approximation of side-force and drag approximations at a design stage. This can then be further analysed with higher fidelity methods or indeed towing tank captive tests, to assess the influence of low-aspect ratios and spacing.

1.2. High fidelity – RANS methods

Various forms of Navier-Stokes solvers are ranked as high-fidelity. They differ mainly by the extent at which they resolve and/or model the turbulence and if they are time-dependent (transient) or quasi-static.

At SSPA the work carried out to this stage on the hydrodynamic side of the hull and appendages was developed with the RANS simulation within ShipFlow XCHAP, as presented below in Section 3. It optimises the solution time with a very time-efficient mesh and solver. Given the necessity to adapt the model herein presented for a large number of appendages-rudder distances, different fin geometries and positions along and across the hull, a high-fidelity option would be to further analyse with XCHAP a chosen set of leeway angles, longitudinal and transverse positions, angles of attack and aspect ratios to then interpolate between the results to find intermediate solutions within SEAMAN. This study is necessary in order to take into consideration all the 3-dimensional flow effects with the large influences on flow speed and angles introduced by different appendages. Two design variations are tested: the first one with two diagonal foils, as seen in Figure 3, and the second one without such foils.

The research herein described brings forward the knowledge and the reliability of the hydrodynamic models developed to ultimately predict the performances of wind powered vessels along a course.

2. EXPERIMENTAL SETUP

The experiments were carried out at SSPA's towing tank. The tank is aligned North-South with slotted roof lights and is spanned by a carriage with maximum speed of 11m/s and is housed in a heated room [10]. The tests were conducted in March 2020.

The main characteristics of the towing tank are:
Length: 260m
Breadth: 10m
Water depth: 5m
Maximum carriage speed: 11m/s.

The represented ship is an early hull iteration of the Wallenius Marine project "The Oceanbird", the concept design can be seen in Figure 2. The geometrical scale factor of the model is 1/25.

Figure 2: The Oceanbird: a wind powered car carrier developed by Wallenius Marine to be launched in 2024.

The model is fitted with two rudders and for the purpose of this study only the towed conditions (i.e. no self-propulsion tests are considered, simulating a condition when the ship is sailing only powered by the wind) captive tests are reported. Two design variations are tested: the first one with two diagonal foils, as seen in Figure 3 and the second one without such foils.

Figure 3: Towing tank model fitted with rudders, brackets, shafts and two diagonal foils.

For each loading condition, the model was ballasted to the corresponding full-scale volume displacement. The draught was verified at forward perpendicular (FP), mid ships (MS) and aft perpendicular (AP), on both sides of the model. The model is marked with 21 equally spaced stations numbered 0 to 20 between the AP and the FP. A continuous waterline was drawn along the entire length of the hull at the corresponding full-scale design draught of 8.5m. A 1 mm trip wire was fitted at station 19 on the bow in order to develop the turbulent boundary layer during testing.

For the purpose of this research, the authors will focus on the results from the captive tests. A dynamometer capable of measuring forces and moments in 4 degrees of freedom was fitted to the model. The model was free to heave and pitch and restricted in the other 4 degrees of freedom, namely surge, sway, roll and yaw, where the forces and moments were measured.
The dynamometer was statically calibrated using in-situ calibration frame. The first run in the morning was a re-run of the hull and inflow settings for specific conditions to ensure repeatability throughout the days.

During the duration of the captive tests, zero (still) measurements were taken at zero speed in different occasions to check consistency of the signals.

The captive tests were carried out at a corresponding full-scale vessel speed of 12 knots. A full uncertainty assessment was carried out and the values are plotted in the figures below as error-bars.

3. NUMERICAL SETUP

The purpose of the model test was to validate CFD-based captive tests, Virtual Captive Test (VCT), which then can be used with confidence for the coming design work. The result from the VCT is used for input to VPP simulations as well as time domain manoeuvring simulation. For the VPP to solve the equilibrium equations, the hydrodynamic forces and moments acting on the underwater surface of the hull and appendages are solved for an extended test matrix that simulates a large number of inflow conditions by the ship hydrodynamics SHIPFLOW solver.

This research not only focuses on the creation of a VCT and the possibility of using the CFD results as a base for a consistent VPP and manoeuvring simulations, but focus has also been applied to assess the influence of scaling effects. The numerical simulations are therefore performed both for full-scale and model-scale hulls.

The numerically investigated conditions range between a Froude Number (Fn) of 0.057 0.17 and a Reynolds number (*Re*) of $4.47x10^8$ and $1.34x10^9$. This range lies in the fully turbulent flow condition. For the purpose of this research only the underwater components of forces and moments of the ship are considered from the RANS simulations with a condition of symmetry plane at the free-surface. However, the free-surface and the corresponding effects are simulated with a three-dimensional potential-flow Rankine-source panel method, when analysing the resistance of the ship, and the total ship resistance is compared to the measured one.

Within SHIPFLOW the Finite Volume Method (FVM) discretisation solver XCHAP has been selected to solve the RANS partial differential equations. With the FVM the flow domain is discretised by a collection of grids. In

SHIPFLOW, the grids are only structured, therefore arranged in a regular pattern. The current research solves a hull completed with appendages, therefore overlapping grids are present. The fluid domain and the underwater part of the hull are represented by a H-O structured grid topology. The appendages, such as the rudders, are represented by a O-O grid topology. In the overlapping grid, the appendages are treated as separate blocks of component grids. The component grids are fitted to the body and then stretched outwards toward the fluid domain, keeping the specified y+ value. The target y+ value for every simulation is specified as 1, fully resolving the boundary layer around the hull and appendages. An example of overlapping grids can be seen in Figure 4.

The SST k-ω turbulence model is used in the simulations. All the simulations are performed with a mesh of approximately 20-25 million cells performed on a Linux Intel(R) Xeon(R) Gold 6142 CPU @ 2.60HHz (32 CPUs), 192 GB RAM. Using 32 cores and depending on the inflow direction the simulations take approximately 7-15 hours for a fully converged solution.
Further details of the numerical setup can be found in [11].

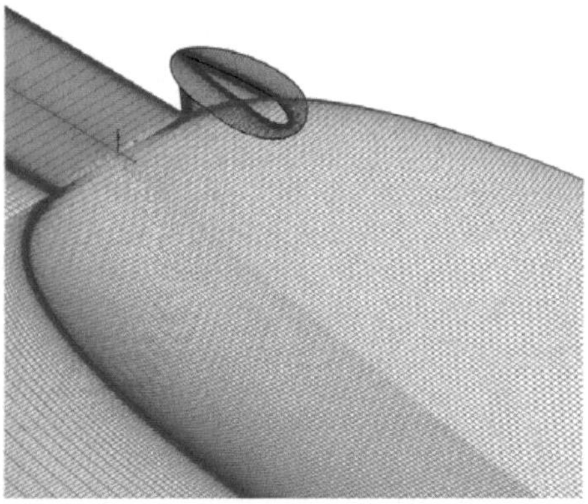

Figure 4: Example of overlapping grid between hull (H-O topology in red) and rudder (O-O topology in blue).

4. VCT AND TOWING TANK COMPARISON

In order to correctly compare the towing tank and the CFD results, the CFD simulations are transposed to the same coordinate system of the measured forces and moments seen in Figure 5.

In the static force model, the forces and moments have a right-handed coordinate system with four degrees of freedom. The origin is set at LPP/2, in the centreline of the hull and vertically in the mean waterline.

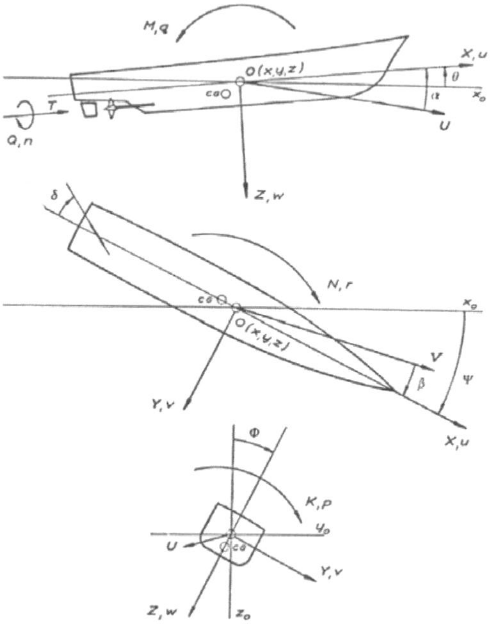

Figure 5: Coordinate system of the towing tank tests as well as the VCT.

Typical static captive model tests include a static drift test with a given drift angle β and a given rudder angle δ [12]. The effective rudder angle, describing the inflow to the rudder, will change both with rudder angle and drift angle. In the current test matrix a range of rudder-drift combined angles have also been included to determine the effective rudder angle for various combinations of rudder angles and drift angles.

The captive test points, shown in Table 1, were selected to give a good representation of the cases that are expected to be experienced in the VPP representing a wide range of flow features.

Table 1: Virtual Captive Test matrix points with overlapping towing tank test matrix points highlighted in red and points only tested in the towing tank highlighted in green.

Test type	F_n	δ [°]	β [°]	φ [°]
Rudder angle	0.137	-2, -5, -10, -15, 5		
Drift angle	0.137		2, 4, 6, 10, -6	
Rudder-Drift	0.137	-5, -7, -9	3,4,6	
Heel Angle	0.137			-5

Initially, in order to account for and understand the increase in resistance given by the diagonal foils, a comparison of the resistance of the towing tank tests and CFD results was performed. For this study in the numerical environment the ship hull is simulated in full-scale. The CFD results, simulating only the underwater bodies, provide results for the viscous resistance (R_V). The total model resistance (R_{TM}) is measured in the towing tank. From the CFD R_V, the viscous resistance coefficient (C_V) is derived. The addition of diagonal foils to the hull increases the viscous resistance, with a maximum increase at the highest investigated speed. This larger value could be mainly associated to the generation of vertical lift from the diagonal foil placed close to the free-surface, that, in ShipFlow software is modelled as a symmetry plane. This could lead to higher uncertainties from the numerical simulation due to a larger flow-complexity. From Figure 6 it is possible to see how the trend for the total resistance coefficient is similar between the CFD and the towing tank results.

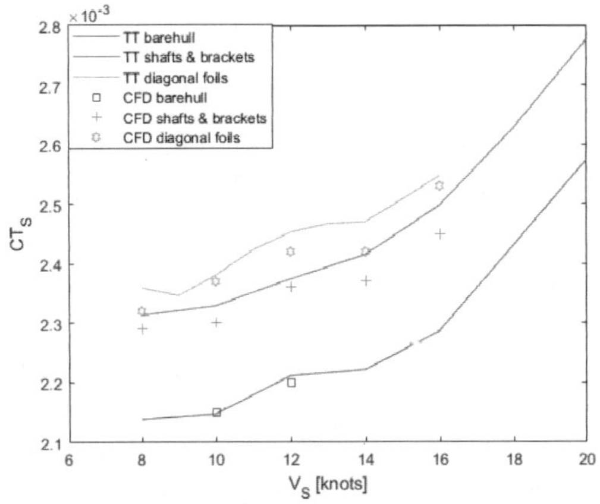

Figure 6: Towing tank and CFD comparison breakdown of resistance coefficient for bare-hull and appended cases estimated with 1978 ITTC Power Prediction Method.

Moreover, the increase in resistance due to the addition of diagonal foils is also well captured by the numerical model. The resistance increases due to the diagonal foils by 2 % at ship speed 12 knots, based on model scale values (orange and yellow line values in Figure 6). The diagonal foils in a resistance tests affect the free-surface only slightly, as seen in Figure 7. The differences in wave profiles are considered insignificant for a ship speed corresponding to V_S=12 knots. However, they could be more significant at higher ship speeds.

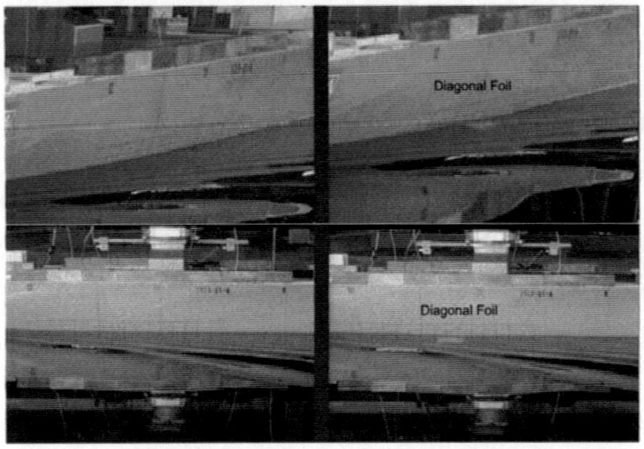

Figure 7: Different views of the free surface in the towing tank to inspect the influence of the diagonal foils on the wave profiles.

For the other degrees of freedom, sway, yaw and roll, the forces and moments from the towing tank are non-dimensionalised to their corresponding coefficients taking into consideration the surface areas and the ship speeds. The figures presented below are therefore results for both the model and full-scale but non-dimensionalised.

Figure 8 shows the side force experienced both in the towing tank and in the CFD simulations when the vessel is subject to leeway angle. From the figure it is possible to see that the CFD closely follows the trend experienced in the towing tank measures and all the CFD results lie within the uncertainty errors. The hull with the diagonal foils presents, as expected, a higher side force compared to the one without. The uncertainties, added as asymmetric error bars to the towing tank results, were measured during the towing tank campaign to assess the hull and rudders alignment compared to the tank. Indeed, during captive tests, great care must be taken when aligning the model with respect to the tank reference axis. The alignment can be checked using pure drift tests at small angles (between +/-2°). During the current towing tank test, both the alignment in leeway and the alignment of the rudder angles have been checked.

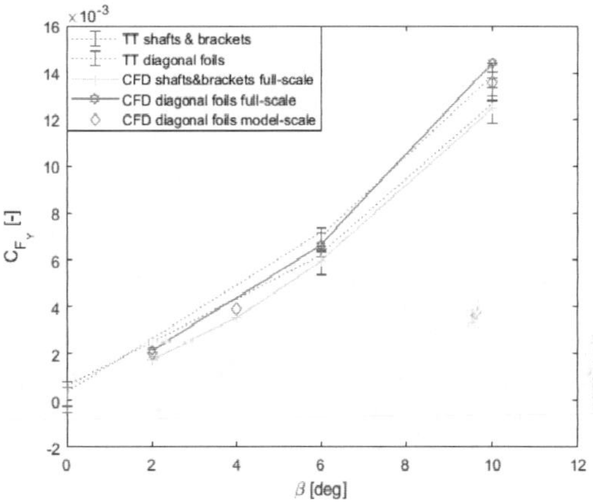

Figure 8: Side force coefficient over leeway angle for both CFD and towing tank results, in model and full-scale for V_s=12 knots.

Figure 9 shows the side force experienced both in the towing tank tests and the CFD environment when subject to a rudder angle. Contrarily from the leeway angle results (Figure 8), where all the results are within the uncertainty errors of the towing tank measures, especially at large negative rudder angles, the full-scale CFD model presents larger side forces, with differences between the model-scale CFD and the full-scale CFD ranging between 20-30%.

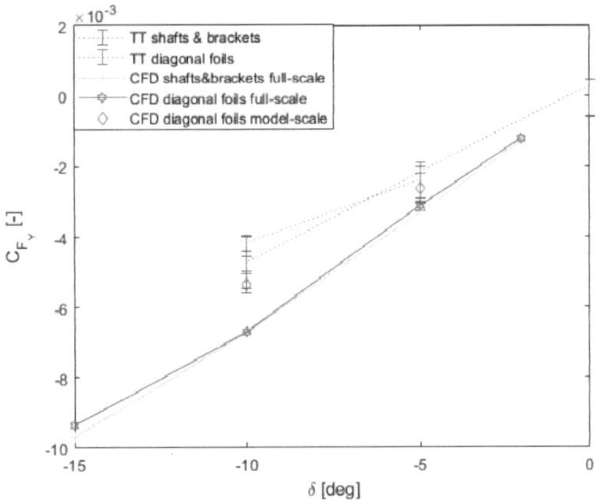

Figure 9: Side force coefficient over rudder angle for both CFD and towing tank, model and full-scale for V_s=12 knots.

As described in [13, 14, 15, 16] the model-scale rudder presents a lower Reynolds number than the full-scale one, leading to errors when estimating the maximum lift and stall angle for example (Figure 10). An increase in *Re* leads to an increase in the lift coefficient and a decrease in the drag coefficient [17]. The relationship of model-scale and full-scale *Re* is as follows:

$$Re_F = \frac{\rho(\lambda^{0.5}V_{RM})(\lambda C_{RM})}{\mu} = \lambda^{1.5}Re_M \qquad (1)$$

where C_R is the rudder chord length, λ is the scale factor, and μ is the dynamic viscosity of water. Subscripts F and M indicate full-scale and model scale. In the current research, at model scale, the *Re* on the rudders is always below 5x10^5, lying in the laminar-flow region.

Furthermore, in [18] the authors experience a similar level of errors ranging between 20-30% comparing the full-scale and the model-scale results for the sway force predictions, leading to errors in the manoeuvring performances of the vessel. Those findings lead to the necessity of simulating the ship in full-scale in CFD for future simulations to obtain the predicted sway forces and to ultimately present an accurate VPP, leading to larger computational times but more accurate results in manoeuvring and performance prediction estimates.

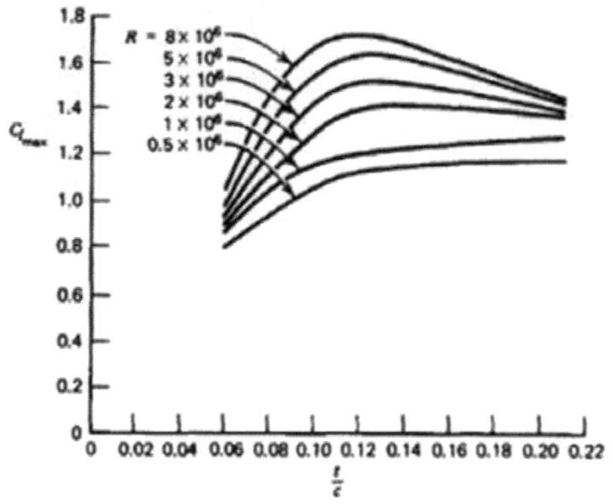

Figure 10: Effects of low Reynolds number on lift coefficients of a NACA24xx series [13].

Figure 11 and Figure 12 present the towing tank and CFD results for the yaw moment. From the figures it is possible to see how the general trend is well represented in the CFD environment, with the model-scale CFD results closely matching the towing tank measures trend. The yaw moment is increased with the addition of the diagonal foils, with the largest influence of the yaw moment given by the rudder angle due to their low aspect ratio and large surface areas and most importantly their position along the hull. The CFD results in full-scale are underpredicting the yaw moment for a change in drift angle compared to the towing tank measures (Figure 11) as the effective rudder angle and the *Re* scaling effects would lead to a reduction in yaw moment. Contrarily they present higher results than the towing tank measures for yaw moment when a negative change in rudder angle is introduced due to associated *Re* scaling issues. When both rudder and drift angles are introduced, the results of the CFD and TT present a very good agreement as the overall effects of scaling and rudder angles would balance. The diagonal foils are introduced slightly forward of LPP/2, so the influence on yaw moment is not large.

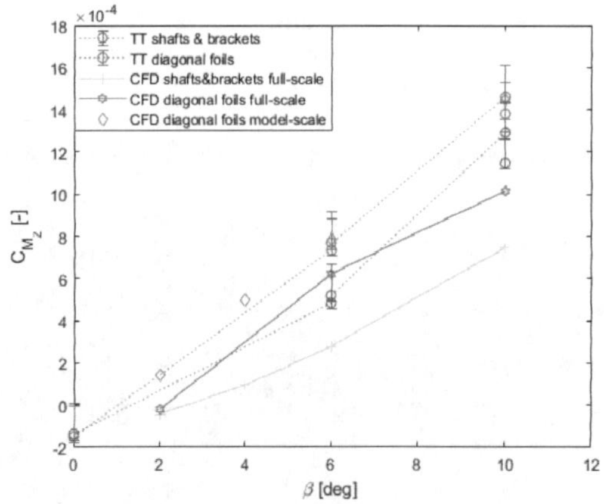

Figure 11: Yaw moment coefficient over leeway angle for V_s=12 knots – towing tank measures and CFD simulations.

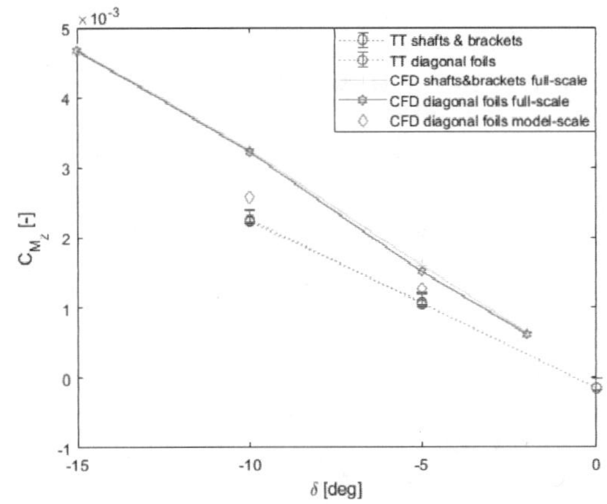

Figure 12: Yaw moment coefficient over rudder angle for V_s=12knots towing tank measures and CFD simulations.

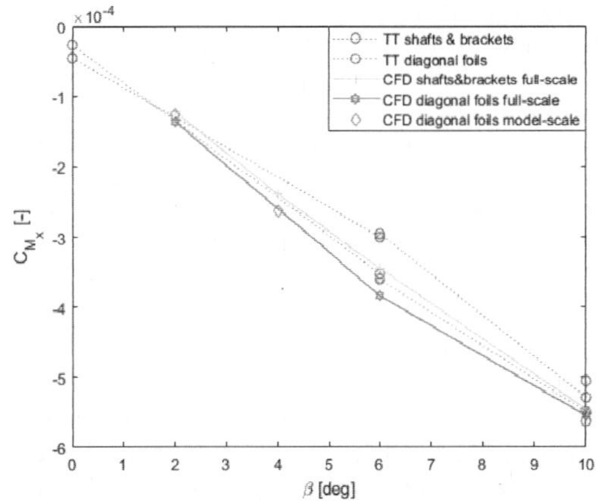

Figure 14: Roll moment coefficient over leeway angle Vs=12 knots towing tank measures and CFD simulations.

The position of additional appendages such as the diagonal foils, when moved longitudinally along the hull will influence greatly the CLR. However, in this configuration, the largest impact on CLR is given by the two rudders. As seen in Figure 13, the creation of side force near the aft perpendicular brings the CLR close to LPP/2, that would contrarily be placed ahead of the forward perpendicular if the bare-hull was to be considered. The results shown present a comparison of outputs from the CFD datapoints and a VPP SEAMAN simulation. More details about the VPP can be found in [11].

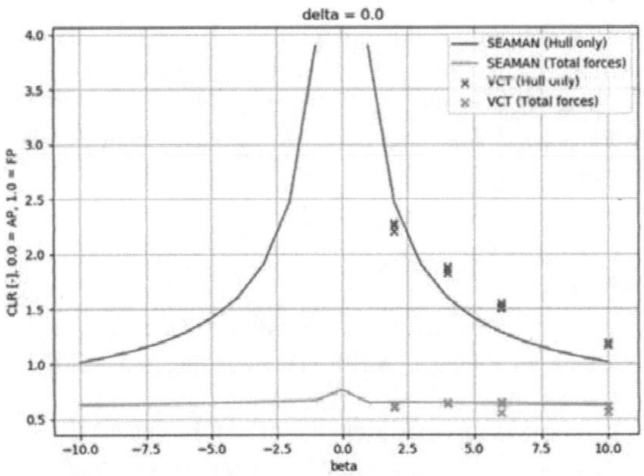

Figure 13: Comparison of CFD and VPP results for the centre of lateral resistance of a bare-hull and with addition of rudders, brackets and shafts.

Roll moment is less subject to scaling effects. As expected, due to the position and geometry of the diagonal foils, the roll moment is decreased with the addition of the appendages (Figure 14).

Figure 15 shows the polar plot of the leeway angle as output from the quasi-static VPP.

From the figure, it is clear how the influence of the diagonal foils (orange line in the plot) on the side force response decreases the overall leeway angle experienced by the vessel, especially in the close-hauled condition, i.e. true wind angles between 40 and 65 degrees. The figure is the response of the vessel sailing in a true wind speed of 10 m/s.

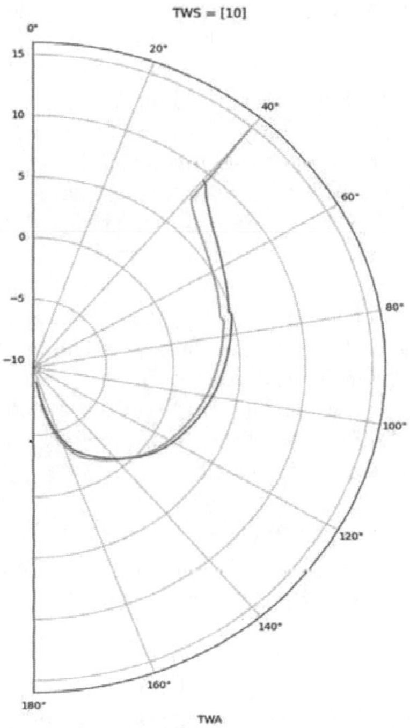

Figure 15: Leeway angle polar plot for the shaft and brackets arrangement (blue line) and the diagonal foils arrangement (orange line) simulated with the forces and moments input from full-scale CFD analysis.

The general trends and increases/decreases in forces and moments due to the addition of the diagonal foils are well captured by the CFD simulations. Analysing the results for yaw and roll moment, as well as for side force, the CFD simulations in model scale are always in line with the towing tank measures but the scaling effects, especially affecting the rudders, are clear in the comparison with full-scale simulations. When transferring the forces and moments to quasi-static VPP or dynamic manoeuvring simulations, it is important to understand the scaling effects and, if possible, minimise them to ultimately correctly predict the performances of the vessels. When testing models, both in numerical and physical environment, it is therefore favourable to test a model as large as possible that leads to sufficiently high *Re* on the lifting surfaces, above which lift and drag coefficients are slightly affected, for instance at *Re* of 6×10^6 where a fully turbulent flow can be expected [19].

5. CONCLUSIONS

The paper presents a thorough comparison of hydrodynamic forces and moments as measured in a towing tank campaign carried out in March 2020 at SSPA and numerical simulations both for model and full-scale vessels. The scaling effects on the lifting surfaces as well as an investigation of a set of appendages is presented and discussed. From the hydrodynamic analysis it is possible to see how the numerical simulations, with the correct inputs, are able to predict the forces and moments acting on the hull and appendages during captive tests. This knowledge gives confidence on the development of the tools to predict at the design phase the future performances of the vessels both for quasi-static and manoeuvring simulations. Whilst separately large commercial vessels and sailing yachts (both cruising and racing) have been extensively studied and their performance predicted, the connection between the two with the advent of wind propelled ships introduces complex scenarios, such as interaction effects of underwater appendages and wing-wing interactions with rigs of the same span and chord. Moreover, the relationship between the CoE and the CLR need to be accurately predicted in order to safely manoeuvre the ship. Therefore, new methods are introduced and herein briefly described.

The tools that SSPA in partnership with Wallenius Marine and KTH are developing form the base of future predictions of vessel performances in a range of different scenarios.

Acknowledgements

The authors would like to acknowledge the collaboration with Wallenius Marine, especially with Mikael Razola, and KTH. Furthermore, the authors would like to thank the technicians within SSPA that have worked relentlessly to be able to deliver the towing tank campaign albeit the global pandemic. The work conducted for this research was supported by the Swedish Transport Agency under grant TRV 2018/96451.

References

1 N. Bialystocki and D. Konovessis, (2016), “On the estimation of ship's fuel consumtion and speed curve: a statistical approach”, *Ocean Engineering,* pp. 157-166.

2 International Maritime Organization, (2017), “Strategic plan for the organization for the six-year period 2018 to 2023”, IMO International Maritime Organization.

3 International Maritime Organization, (2019), “Strategy, planning and reform”, IMO International Maritime Organization.

4 G. M. Atkinson and J. Binns, (2018), “Power profile for segmented rigid sail”, *Journal of Marine Engineering & Technology,* pp. 99-105.

5 T. Smith, P. Nweton, G. Winn and A. L. Rosa, (2013), “Analysis techniques for evaluating the fuel savings associated with wind assistance”, in *Low Carbon Shipping Conference*, London.

6 F. Smulders, (1985), “Exposition of calculation methods to analysie wind-propulsion on cargo ships”, in *International Symposium on Windship Technology (WINDTECH'85)*, Southampton.

7 I. M. Viola, M. Sacher, J. Xu and F. Wang, (2015), “A numerical method for the design of ships with wind assisted propulsion”, *Ocean Engineering,* vol. 105, pp. 32-42.

8 Offshore Racing Congress, (2016), “ORC VPP Documentation”.

9 J. A. Keuning, M. Katgert and K. J. Vermeulen, (2007), “Further analysis of the forces on keel and rudder of a sailing yacht”, Annapolis.

10 M. Brown, I. Campbell and J. Robinson, (2002), “The accuracy and repeatibility of tank testing from experience of ACC yacht development”, in *High Performance Yacht Design Conference*, Auckland, NZ.

11 L. Marimon Giovannetti, F. Olsson, M. Alexandersson, S. Werner and C. Finnsgård, (2020), “The effects of hydrodynamic forces on maneuvrability coefficients for wind-assisted ships”, in *Proceedings of the ASME 2020 39th International Conference on Ocean, Offshore and Arctic Engineering OMAE 2020*, For Lauderdale, FL, USA.

12 Y. Liu, L. Zou, Z. Zou and H. Guo, (2018), “Predictions of ship maneuvrability based on virtual captive model tests”, *Engineering Applications of Computational Fluid Mechanics,* vol. 12, no. 1, pp. 334-353.

13 B. W. McCormick, (1995), “Aerodynamics, Aeronautics, and Flight Mechanics”, John Wiley & Sons, Inc.

14 N. Sasaki and M. Atlar, (2019), “Scale effect of Gate rudder”, in *Sixth international symposium on Marine Propulsors smp. 19*, Rome.

15 N. Sasaki, S. Kuribayashi, M. Fukazawa and M. Atlar, (2020), "Toward a realistic estimation of the powering performance of a ship with a gate rudder system", *Journal of Marine Science and Engineering,* vol. 8.

16 A. F. Molland and S. R. Turnock, (2007), "Marine rudders and control surfaces", Oxford: Butterworth-Heinemann.

17 J. Liu and R. Hekkenberg, (2017), "Mesh properties for RANS simulations of aerofoil-shape rudder hydrodynamics", in *10th International Workshop on Ship and Marine Hydrodynamics (IWSH 2017),* Keelung, Taiwan.

18 Y. Jin, S. Chai, J. Duffy, C. Chin and N. Bose, (2015), "Scale effects on hydrodynamic manouvring force prediction", in *Twenty-fifth Intranational Ocean and Polar Engineering Conference*, Kona, Hawaii, USA.

19 J. Liu and R. Hekkenberg, (2017), "Sixty years of research on ship rudders: Effects of design choices on rudder performance", *Ships and Offshore Structures,* vol. 12, no. 4, pp. 495-512.

EFFICIENT JIB-MAINSAIL FLUID-STRUCTURE INTERACTION MODELLING BASED ON AN EXTENDED *LIFTING-LINE THEORY* – VALIDATIONS WITH SEMI-RIGID SAILS EXPERIMENTS

Antoine Morvan[1], antoine.morvan@ensta-bretagne.org, Matthieu Sacher[1], matthieu.sacher@ensta-bretagne.fr
Alain Nême[1], alain.neme@ensta-bretagne.fr, Jean-Baptiste Leroux[1], jean-baptiste.leroux@ensta-bretagne.fr
Christian Jochum[1], christian.jochum@ensta-bretagne.fr, Nicolas Abiven[2], Nicolas.Abiven@chantiers-atlantique.com

Abstract. The present work takes place in the framework of the "Jib Sea" project. The main purpose of the project is to develop a new sail design made of articulated composite panels, for large merchant ships. The French laboratory "ENSTA Bretagne, IRDL" is involved in this industrial project to provide its expertise on fluid-structure interactions modelling. A fast and robust approach to model fluid-structure interactions for yacht sails is presented. Specifically, interaction effects between the jib and the mainsail are taken into account in the flow model presented. This is achieved using the lifting-line theory combined with a discrete vortex method, involving distributions of lumped-vortex elements along sail sections. The flow model is coupled with a structural finite element software, using shell elements for the modelling of sail membranes, beam stringers for battens modelling and a quasi-static resolution based on a dynamic backward Euler scheme. Numerical comparisons with experiments are conducted on a 50 m² composite mainsail prototype and a conventional jib , built and hoisted on an onshore balestron rig. Measurements, such as strain gauges or cable tensions, are synchronized with a wind sensor. These data collected together enable both global and local numerical-experimental comparisons for forces and moments, providing a validation of the proposed fluid-structure interactions modelling of yacht sails. A good matching between experimental and numerical modelling is observed on local comparisons. Indeed, relative differences are all less than 25% for $TWA \in [-20; 20]$ and $TWS < 10$ kn. Global comparison results exhibit validations with experiments for $|TWA| < 10$ deg and $TWS <$ 10 kn, where numerical-experimental relative differences are less than 10%.

1. NOMENCLATURE

C_{var}	Variation coefficient $[-]$
c	Chord length $[m]$
ds	Integration element $[m]$
e	Euler number $[-]$
F	Force $[N]$
h	Batten height $[m]$
i	Imaginary number $[-]$
j,k	Index $[-]$
L	Length $[m]$
M	Moment $[N.m]$
N	Number of element $[-]$
Q	Element for DVM $[-]$
U	Wind velocities $[m.s^{-1}]$
V	Flow velocities $[m.s^{-1}]$
w	Complex velocity $[m.s^{-1}]$
z	Complex number $[-]$
α	Angle for LLT $[deg]$
Γ	Global circulation $[m^2.s^{-1}]$
γ	Local vortex circulation $[m^2.s^{-1}]$
ΔP	Pressure variation $[Pa.]$
Δl	Elongation $[m]$
ε	Relative error $[-]$
θ	Local angle per element $[deg]$
μ	Mean $[N]$ or $[N.m]$
σ	Standard deviation $[N]$ or $[N.m]$
ρ_0	Density $[kg.m^{-3}]$
ω_i	Induced velocity $[m.s^{-1}]$
$\langle EI \rangle$	Out-of-plane bending stiffness $[N.mm^2]$
LE	Leading edge
TE	Trailing edge
x,y,z	Spatial coordinates $[m]$

2. INTRODUCTION

The present work is part of a substantial industrial project, called "Jib Sea", bringing together several industrial companies and ENSTA Bretagne as the academic partner. The main purpose of this project is to develop new rigid sails for use on large merchant ships. Such sails are made of composite panels that hoist and lower like an accordion. The main benefit of this kind of articulated composite sails stays in the increase of the lifetime, and the possibility to significantly increase sail areas to face values of more than 1000 m².

The ENSTA Bretagne laboratory is involved in this industrial project to provide its expertise on Fluid-Structure Interactions (FSI) modelling. In most cases, sails are manufactured with thin textile fibers [1]. The aerodynamic loading on thin sail membranes usually involves large displacements and deformations [2], that directly affect the flying shape of the sail and consequently pressure fields. It is therefore necessary to consider a nonlinear coupled approach to correctly capture the sail FSI phenomenon [3]. However, because of the complex coupled aero-elastic behaviour of yacht sails, nonlinear FSI computations are rather difficult to implement. Furthermore, the modelling strategy must be able to couple both structure and flow solvers.

[1] ENSTA Bretagne, CNRS UMR 6027, 2 Rue François Verny, F29200 Brest, France
[2] Chantiers de l'Atlantique, Avenue Antoine Bourdelle, F44600 Saint-Nazaire

In most of cases, structure solvers are based on discretized method, such as the well-known Finite Element Method (FEM). For flow solvers, resolution methods are usually based on viscous or inviscid flow models [4]. The fairly simple implementation and the fast computation time of inviscid models makes them very popular and used today. Viscous flow approaches start also to be used, but they are usually limited to large industrial projects with adequate time and money. In most of industrial projects, these computations are still very expensive, in addition to convergence or mesh deformation problems that may happen. Consequently, there is a real need nowadays for the design departments of yacht companies to work with fast and robust FSI solvers, allowing a fast and easy to use accurate enough modelling.

The FSI computation approach of the present paper follows a strategy very similar to the work of Sacher et al. [5]. However, the method is extended to the flow solver part. Especially, the proposed flow model considers now interaction effects between jib and mainsail. The structural and flow solvers, respectively FEM and Lifting-Line Theory (LLT), are the same, considering a steady solution of the nonlinear FSI problem. The LLT model from the Prandtl [6] theory is applied as in [5], but pressure fields are now computed using the Discrete Vortex Method (DVM) from Katz and Plotkin [7]. The structural solver part of, the presented method is performed using *Abaqus™ 2017* finite element software. As in [5], a quasi-static approach is involved with a dynamic backward *Euler* scheme [8] to improve the convergence robustness using a standard *Newton-Raphson* method. The mainsail membrane and the reinforcements are modelled with shell elements, battens are modelled with stringers elements connected to shell nodes. The cables connecting the panels are modelled with spring connecters. Each panel luff is pinned to the mast and is free to move along it. In order to simplify the FSI problem, the solving of the jib structure response was not conducted even if it could have been considered by the strategy presented in this work. However, the jib was modelled from an aerodynamic point of view based on a rigid geometry assumption to be considered in the flow solver.

Numerical comparisons with experiments are furthermore provided in the second part of this work. As for experimental setup ,a 50 m² composite mainsail prototype and a conventional jib have been built and hoisted on an onshore balestron rig (see Figure 1) to provide several measurements, such as strain gauges or cable tensions, synchronized with a wind sensor. These data collected together enable both local and global numerical-experimental comparisons, to validate the proposed FSI modelling of yacht sails.

In the following, Section 3 of the paper outlines the FSI modelling strategy proposed. Section 4 presents the composite mainsail prototype and its dedicated sensors. In Section 5, numerical to experimental comparisons on the composite mainsail prototype are presented and results observed allow a reasonable validation of the sail FSI modelling strategy. Finally, main conclusions on this work are given in Section 6.

Figure 1 Jib Sea prototype at 1:5 scale.

3. NUMERICAL MODEL

Compared with a previous work [5], where only the mainsail was considered, the present study is extended to account for the jib. In this work, the mast is still considered rigid and we assume that the FSI problem has a steady solution, which allows to couple flow and structural solvers with a quasi-static approach.

In the following, the flow and the structural solver are respectively described in Part 3.1 and Part 3.2. Part 3.3 presents the FSI coupling technique and Part 3.4 summarizes the major uncertainties and modelling strategies.

3.1. Aerodynamic Models

In this part, the flow model for wind aerodynamic loads on sails is derived.

As a hypothesis, we consider that the wind acting on sails can be modelled by an inviscid flow. Paragraph 3.1(a) provides details on the computation, of flow characteristics on sails along the span (mainsail and jib), based on the *Lifting-Line Theory* modelling approach. Paragraph 3.2(b) describes the calculation of the aerodynamic pressure fields along the chord of the mainsail and the jib, using the *Discrete Vortex Method*.

3.1(a) Flow characteristics

Using the *Lifting Line Theory* (LLT) and similarly to Sacher et al. [5] the two sails are modelled as a single straight lifting line and an associated vortex sheet. The main difference with [5] is that here the sections are composite, as being made up of two sub-sections: one for the jib and one for the mainsail. This is illustrated in Figure 2, considering at a given z coordinate a two-dimensional (2D) section subjected to a uniform flow $\vec{U}_0$ of angle α with the x-axis defined by the lifting sections of the jib and the mainsail of chord lengths $c_{jib}(z)$ et $c_{main}(z)$:

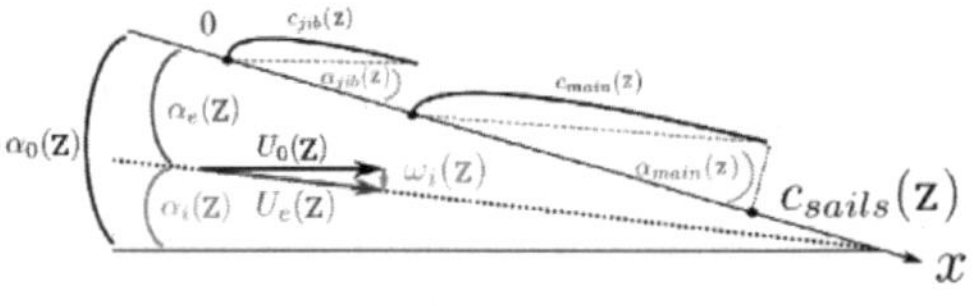

Figure 2 Induced velocity and induced angle in the vicinity of the composite section of the lifting line model.

The LLT allows to compute the induced velocity field along the span, $\omega_i(z)$ and $\alpha_i(z)$, as defined in Figure 2 (see references [5], [6], [7]). The effective angle $\alpha_e(z)$ and the effective velocity $U_e(z)$ of the flow at a given height z are then obtained, according to equations (1):

$$\begin{cases} \alpha_e(z) = \alpha(z) - \alpha_i(z) \\ U_e(z) = U_0(z) + \omega_i(z) \end{cases} \quad (1)$$

In the following, aerodynamic pressure fields along the sails chord are computed section by section using $\alpha_e(z)$ and $U_e(z)$

3.1(b) Aerodynamic pressure field

The LLT method alone cannot compute local aerodynamic sail pressure loads, but provides chord- integrated loads along the span. An additional aerodynamic formulation is required for this purpose. In the previous work [5], a semi-analytical model from the well-known linearized thin wing sections theory was used. In this work, it has been replaced by the use of the *Discrete Vortex Method* (DVM) [7], more suitable to deal with two-element sections. Considering at a given z coordinate, the section (jib + mainsail), in the effective velocity field ($U_e(z)$, $\alpha_e(z)$) calculated from the LLT, the DVM allows to model the lifting potential flow subjected to the boundary conditions $(2.a)$ and $(2.b)$:

$$\begin{cases} \vec{V}_{jib}\cdot\vec{n}_{jib}|_{L_{jib}} = 0 \; and \; \vec{V}_{main}\cdot\vec{n}_{main}|_{L_{main}} = 0 & (2.a) \\ \Delta P|_{TE_{jib}} = 0 \; and \; \Delta P|_{TE_{main}} = 0 & (2.b) \end{cases}$$

The circulation Γ, defined in equation (3), where $\vec{V}_{jib}$ and $\vec{V}_{main}$ are respectively the flow velocity around the jib and the mainsail section, is generated using a discrete vortex distribution.

$$\Gamma = \oint_{L^-_{jib}\cup L^+_{jib}} \vec{V}_{jib}\cdot\vec{t}_{jib}\,ds + \oint_{L^-_{main}\cup L^+_{main}} \vec{V}_{main}\cdot\vec{t}_{main}\,ds \quad (3)$$

The two-element lifting section is discretised into N elementary facets $[Q_{k-1}, Q_k]$ of length ds_k (see Figure 3).

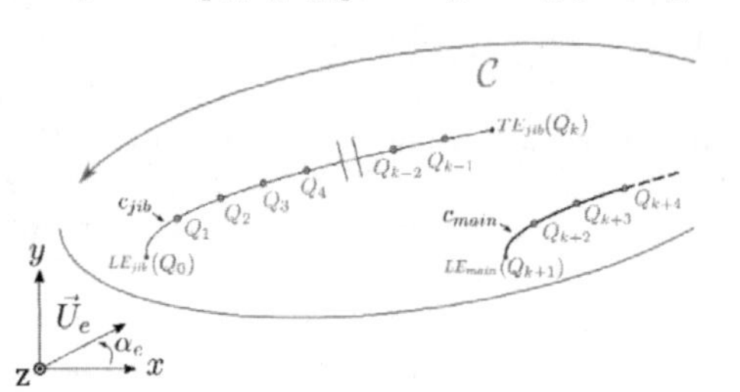

Figure 3 Two-element lifting section discretization. Here only a focus on the jib section has been represented.

The number of N elementary facets is about 2500. The circulation contribution of each ds_k facet is modelled by a discrete vortex of intensity γ_k located at one quarter of the leading edge of the facets.The flow tangency boundary condition is applied at one quarter of its trailing edge (see Figure 4 and [7]).

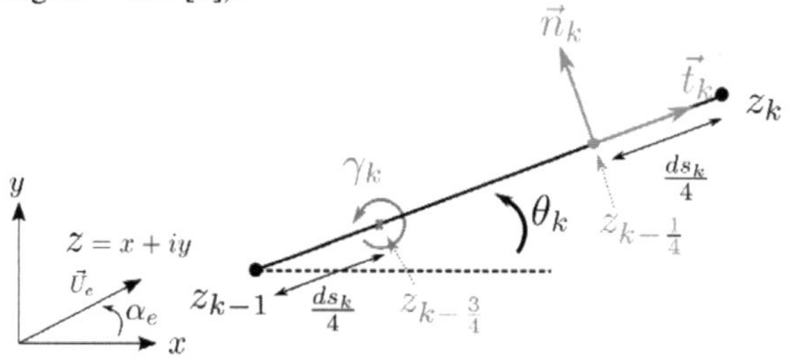

Figure 4 Local parametrization of a ds_k facet.

For each section, the flow tangency boundary condition (4) is expressed in the complex number space as presented in equation (5):

$$Im\left[w_{j-\frac{1}{4}}e^{i\theta_j}\right] = 0 \quad (4)$$

Where:

$$w_{j-\frac{1}{4}} = U_e e^{-i\alpha_e} - \sum_{k=1}^{N_{jib}} \frac{i\gamma_k}{2\pi\left(z_{j-\frac{1}{4}} - z_{k-\frac{3}{4}}\right)} - \sum_{k=1}^{N_{main}} \frac{i\gamma_k}{2\pi\left(z_{j-\frac{1}{4}} - z_{k-\frac{3}{4}}\right)} \quad (5)$$

Three-dimensional effects being considered through the couple $(\alpha_e(z), U_e(z))$ computed by the LLT.
The writing of the $N = N_{jib}+N_{main}$ flow tangency boundary condition leads to a system of N linear equations for the N unknowns $s\gamma_k$. The solving of this linear system provides all vortex intensity γ_k per lumped-vortex elements for the jib and the mainsail.
Finally, for each element of the discretisation of the jib and mainsail section, the pressure load (6) is approximated constant per element from the Kutta formula:

$$\Delta P_k = -\frac{\rho_0 U_e \gamma_k}{ds_k} \quad (6)$$

The FSI coupling strategy is presented.in following Part 3.2 and Part 3.3,

3.2. Structure model

The structure modelling approach is performed using the *Abaqus*™ finite element analysis software and is based on a discretisation of the mainsail only, by means of several finite element panels. Main leech and luff connections between the 12 mainsail panels are modelled with nonlinear spring elements, having following stiffness law:

$$F = 6310\Delta l^{1.19} \quad (7)$$

This stiffness law has been determined from experiments, where F denotes the tension in the loop in Newton and Δl the corresponding elongation in millimetres. The secondary connections between panels along the battens are modelled with two-dimensional linear spring connections.

Sail battens are modelled with equivalent beam stringer elements "B33" and "STRI3" shell *Kirchhoff* elements are used for the modelling of thin composite panel membranes (see Figure 5). For more information, the reader may refer to the *Abaqus*™ User's Manuel [9]. The whole number of shell element at mesh convergence is about 200 000 [5].

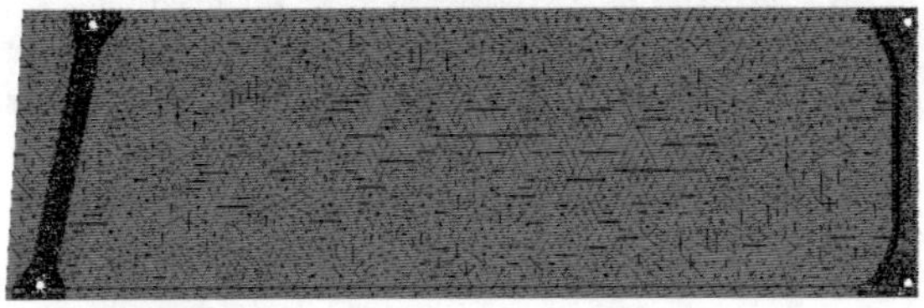

Figure 5 GP1 panel mesh.

Structural computations of thin sail membranes are often complex because of buckling issues due to possible appearance of wrinkles. In this work, respectively shell *Kirchhoff* elements and beam *Euler-Bernoulli* elements are suitable for thin plate and slender rod, since they are not subjected to transverse shear locking phenomena. A specific shell thickness approach is presently implemented [5] to catch sail extension deformations by neglecting small wrinkles. Each panel consists of glass-epoxy membranes and carbon-epoxy reinforcements. The classical laminate theory [10] is applied for the behaviour of the shell *Kirchhoff* elements. Note that the jib structure is not presently computed. Many tied contacts have been introduced in the finite element model to reproduce the bonding between the reinforcement parts (see Figure 6).

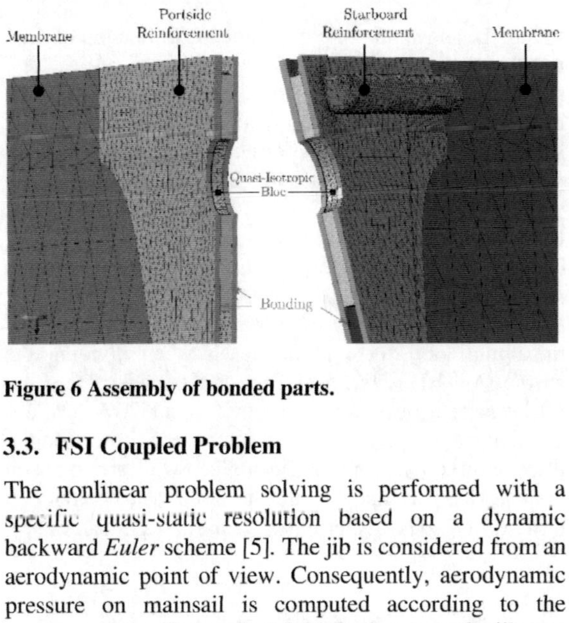

Figure 6 Assembly of bonded parts.

3.3. FSI Coupled Problem

The nonlinear problem solving is performed with a specific quasi-static resolution based on a dynamic backward *Euler* scheme [5]. The jib is considered from an aerodynamic point of view. Consequently, aerodynamic pressure on mainsail is computed according to the presence of the jib (see Part 3.1). In that way, the jib acts as a flow disturbance seen by the mainsail, and an example of mainsail pressure field, with and without jib showing jib effects, is given in Figure 7. After a *Delaunay* triangulation of the sails (Figure 8), the aerodynamic pressure field is applied on each mainsail structural shell element centroids by using bi-linear interpolations with *Lagrange* polynomials.

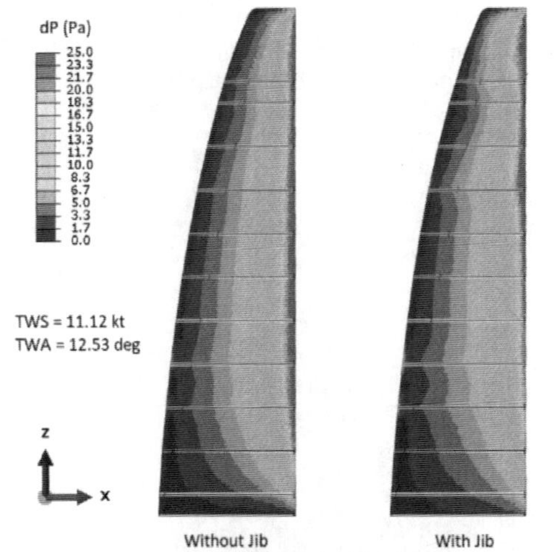

Figure 7 Pressure field on the mainsail with (left) and without (right) jib.

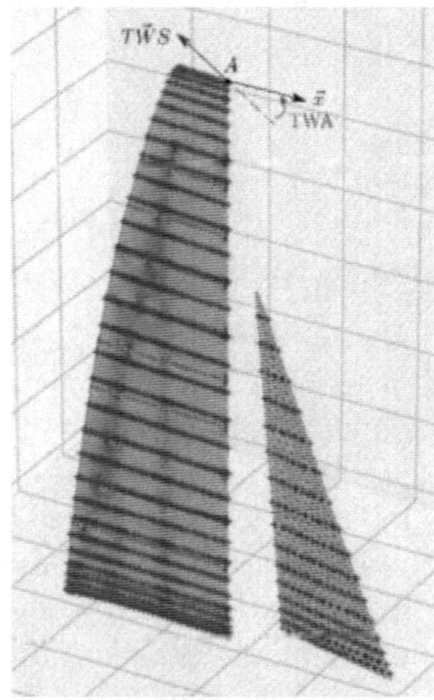

Figure 8 Delaunay mesh triangulation.

An overview of the FSI coupling algorithm scheme is presented in Figure 9

input : Non-deformed structural mainsail mesh
output : Deformed and loaded mainsail mesh
compute # Mainsail and Jib pretension
 # FSI main loop
while Not Converged **do**
- ***# Fluid solving***
- **compute #** LLT
  - Delaunay triangulation of structural mesh
  - Sail sections along $\vec{z}$ using $(\vec{x}, \vec{y})$ plane intersections with triangulation
  - From $U_0(z)$ and $\alpha(z)$ solve Prandtl equation see Sacher et al. [5] and Katz and Plotkin [7]
  - Use Equation (1) to get $U_e(z)$ and $\alpha_e(z)$
- **compute #** Pressure loads DVM
  - From $U_e(z)$ and $\alpha_e(z)$ linear system implementation through Equation (4)
  - $\gamma(x)$ on each mainsail and jib section with (5)
  - $\Delta P(x)$ on each mainsail section with (6)
  - Bi-linear interpolations of ΔP on structural shell element centroids
- ***# Structure solving → compute Abaqus ™***

Figure 9 FSI Algorithm.

An example of nonlinear FSI results is provided in Figure 10, where the transverse (y-axis direction) mainsail displacement $[m]$ is illustrated with colored contours.

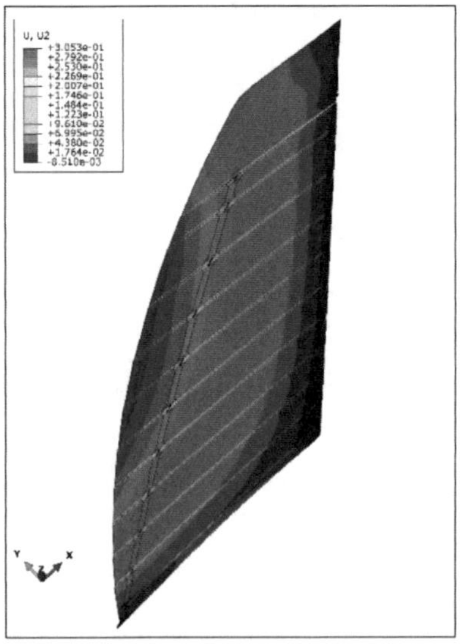

Figure 10 Example of FSI results.

3.4. Numerical uncertainties and investigations

Part 3.4 gathers several important remarks related to the development of the numerical model and exposes relevant strategies to solve modelling issues.

The question of the meshes convergence is not presented in this work. To determine the accurate mesh size, the previous study [5], which contains a study of meshes convergence onto a similar semi-rigid sail, is used. The accurate average size of elements has been scaled for the "Jib Sea" sails and set to 0.125m.

The stopping criterion of the FSI calculation presented in Figure 9 is reached as soon as relative differences between consecutive iterations of several relevant mechanical parameters is below a fixed threshold. The threshold is set at 10^{-4} for following mechanical parameters: mainsheet load, cunningham load, halyard load, side force, global strain energy and maximum displacement. In practice, the FSI calculation converges after six iterations.

To validate the structural numerical model and the calculation process, an experimental study on GP5 regarding its bending behavior is carried out. Indeed, for a fully modelled panel, the bending behavior of the panel is mainly controlled by the stiffness of the battens (Figure 11).

This study has the aim to investigate the mechanical response of the battens through the height of the carbon epoxy UD part inside of the composite battens section (Figure 13).

Figure 11 Picture of GP5 panel under loading: 14.26 kg per battens.

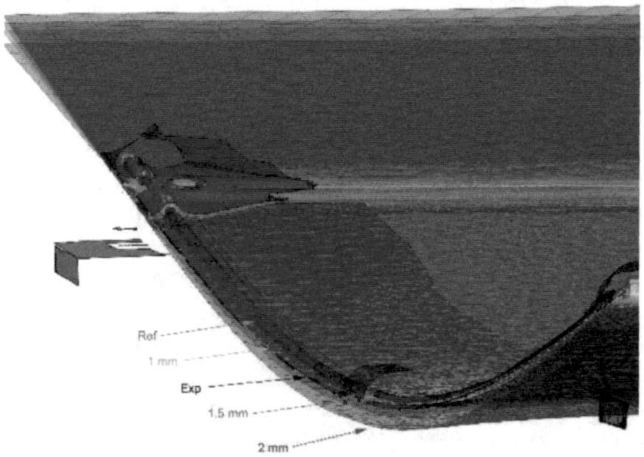

Figure 12 Experimental deflection of the GP5 (Exp) and numerical deflections for 4 carbon UD locations (Ref, $1mm$, 1.5mm and 2mm) under pure bending loading of 28.52 kg.

Figure 12 shows that the experimental deflection of GP5 is between the numerical deflections results respectively for a carbon UD offset of 1mm and 1.5mm. Thus, for all mainsail battens in the numerical model, carbon UD was shifted by 1.5 mm (y-direction) compared to the original geometry of the battens (see Ref from Figure 12).

In a previous work, the structural model of synthetic connectors between panels (i.e. Loop in Figure 18), was a simple axial spring having a stiffness of $10^7 N/m$. A sensibility analysis has been conducted to compare the maximum loop force at the leach when dividing the stiffness by 10 or 100. A division of the axial stiffness by 10 leads to a decrease in force of about 26%, while a division by 100 decreases the force by 61%. According to these observations, an experimental test was carried out on a single loop to characterize the non-linear behavior in terms of force and displacement as shown by equation (7).

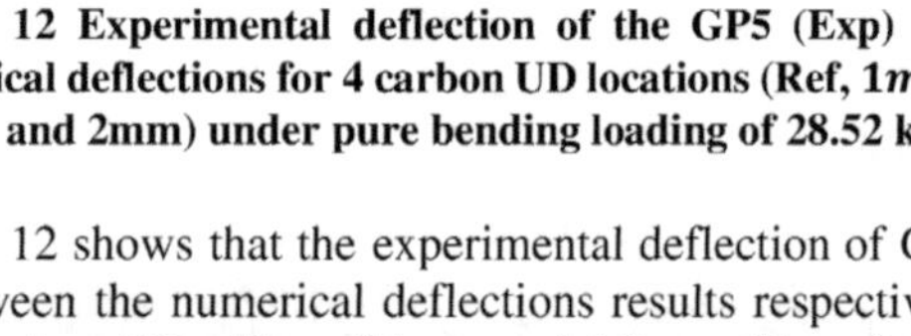

Figure 13 Side view of batten CAD.

4. EXPERIMENTAL SETUP

In this section, the experimental setup of the 1:5 composite mainsail prototype is presented. Part 4.1 deals with the materials used for the manufacture of the prototype, while Part 4.2 presents the sensors that were set on the mainsail prototype. Part 4.3 summarizes relevant uncertainties linked to sensors measurements and highlights important information on sensors uncertainties. Due to financial restrictions, we did not measure the deformation and displacement of the sail. This would have been an additional means of validating the simulations. During the tests, we did not visually encounter any major instability effect, such as membrane large wrinklings , batten buckling or flapping of the sail for instance. We therefore assume the uniqueness of the mechanical solution. Thus, only force or moment measurements allow us to validate the whole numerical model which includes sail deformation. Moreover, force or moment are physical quantities that triger the design of the sail.

4.1. Prototype overview

The 1:5 scale mainsail prototype is made of 12 composite panels connected to each other with textile loops made of "Dyneema SK78", as shown in Figure 16. The mainsail area is about 50 m². The jib of about 13 m² is a classical one made of Hydranet (polyethylene+polyesther) and manufactured by XVOILES. Mainsail membranes are made as a fiberglass composite, having a thickness of about 0.2 mm. The panel reinforcements and battens are manufactured using carbon epoxy laminates

All components of the mainsail are held up on a super-structure called a balestron rig (see Figure 14), which is made of aluminium alloy. The major difference with a standard rig is that there is no boom per se.

Figure 14 Balestron rig full view.

The balestron rig supporting the composite mainsail and the classical jib has been installed onshore at Pornichet in France, and was fitted with a motor allowing it to rotate around the z-axis (vertical axis) of the ground coordinate system.

This feature is very useful to target a true wind angle (TWA) during testing. The TWA is obtained by an anemometer located at the top of the mast as shown in Figure 8. For this configuration, the x-vector is unitary and parallel to the vector connecting the clew of the main sail and the mast foot. TWA is positive when the wind comes from starboard.

4.2. Prototype sensors

The 1:5 scale prototype (see Figure 1) is fitted with several sensors to study the mechanical behaviour of the mainsail under aerodynamic loadings. The jib is not instrumented. The mainsheet, the cunningham and the halyard are fitted with snap hook with load cell sensor (see Figure 15).

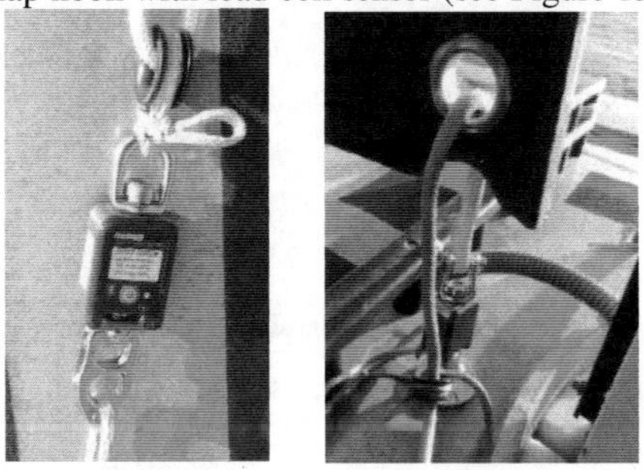

(a) (b)

Figure 15 (a) mainsheet load cell sensor (b) cunningham load cell sensor.

Most of the sensors are set on the first large panel, named GP1 as illustrated in Figure 18. A homemade load cell sensor, *LoopBckUp,* is set on the textile loop located at the bottom leech of the GP1 panel (see Figure 16) in order to measure tension forces between the two composite panels, PP1 and GP1.

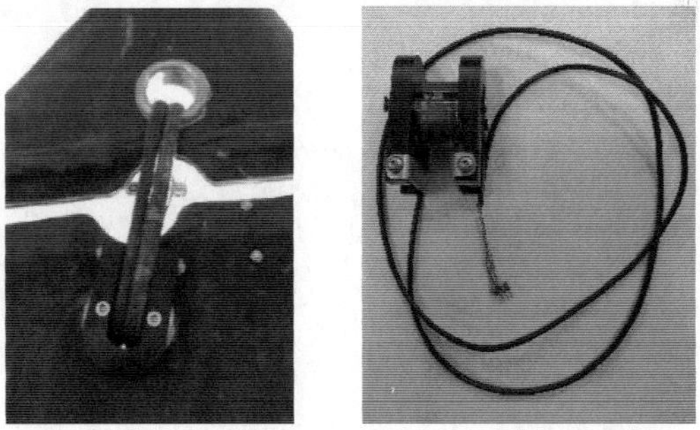

(a) (b)

Figure 16 (a) *LoopBckUp* close-view (b) homemade load cell

The link between the GP1 panel and the rigid mast is instrumented through the luff rail (Figure 18). These rail sensors, called *RailTransver,* are made of strain gauges and are able to measure starboard and portside components of the GP1 forces. Other gauges, called *RailAxial* are installed on the luff rail and are able to measure tension and compression forces generated by GP1 on the mast (see Figure 17).

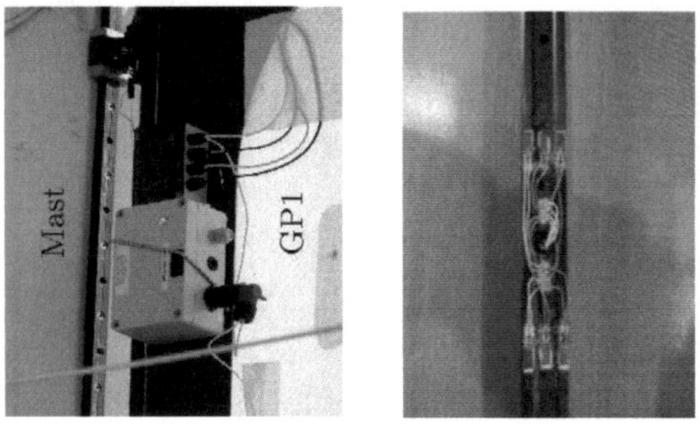

(a) (b)

Figure 17 (a) Instrumented luff rail (b) strain gauges installed on the luff rail.

The battens of GP1 are also instrumented (see Figure 18) to measure the bending moment due to the aerodynamic loading.

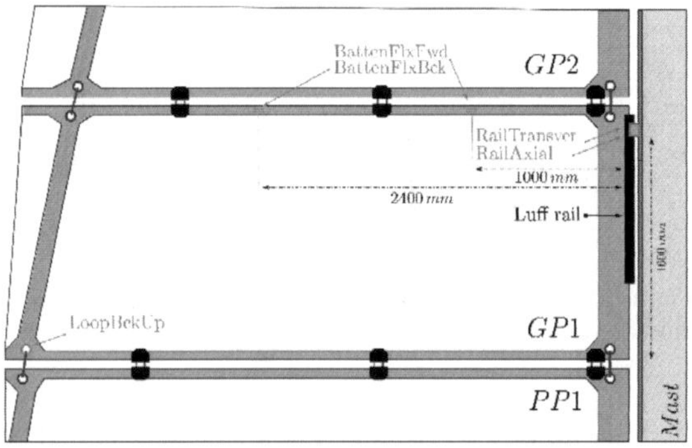

Figure 18 Sketch of the GP1 with sensor locations.

For these measurements, two sets of strain gauges, respectively called *BattenFlxFwd* and *BattenFlxBck*, have been bonded on the upper batten of GP1 (see Figure 18).

4.3. Sensors Uncertainties

The quantification of measurement noise in the data recorded by the sensors was performed when the sails are in the eye of the wind. For the load cells, the highest contribution of noise is about 23 N for the *Halyard*. For the *RailTransver* and *RailAxial*, the contribution of the noise is about 5 N. and for the *BattenFlxFwd* and *BattenFlxBck* less than 1 $N.m$.

The calibration of all the sensors except those of the Rail has been carried out indoor by means of a calibration bench. *RailTransver* and *RailAxial*, sensors calibration was carried out on the prototype using a calibration bench. All data are recorded in real time at a sampling rate of 1Hz and are stored directly in a data control center "onboard".

5. RESULTS: NUMERICAL VS EXPERIMENTAL

Section 5 provides several numerical and experimental results related to the experimental set-up presented in Section 4. The numerical FSI model presented in Section 3 is applied. Two types of experimental vs numerical comparisons are presented. Specifically, Part 5.2 deals with global comparisons based on metamodeling, while Part 5.1 presents local comparisons. All following comparisons are carried out with the jib.

5.1. Local comparisons

In this part numerical to experimental local comparisons results are provided. This type of comparisons consists of comparing experimental data with the corresponding *Abaqus*™ calculations for a specific almost "stable" TWA and TWS aerodynamic loading. Actually, due to the onshore experimental system, it is really challenging to obtain stable wind conditions allowing almost constant sail responses to the loads.

However, a moving average approach, with a six seconds window was implemented and set with a criterion on the variation coefficient.

The acquisition frequency of the sensors is one *Hertz*. This method allows to keep relevant only data where TWA and TWS are nearly stable. The variation coefficient is defined as follows:

$$C_{var} = \frac{\sigma}{|\mu|} \tag{8}$$

Where σ and μ are respectively the standard deviation and the mean of the TWA or the TWS, over a given time period. In practice, a time period is arbitrary defined to be stable if the variation coefficient is less than 15%. Figure 19 shows an example of experimental data, on a selected time period, that is presently used for the following local comparisons.

The decrease of the *Mainsheet* tension, during the selected time period is probably due to the viscoelastic relaxation of the loops. A sensitivity analysis on the simulation results showed that around the average parameters (Eq. 9) a 15% variation of TWS and TWA generates a standard deviation of 52 N on the values of the mainsheet tension. This standard deviation is relatively small compared to the decrease of about 300 N observed in the selected experimental window (Figure 18) and confirms viscoelastic relaxation of the loops..

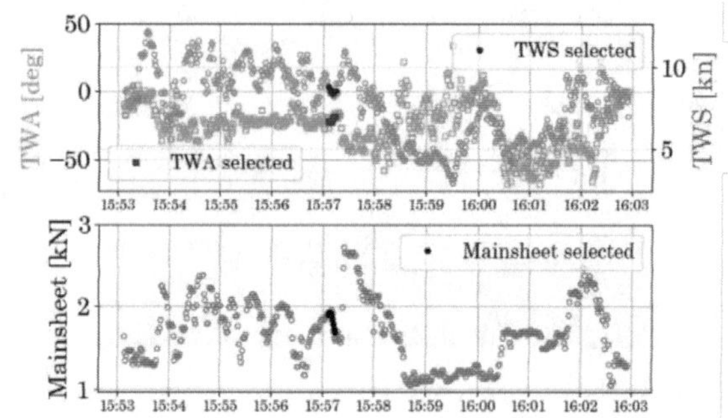

Figure 19 Selected experimental data for local comparisons. TWS and TWA are respectively plotted with blue and red markers. Black markers correspond to the *Mainsheet* tension. Bolded markers represent the stable time period.

The aerodynamic loading is determined by averaging the selected stable time period. In that case (see Figure 19), it leads to following reasonable set:

$$\begin{cases} TWA = & -20.76\ [deg] \\ TWS = & 8.53\ [kn] \end{cases} \quad (9)$$

Each local experimental quantity (see sensors descriptions in Section 4), to be compared with the *Abaqus*™ computation, is obtained by averaging the almost stable selected time period. The relative difference is used to assess numerical to experimental local comparisons:

$$\varepsilon[\%] = 100 \times \left|\frac{X_c - X_e}{X_e}\right| \quad (10)$$

where X_c and X_e are respectively numerical and experimental quantities to be compared. Bar charts are used to highlight comparisons results. The black item denotes the measurement dispersion at more or less 2σ.
Figure 20 shows that local comparisons for *Mainsheet*, *Cunningham*, *Halyard* and *LoopBckUp* sensors are overall satisfactory.

The highest relative difference is around 25% for the *Halyard*, while all the others are less than 15%. As for measurement dispersion, only the numerical *LoopBckUp* may match the simulation.
For these comparisons, related to the mechanical behaviour of textile fibre ropes, it is not obvious to explain where the differences come from. However, for the *Mainsheet* and the *Cunningham*, numerical to experimental differences can be explained by the mechanical relaxation of the polymer textile fiber.

In the current numerical model, the input pre-tensions of the *Mainsheet* and *Cunningham* are determined with experimental measurements evolving over time. In order to determine input pre-tensions, a linear approximation between 3 points was performed. In order to better determine these input pre-tensions, it would be interesting to use a decreasing exponential [11] approximation.

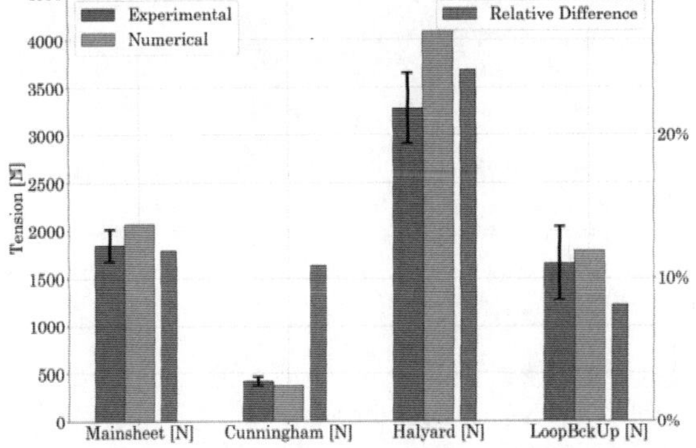

Figure 20 Local comparisons and relative difference for the *Mainsheet*, *Cunningham*, *Halyard* and *LoopBckUp*.

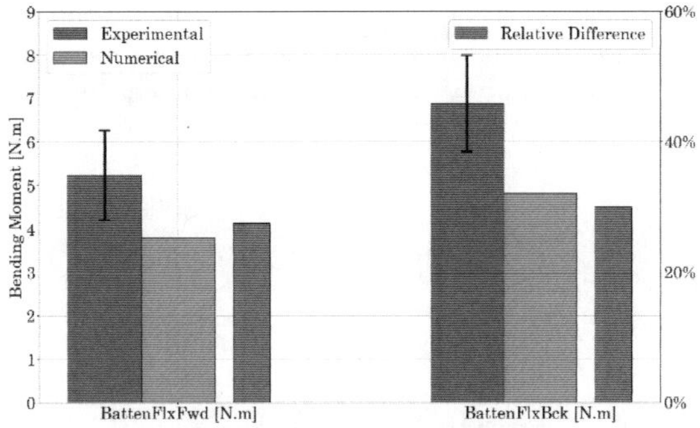

Figure 21 Local comparisons and relative difference for the *BattenFlxFwd* and *BattenFlxBck*.

Figure 21 shows that numerical results underestimate experimental measurements of about 30%. By considering measurement dispersions at $\pm 2\sigma$, for both battens, relative differences drop to 10%. Regarding these last observations, comparisons for battens are not so bad for bending moments. Nevertheless differences might be explained by analyzing the data of the formula used to compute the batten bending moments:

$$M_{ben} = \frac{2\langle EI\rangle}{\kappa h}(x_{mes\ batten} - x_{0\ batten}) \quad (11)$$

Where $\langle EI\rangle$ is the out-of-plane bending stiffness of the batten, h its height and κ the gauge factor.

Some discrepancies may subsist between the real value of $\langle EI\rangle$ and h from the manufacturing and the theoretical numerical value set in the FEM *Abaqus*™ model. Moreover, secondary connecting systems between panels along the battens are complex and difficult to model well. Two dimensional linear spring connectors are used. This approach should be improved to obtain a better correlation with experiment regarding bending moment along the batten.

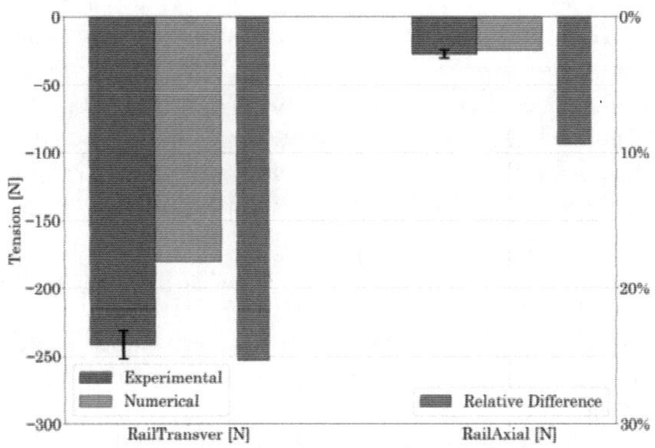

Figure 22 Local comparisons and relative difference for the *RailTransver* and the *RailAxial*.

Figure 22 shows similar results to those in Figure 21. From a magnitude point of view, numerical results underestimate experimental results of about 25% for the *RailTransver* and of about 10% for the *RailAxial*. By estimating measurement dispersions, numerical results can match experimental measurements for the *RailAxial*.

As previously said, improvement of the secondary connecting system modelling should enhance numerical bending behavior of the battens and consequently the transversal reaction force of the mast.

Since the luff rail can undergo a large rotation, a fictive new local coordinate system $(O, \vec{X}_R, \vec{Y}_R, \vec{Z}_R)$ is built from the deformed coordinates of points HR, LR and MC (see Figure 23).

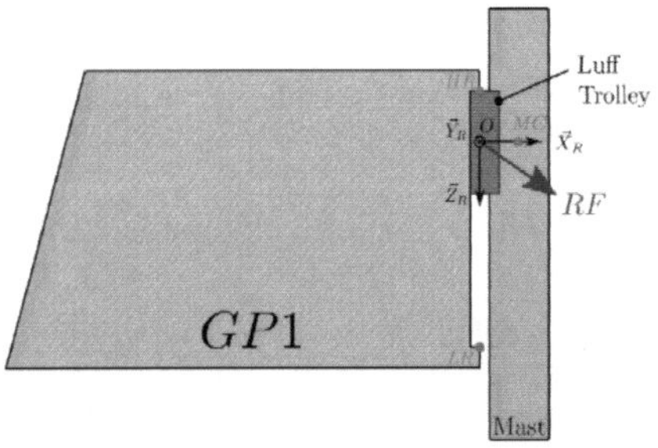

Figure 23 Sketch of the new local coordinates system

Afterwards, in order to perform comparisons with experimental sensors measurements, *Abaqus*™ output results for the rail, as vector $\overrightarrow{RF}$, at point O is projected along vector $\vec{X}_R$ for the *RailAxial* and along $\vec{Y}_R$ for the *RailTransver*. Consequently, differences may come from the projection of $\overrightarrow{RF}$ along $\vec{X}_R$ and $\vec{Y}_R$ through the building of the fictive local coordinates system. Because the mast is considered as rigid, coordinates of MC are unchanged after the aerodynamic loading contrary to the point HR and LR.

5.2. Global comparisons

Global comparisons were carried out in three steps. Firstly, to compare experimental data collected by a single sensor with numerical results for a wide range of *TWS* and *TWA*. In this way, two 3D scatter plots are superimposed: denoting respectively experimental data and numerical results. Secondly, based on these two scatter plots, two response surfaces are computed using a *Gaussian processes* [12]. Thirdly, in order to quantify more precisely differences between numerical results and experimental measurements, coloured "iso-contours" plots are used. Differences between numerical and experimental response surfaces are then analysed to validate the proposed numerical modelling of sails FSI. This numerical to experimental comparison allows a validation of the numerical FSI model in terms of trend and relative differences on a large range of sails operating points. This three-step process was carried out for the *Mainsheet*, the *RailTransver* and the *BattenFlxBck*.

Figure 24 shows numerical and experimental response surfaces of the $Mainsheet$ tension in $[N]$. It can be noticed that the numerical model (*blue*) overestimates the measurements (red) for large values of $|TWA|$. However, for $TWA \in [-10; 10]$ deg, the numerical model response almost fits the experiments.

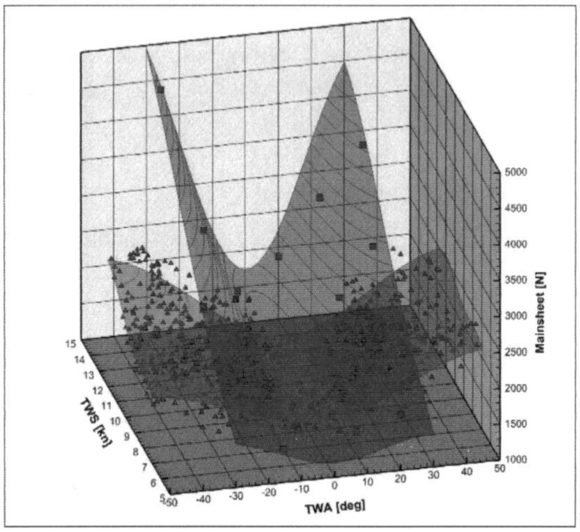

Figure 24 Experimental (red) and numerical (blue) response surfaces of the *Mainsheet tension* built from experimental measurements (red triangles) and numerical results (blue squares).

Figure 25, where relative differences between numerical and experimental responses are shown, confirms the observations made in Figure 24. Relative differences are less than 5% for $TWA \in [-10; 10]$ deg. It should be noted that even for large $TWA \in [-20; 20]$, relative differences remain less than 25% for $TWS < 10$ kn. These observations are in accordance with the limits of an inviscid flow model, which computes an attached sail flow and therefore overestimates sail forces for large TWA values.

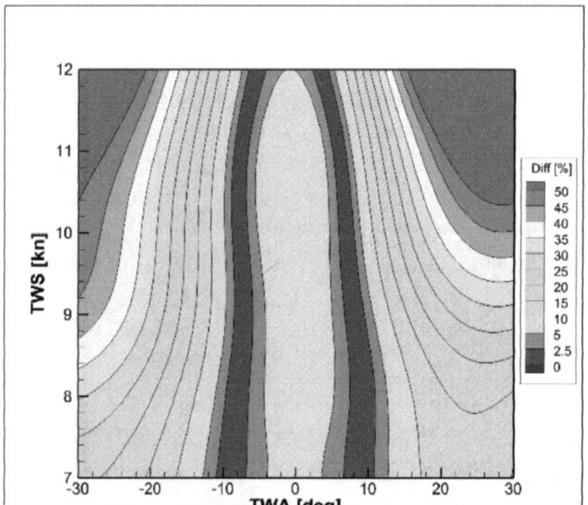

Figure 25 Relative difference between experimental and numerical results for the *Mainsheet* tension.

Figure 26 shows numerical and experimental response surfaces of the bending moment in the $BattenFlxBck$. Response surfaces are very close for $TWS < 10$ kn, while for large TWS values, the numerical model seems to overestimate the experiments. This trend is confirmed by absolute differences of response surfaces that are provided in Figure 27. Absolute differences of the bending moment appear to be less than 1.5 N.m for $TWS < 10$ kn.

Regarding experimental values of the bending moment, relative numerical to experimental differences are about 5 to 10%. Higher absolute differences can be observed for a starboard loading when TWS is greater than 11 kn and 25 deg. of TWA. This asymmetrical response of the batten bending moment regarding the TWA comes from its starboard to portside asymmetrical mechanical properties as battens are located on the starboard side of the panel. Comparing the $Mainsheet$ tensions, differences observed here between numerical predictions and measurements are less pronounced for large TWA. This may be explained by the fact that at low or moderate TWS, the GP1 panel is lightly loaded.

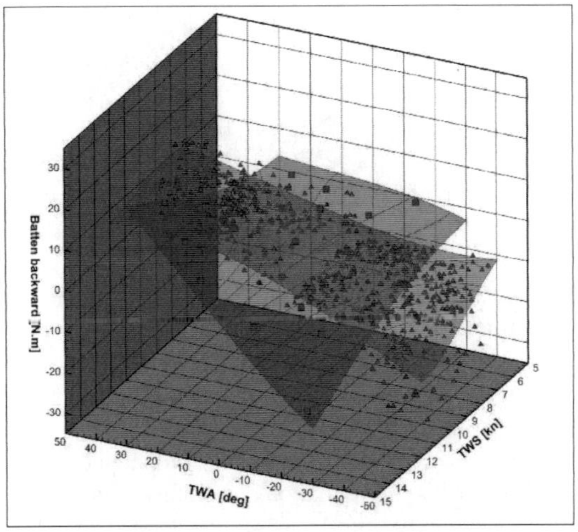

Figure 26 Experimental (red) and numerical (blue) response surfaces of the bending moment in *BattenFlxBck* built from experimental measurements (red triangles) and numerical (blue squares).

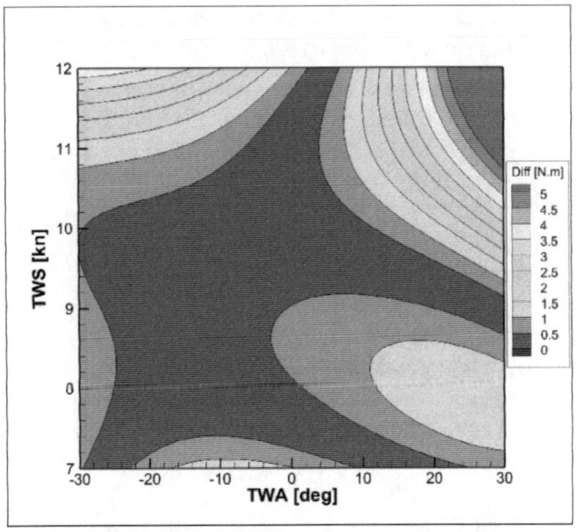

Figure 27 Absolute difference between experimental and numerical results for the *BattenFlxBck* moment.

Figure 28 is related to the force in the $RailTransver$, and is similar to results provided in Figure 26. For the common $[TWA, TWS]$ range both response surfaces are evolved in the same way and seem to be very close. Figure 29 confirms that both response surfaces are evolving in the same way within $3/4$ of their common $[TWS, TWA]$ pair. In this area the absolute difference is less than 6 N, that corresponds to relative differences of again 5 to 10%.

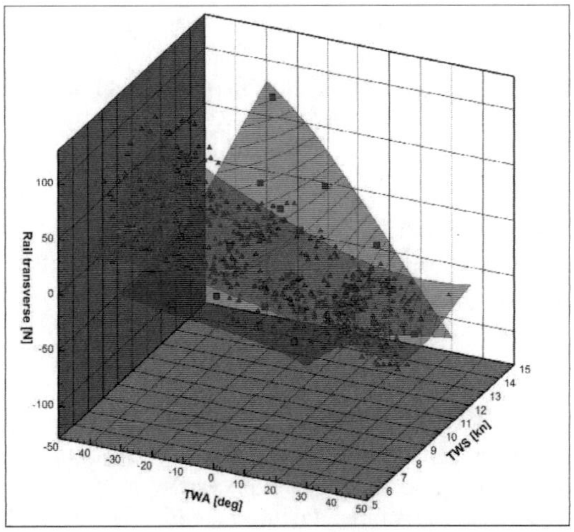

Figure 28 Experimental (red) and numerical (blue) response surfaces of the force in *RailTransver* built from experimental measurements (red triangles) and numerical (blue squares).

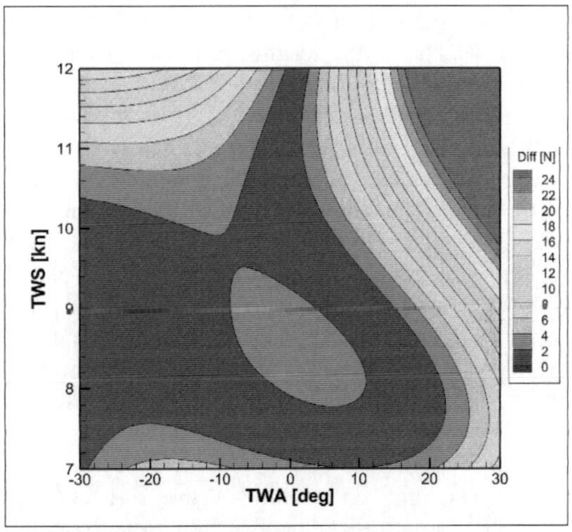

Figure 29 Absolute difference between experimental and numerical results for the *RailTransver* force.

According to these last observations and more specifically numerical to experimental differences, it has been shown that numerical FSI results are very close to the experiments for moderate values of TWA. This is particularly true for the $Mainsheet$ tension that is directly related to the global mainsail lift force. As already mentioned, this overestimation of the mainsail lift for large TWA is explained by the limits of the presently implemented inviscid flow model, that does not compute possible sail flow separations. Regarding at more local quantities, such as the rail transverse force and the batten bending moment of the GP1 panel, numerical to experimental differences are also increasing for large TWA values, but only for $TWS > 10$ kn.

At low and moderate TWS, the GP1 panel appears to be not enough loaded to see numerical to experimental differences. Nevertheless, it should be noticed that even if the numerical FSI model is overestimating the magnitude of experimental results in cases of both large TWA and TWS, it stays conservative and can therefore be considered as a safety factor in the industrial design context.

6. CONCLUSION

A robust method to solve nonlinear FSI yacht jib and mainsail has been presented in this paper. This study is part of an industrial project which aims to develop new designs of rigid sails of about 1000m² for cruise ships. A 50m² semi-rigid composite mainsail prototype has been built onshore on the Pornichet's dyke in France. The prototype is fitted with several sensors in order to study its mechanical behavior under aerodynamic loadings. The prototype has been modelled numerically to simulate the FSI on the structure. In this study, two types of comparisons were carried out in order to validate the FSI algorithm proposed here. The first type of comparisons deals with local comparisons between experimental measurements from the sensors and FSI results for a given TWA and TWS set. It has been shown that for local comparisons, the FSI algorithm is overall valid for all sensors as relative differences are less than 30%. The remaining differences can be partly explained by modelling and manufacturing discrepancies. The second type of comparisons is about global comparisons for three specific sensors and for a wide range of TWA and TWS. It was observed that for these global comparisons, the FSI algorithm is validated only for a specific range of TWA and TWS i.e. when $|TWA| < 10\, deg$ and $< 10\ \ kn$. Beyond these ranges, FSI results overestimate the magnitude of experimental measurements, but it can be seen as a safety factor for an industrial and conservative point of view. Moreover, dissimilarities might be explained by the fact that the dynamic effect of the structure was not considered in the FSI algorithm. Differences can also be explained by weather conditions during experimental campaigns. Indeed, during tests, the $[TWA, TWS]$ set was very unstable and therefore made it difficult to isolate stable and relevant time periods for comparisons. Future developments in order to improve FSI algorithm, should also model the jib's structure. In the light of these first results, the next consistent step related to the « Jib Sea » project is to perform FSI computations by using a viscous flow model. This will allow to better quantify the impact of using an inviscid flow model for the design stage of composite sails.

Acknowledgements

This study began in the "Solid Sail 2.0" project and now takes part in the "Jib Sea" project funded by the "Région Bretagne", and approved by the "Pôle EMC2" and the "Pôle Mer Bretagne Atlantique". This project involves a consortium of three industrial partners: "Chantiers de l'Atlantique", "GSea Design", "Multiplast" in collaboration with "ENSTA Bretagne" a graduate and post-graduate school of engineering as an academic partner. "Ocean Data System", subcontractor of "ENSTA Bretagne", has provided almost all components of the experimental setup for the mainsail. "Capacités SAS" carried out the instrumentation of the strain gauges. The authors gratefully acknowledge the four industrial partners for their technical support and collaboration and the Brittany region for its financial support.

Références

[1] S. Adanur, Wellington sears handbook of industrial textiles, CRC Press, 1995.

[2] I. Viola et R. Flay, «Sail aerodynamics: understanding pressure distributions on upwind sails,» *Experimental Thermal and Fluid Science,* vol. 35, n° %18, pp. 1497-1504, 2011.

[3] S. Bak et J. Yoo, «FSI analysis on a sail performance of a yacht with rig deformations,» *International Journal of Naval Architecture and Ocean Engineering,* vol. 11, n° %12, pp. 648-661, 2019.

[4] M. Sacher, H. Frédéric, P. Bot et M. Durand, «Sail trimming FSI simulation-comparison of viscous and inviscid flow models to optimise upwind sails trim,» *HPYD5,* 2015.

[5] M. Sacher, J.-B. Leroux, A. Nême et C. Jochum, «A fast and robust approach to compute nonlinear fluid-structure interactions on yachts sails - application to semi-rigid composite mainsail,» *Ocean Engineering,* vol. 201, pp. 107-139, 2020.

[6] L. Prandtl, «Theory of lifting surfaces. Technical Report. NASA,» 1920.

[7] J. Katz et A. Plotkin, Low-Speed Aerodynamics, Cambridge Aerospace Series 2nd edition: Cambridge University Press, 2001.

[8] E. Hairer, Norsett S.P. et G. Wanner, «Solving ordinary differential equations I: Nonstiff problems,» *Springer Science & Business Media,* vol. 8, 2008.

[9] Simulia, «Abaqus 2019 User's Manual,» 2019.

[10] D. Gay, Composite Materials : Design and Applications, 3rd éd., T. &. F. Group, Éd., CRC Press, 2015, pp. 263-285.

[11] A. Schiavi et A. Prato, «Non-linear stress relaxation of polymers: experimental evidences and stretched-exponential model,» chez *23rd International Congress on Sound & Vibration,* Athens, 2016.

[12] M. Sacher, H. Frédéric, R. Duvigneau , O. Le Maître, N. Aubin et M. Durand , «Efficient optimization procedure in non-linear fluid-structure interaction problem: Application to mainsail trimming in upwind conditions,» *Journal of Fluids and Structures,* vol. 69, pp. 209-231, 2017.

ON THE EFFECT OF THE LEADING-EDGE SEPARATION BUBBLE ON THE AERODYNAMICS OF SPINNAKERS

Jean-Baptiste R. G. Souppez[1], j.souppez@aston.ac.uk
Patrick Bot[2], patrick.bot@ecole-navale.fr
Ignazio Maria Viola[3], i.m.viola@ed.ac.uk

Abstract. The spinnaker is the most powerful and one of the most used sails both in racing and cruising - yet its complex aerodynamics governed by flow separation is still not fully understood. While the flow around a spinnaker is unsteady and highly tridimensional, locally the governing fluid mechanics may be represented by the quasi-steady bidimensional flow around a cambered circular arc with a sharp leading edge. The spinnaker is typically trimmed such that the stagnation point is at the leading edge with the sail streamline separating on the suction side and reattaching within the first 10% of the chord length, forming a leading-edge separation bubble (LESB). This flow feature sets the beginning of the boundary layer, whose separation further downstream is paramount for the global aerodynamic forces on the sail. This study investigates the effect of the LESB on the boundary layer regime and downstream flow separation through particle image velocimetry on a circular arc. The existence of the combination of a critical Reynolds number and a critical angle of attack to trigger turbulent separation is demonstrated. A turbulent LESB followed by a laminar boundary layer is observed in sub-critical regime. Conversely, in a post-critical condition, a turbulent LESB ensued by a turbulent boundary layer is detected, the latter continuing all the way to trailing-edge separation. This behaviour ultimately yields a sharp lift increase and drag reduction. These findings reveal the critical effect of the leading-edge vortical structures on the global flow field and forces experienced by cambered wings with leading-edge separation, including high performance spinnakers. It is envisaged that these results will contribute to improve the design and performance of downwind yacht sails.

1. NOMENCLATURE

AR	Aspect ratio (s/c)
c	Chord length (m)
C_D	Drag coefficient
C_L	Lift coefficient
D	Drag (N)
f	Chordwise maximum draft location (m)
L	Lift (N)
Re	Reynolds number
s	Span (m)
t	Thickness (m)
u	Streamwise flow velocity (m.s^{-1})
u'	Streamwise flow velocity fluctuation (m.s^{-1})
U_∞	Freestream flow velocity (m.s^{-1})
v'	Streamnormal velocity fluctuation (m.s^{-1})
x	Streamwise coordinate (m)
x_s	Streamwise position of the separation point (m)
y	Streamnormal coordinate (m)
y_c	Sail camber (m)
α	Angle of attack (°)
Γ	Circulation (m^2.s^{-1})
κ	Nondimensional turbulent kinetic energy
ρ	Fluid density (kg.m^{-3})
LESB	Leading-edge separation bubble
PIV	Particle image velocimetry
RANS	Reynolds-averaged Navier-Stokes
TKE	Turbulent kinetic energy

2. INTRODUCTION

Modern symmetric spinnakers appeared in the 1960s and further developed in the 1970s, before asymmetric spinnakers were introduced in the 1980s in the 18ft fleet in Sydney, and then popularised on offshore racing yachts in the 1990s. These new sails were promptly adopted in many prominent sailing events, from offshore races to the America's Cup [1, 2, 3]. Indeed, the 1995 edition of the America's Cup gave greater importance to downwind legs [1], thus prompting further developments. Because of the complexity of the flow around downwind sails, the necessity to undertake experimental tests was perceived [4], ultimately resulting in dedicated experimental facilities, namely twisted flow wind tunnels [5]. On the other hand, the 1990s also coincide with a fast increase in accessible computational power, allowing advanced numerical methods [6]. As downwind sails feature a high camber and the flow is largely separated, the use of Reynolds-averaged Navier-Stokes (RANS) simulations is necessary [7], with the first instance of RANS models occurring in 1996 for spinnakers [8]. This is in contrast with upwind sails, exhibiting lower camber and largely attached flow. As a result, these have been successfully analysed using inviscid codes since the 1960s [9, 10], methods that have been extensively utilized in America's Cup sails development [11].

The flow around spinnakers is unsteady and highly tridimensional, owing to the complex 3D geometry of the sail. However, locally, the governing fluid mechanics may be represented by the quasi-steady bidimensional flow around a highly cambered, thin circular arc with a sharp leading edge [12, 13, 14].

[1] Senior Teaching Fellow, College of Engineering and Physical Sciences, Aston University, UK.
[2] Associate Professor, Naval Academy Research Institute, Brest, France.
[3] Reader, School of Engineering, Institute for Energy Systems, University of Edinburgh, UK.

Spinnakers are typically trimmed such that the stagnation point is at the leading edge with the sail streamline separating on the suction side and reattaching within the first 10% of the chord length, forming a leading-edge separation bubble (LESB), also referred to in the literature as a leading-edge bubble. This flow feature sets the beginning of the boundary layer, whose separation further downstream is paramount for the global aerodynamic forces on the sail.

This study investigates the effect of the LESB on the boundary layer regime and downstream flow separation through particle image velocimetry (PIV) on a circular arc. First, the literature inherent to low and high camber circular arcs is reviewed. Then, the experimental setup is presented, and subsequently validated to ensure flow bidimensionality and measurement accuracy. The results eventually present novel findings on the effect of the leading-edge separation bubble on the aerodynamics of modern asymmetric spinnakers, including the flow fields and global forces.

3. BACKGROUND ON CIRCULAR ARCS

3.1. Low camber ($y_c \leq 0.2c$) circular arcs

Early instances of experiments on circular arcs can be traced back to Wallis [15]. Later, tests on bent plates were undertaken by Maekawa & Atsumi [16], that refined the idea of a critical Reynolds number (Re) originally introduced by Tani [17], and eventually resulting in the definition of the O-K criterion [18]. A circular arc with a camber-to-chord ratio (y_c/c) of 7.5% was tested by DeLaurier & Harris [19] for an atypical application, namely paper model airplanes. The work of Buehring [20] on windmills, later expanded by Bruining [21] tackled a circular arc with y_c/c = 10%, while Pandey et al. [22] undertook wind tunnel testing of various circular arcs up to y_c/c = 14%.

Additional investigations into low camber circular arcs were performed by Tse [23], who tested 5%, 7.5% and 10% camber-to-chord ratios, subsequently complemented by Cyr [24], with the addition of an 18% thin cambered aerofoil with a sharp leading edge. A larger y_c/c = 20% was considered in Sunada et al. [25] as part of a parameter study, including camber-to-chord ratios of 5%, 10% and 20%. The results highlighted the drastic changes in behaviours with greater camber, such as increased lift and delayed stall, supported by the later findings of Okamoto & Azuma [26] on 3D circular arcs.

The geometries previously discussed had a thickness-to-chord ratio (t/c) circa 5%, i.e. similar to the experiments of Velychko [27], Flay et al. [12] and Bot [13], with a value close to 4%. These remain relatively high when compared to membranes or sails, and thus a smaller value should be targeted to achieve as thin a wing as possible while structurally suitable for experimental campaigns.

Furthermore, Sunada et al. [25] eventually concluded that a thin aerofoil with a sharp leading edge is paramount for performance, based on the comparison of force coefficients, and noted that performance is strongly affected by leading edge vortices. This implies that promoting the appearance of the leading-edge separation bubble with a thin and sharp leading edge improves performance. Moreover, Okamoto & Azuma [26] revealed that sharpening the leading edge does provide benefits analogous to those of a higher camber profile. Finally, the work of Brault [28] also supports these findings, demonstrating the benefits of a sharp leading edge over a chamfered one.

Consequently, it appears that only very few historical studies tackled y_c/c in excess of 20%, with further limitations in terms of both thickness and leading edge shape. While the original study of a 20% camber sail by Milgram [29] is worth mentioning, it is the recent literature on modern asymmetric downwind yacht sails that proves most relevant. Indeed, highly cambered (22% < y_c/c < 25%), thin circular arcs with a sharp leading edge were adopted as a 2D simplification of the more complex 3D shape of downwind sails. A greater focus on flow visualisation and flow features are also core elements of the newer literature, thereby extending beyond comparative force measurements.

3.2. High camber ($y_c > 0.2c$) circular arc for downwind sails application

The pressure distribution of membranes, equivalent to circular arc sails with high camber-to-chord ratios ranging from 19.3% to 25.7% was the focus of Cyr & Newman [30], thereby representing one of the earliest studies into higher camber. Although the exact Reynolds number or angle of attack (α) are not specified, the pressure distribution of a similar high camber arc was tested by Flay et al. [12], yielding a comparable qualitative behaviour.

Since 2006, highly cambered circular arcs have been extensively studied [12, 13, 14, 31, 32, 33, 34, 35]. Sharp transitions in lift forces were identified and related to variations in flow patterns, as summarized by Bot [13, 36] while additional correlation between forces and flow fields were provided by Souppez et al. [14, 37].

From the published experimental results of Bot [13], depicted in Figure 1, it is possible to observe an offset in the lift coefficients (Figure 1a) up to α = 14°. At this point, the lower Reynolds number arc suddenly produces more lift, and its trend merges with that of the arc at a Reynolds number of 218,000, which is known to have undergone transition [36]. At the exact same angle of attack, an abrupt reduction in drag can be noticed (Figure 1b). Furthermore, this coincides with a sudden downstream shift in the location of the trailing edge separation point (Figure 1c), in line with that of the arc experiencing turbulent flow.

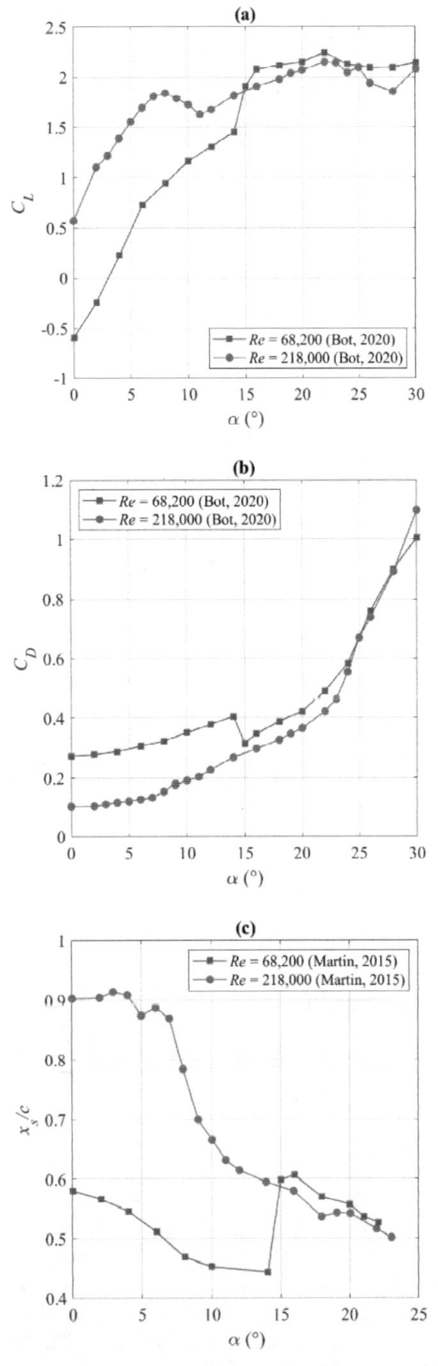

Figure 1. Lift (a) and drag (b) coefficients [13] and location of the trailing-edge separation point [34] (c) at a Reynolds numbers of 68,200 and 218,000.

These observations would indicate that the abrupt change in behaviour is the result of laminar-to-turbulent transition, occurring due to a critical angle of attack being reached at this particular value of the Reynolds number. This implies that a combination of a critical Reynolds number and a critical angle of attack are required to trigger transition, thus suggesting that, even following a leading-edge separation bubble, there can be a laminar boundary layer leading to laminar separation. Hence, this paper endeavours to assess the role of the leading-edge flow on boundary layer regime and turbulent separation, as well as the correlation between flow fields and global forces.

4. EXPERIMENTAL SETUP

4.1. Geometry

The geometry of the highly cambered, thin circular arc with a sharp leading edge under study is the same as that of Bot [13], with y_c/c = 22.32%, over chord lengths of 100 mm, 150 mm and 200 mm for the three arcs manufactured. The requirement for multiple chord lengths arose from the necessity to achieve a wider range of Reynolds number within the given flow speeds that could be achieved in the experimental facilities.

Carbon fibre prepreg was employed to achieve the required strength and stiffness at the lowest possible thickness, namely 1.8 mm. This is primarily motivated by the will to achieve as thin as possible a wing, as per yacht sails, but also prevent any upstream pressure increase and associated alteration in effective angle of attack that can arise from a thicker section. The relevant specifications are presented in Table 1, and a representative geometry is depicted in Figure 2.

Table 1. Geometric definition of the circular arcs employed.

Circular Arc	100 mm Arc	150 mm Arc	200 mm Arc
Chord, c (mm)	100	150	200
Span, s (mm)	370	370	370
Camber, y_c (mm)	22.32	33.48	44.64
Camber-to-chord ratio, y_c/c	22.32%	22.32%	22.32%
Chordwise draft location to chord ratio, f/c	50%	50%	50%
Leading edge and trailing edge angles (°)	48	48	48
Thickness, t (mm)	1.80	1.80	1.80

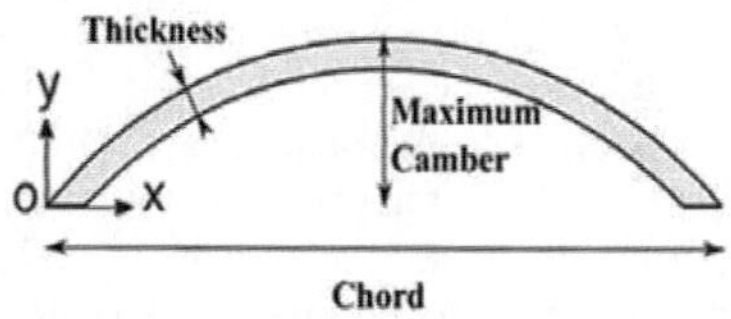

Figure 2. Experimental geometry, adapted from [13].

4.2. Towing tank testing

Part of the force measurements were undertaken in a towing tank, having a length of 60 m, breadth of 3.7 m, depth of 1.8 m and top speed of 4.6 $m.s^{-1}$ [39]. The circular arc was fitted between end plates to achieve a 2D section. The lift and drag were recorded at 1000 Hz for six seconds once the desired speed was reached. Forces created by the test rig including end plates were subtracted, and resulting forces from the circular arc section were converted into force coefficients for lift and drag.

Considering the fluid's density ρ, the surface area given as the product of the span s and chord c, and the free stream velocity U_∞, the lift L and drag D yield the associated coefficients C_L and C_D, using Eq. 1 and Eq. 2 respectively:

$$C_L = \frac{L}{\frac{1}{2}\rho c s U_\infty^2}, \quad (1)$$

$$C_D = \frac{D}{\frac{1}{2}\rho c s U_\infty^2}. \quad (2)$$

4.3. Water tunnel testing

The water tunnel at the University of Edinburgh provided flow visualisation, as well as the ability to conduct additional force measurements. The 12 m long water tunnel features a breadth of 0.4 m, a depth of 0.9 m [39] with a water level at 0.34 m in this instance, and an ideal operating speed range between 0.15 m.s^{-1} (below which high turbulence intensity can interfere with the measurements) and 0.4 m.s^{-1} (after which wave interference begins). Force measurements were realised by employing a six-axis force/torque sensor, allowing for forces to be recorded up to ±35 N for the lift, and ±25 N for the drag, with a resolution of 1/160 N [40]. Moments could be quantified up to ±250 N.mm around all three axis; this proved a limiting factor on the 200 mm chord length geometry at higher flow speeds, thus further justifying the need for varying arc sizes. The data was sampled at 1000 Hz for 45 seconds. This is substantially longer than the record time in the towing tank, owing to both the limited run length and therefore run time in the towing tank, but also the higher expected streamwise turbulence intensity in the water tunnel, measured using laser doppler velocimetry as 3.63%.

4.4. Particle image velocimetry

A 200 mJ Nd:YAG pulsed laser at a wave length of 532 nm was employed to illuminate silver coated hollow glass spheres, with a nominal 14 μm diameter and specific gravity of 1.7, and images having a 2056 px by 2060 px resolution. The laser sheet, of thickness lesser than 2 mm, was directed parallel to the onset flow, illuminating the upper surface of the arc at midspan. The work of Bot et al. [36] having previously characterised the flow field around both the pressure and suction side of the same geometry, it is expected only the flow around the suction side will be of interest for the speeds and angles of attack investigated, hence no flow visualisation of the underside of the arc was undertaken. Indeed, it is the LESB, boundary layer and trailing-edge separation on the suction side that are of primary relevance in the present research.

Each particle image velocimetry flow visualisation featured a minimum of 100 pairs of images sampled at 7.5 Hz. A multi-pass (decreasing size) cross-correlation was adopted, with two initial passes having a 96 px by 96 px interrogation window and 50% overlap, before a final 32 px by 32 px pass with a 75% overlap.

5. VALIDATION AND EQUIVALENCE OF EXPERIMENTAL SETUP

5.1. Flow Bidimensionality

Firstly, the effectiveness of the end plates was ascertained, a vital process considering the low aspect ratio (AR) of the 200 mm chord test geometry ($AR = s/c = 1.85$). Lift and drag where recorded in the towing tank, for angles of attack ranging from 10° to 25° in one degree increment at Re = 53,530. The results were evaluated against the wind tunnel tests of Velychko [27], with an identical cross section, but an aspect ratio of 10. The comparison is shown in Figure 3.

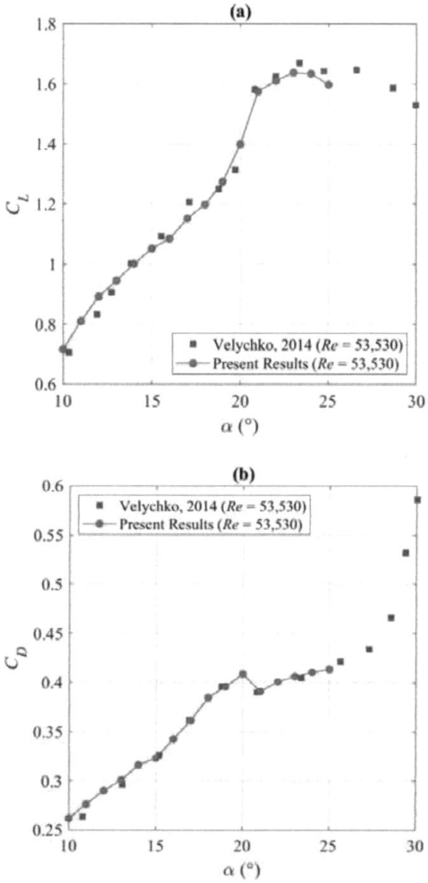

Figure 3. Lift (a) and drag (b) coefficients comparison between wind tunnel [27] (AR = 10) and towing tank (AR = 1.8) on identical circular arcs (y_c/c = 22.32%), both fitted end plates, at Re = 53,530.

Due to the difference in achievable flow velocities between the towing tank and water tunnel, different chord lengths had to be used to achieve the same Reynolds number range, as well as widen it. In order to guarantee the reliability of testing different chord length geometries, a repeatability experiment at Re = 68,200 was conducted, with arcs of 100 mm and 150 mm chord length respectively. The lift and drag coefficients are presented in Figure 4.

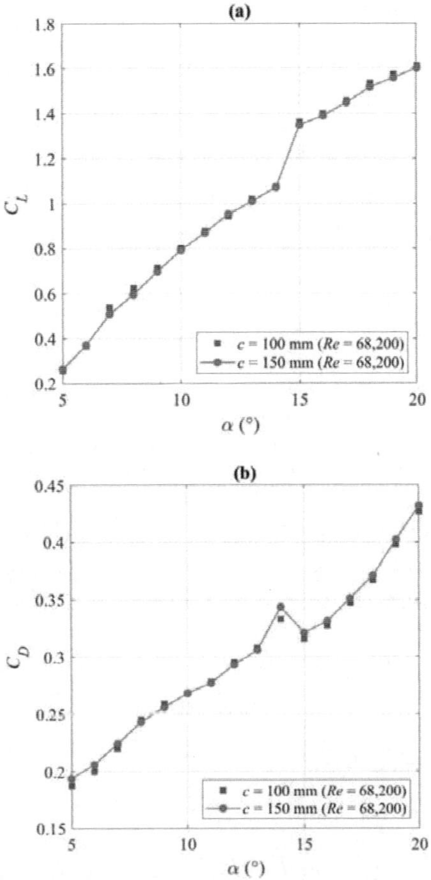

Figure 4. Lift (a) and drag (b) coefficients for arcs of 100 mm and 150 mm chord length respectively, at *Re* = 68,200.

5.2. Towing tank force measurements

Towing tank tests were performed at Reynolds numbers comparable to that of the literature, namely 53,530 [27], 68,200 [32, 34], 150,000 (arbitrarily chosen to provide intermediate data) and 218,000 [32, 34]. The angles of attack ranged from 5° to 25°. The results are depicted in Figure 5.

For increasing Reynolds number, a shallower critical angle of attack is needed for transition to occur and thus induce a sharp lift increment. It also appears that a Reynolds-number-independent pre-critical trend exists. In the case of Re = 218,000, the increase in lift coefficient occurs at zero degrees angle of attack, an effect known as the lift crisis [36]. Moreover, in the post-critical range, a common trend is also present. These observations are in favour of the hypothesis that the abrupt change in lift and drag coefficient is due to laminar-to-turbulent transition.

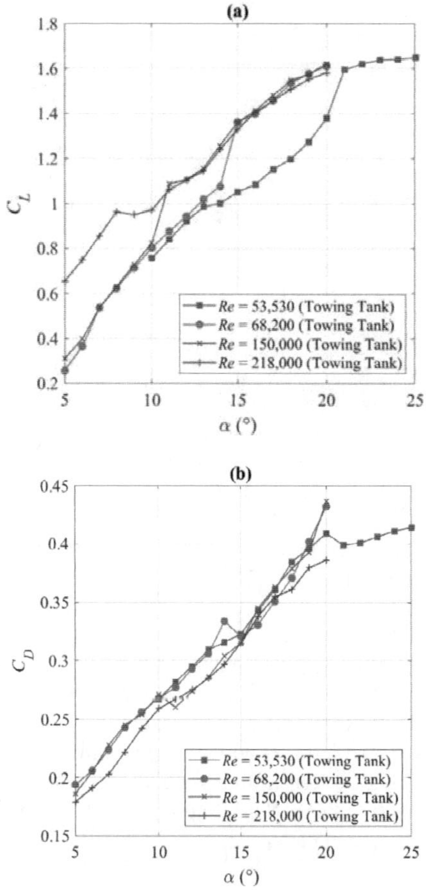

Figure 5. Lift (a) and drag (b) coefficients recorded in the towing tank, at *Re* = 68,200.

6. RESULTS

6.1. Ideal angle of attack and associated critical Reynolds number

The ideal angle of attack, defined as the angle of incidence when the stagnation point is on the leading edge, was previously suggested at a value of either 8° [34] or 9° [31], likely under the misconceived assumption that the idea angle of attack can be identified by a local maximum in the lift, and without supporting flow visualisation evidence of a leading-edge separation bubble until α = 12° [35], though the ideal angle of attack may depend on Re [13]. The present study employed particle image velocimetry to locate the stagnation point on the arc, starting at the lowest suggested value of α = 8°, and increasing until the ideal angle of attack was reached, and then exceeded. Thus, the flow field and its relation to the lift coefficient's trend at the idea angle of attack can be characterised. Knowledge of the ideal angle of attack is important in the case of a soft membrane, as it constitutes the minimum angle of attack required for inflation. In the case of downwind yacht sails, a lower angle of incidence would not inflate the sail, thus rendering it unusable. As illustrated in Figure 6, the ideal angle of attack was ascertained to be 11°.

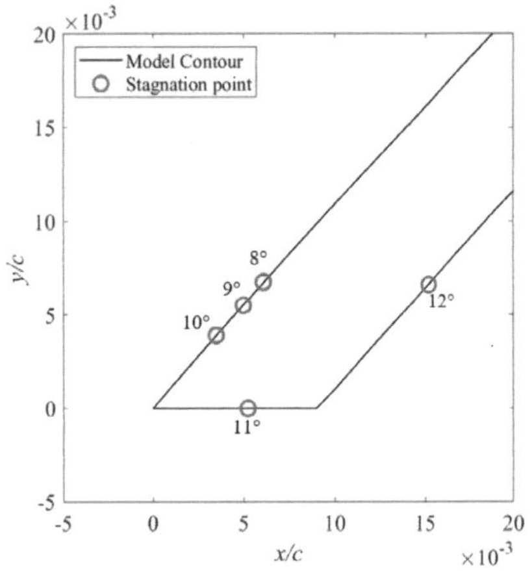

Figure 6. Location of the stagnation point near the leading edge for different angles of attack at *Re* = 68,200.

The circular arc was then tested at increasing flow speeds to determine the critical Reynolds number at $\alpha = 11°$, for $y_c/c = 22.32\%$ at $Re = 68{,}200$. Results for lift and drag coefficients, presented in Figure 7, clearly show the change in behaviour at $Re = 144{,}000$ (+/- 2,000). This particular Reynolds number, i.e. the critical Reynolds number at the ideal angle of attack (for $y_c/c = 22.32\%$ at $Re = 68{,}200$), therefore represents the minimum required value during model scale testing of downwind sails to ensure a turbulent flow, representative of full-scale behaviour, is replicated.

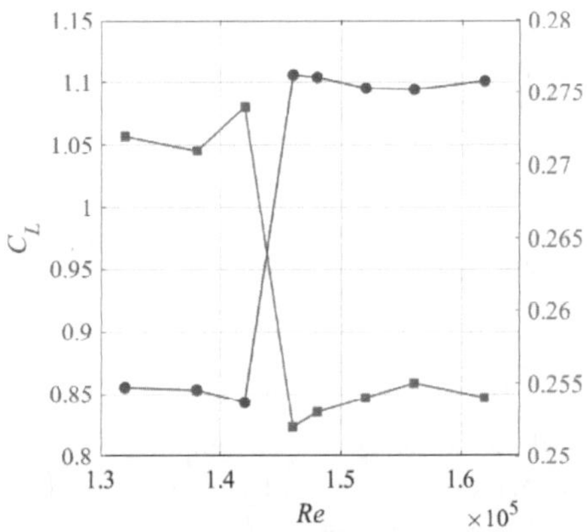

Figure 7. Lift and drag coefficient for increasing Reynolds number at the ideal angle of attack.

6.2. Separation point

The behaviour of the force coefficients, namely an increase in lift with an associated drop in drag, is consistent with the laminar-to-turbulent transition. This is further strengthened by the downstream shift in the separation point observed in the time-averaged PIV flow fields. Indeed, separation occurs further upstream prior to the critical angle of attack, and reaches a similar location as that of a higher Reynolds number, where the flow is turbulent, when surpassing the critical angle of attack. The location of the separation point over a range of pre and post-critical incidences can be found in Figure 8, yielding a similar behaviour as found in the literature [34].

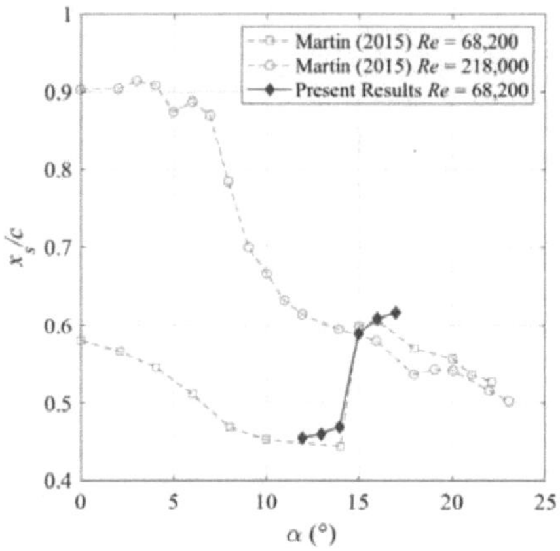

Figure 8. Streamwise location of the separation point at *Re* = 68,200 and comparison with Martin [34].

6.3. Turbulent kinetic energy at *Re* = 68,200

To further describe the flow field and the role of the leading-edge flow on trailing-edge separation as well as boundary layer regime, the nondimensional turbulent kinetic energy (TKE), κ, given by Eq. 3, was calculated.

$$\kappa = \frac{\overline{(u')^2} + \overline{(v')^2}}{2U_\infty^2} \tag{3}$$

This is a function of the streamwise and streamnormal velocity fluctuations u' and v', and free stream velocity U_∞. TKE is commonly employed to characterise turbulent flow on flat plates [41], thick airfoils [42, 43, 44], low camber membranes [45, 46], as well as high camber wings [47, 48, 49]. The literature further reveals comparable magnitudes of κ on cambered profiles (*e.g.* $0 < \kappa < 0.2$ [43]). Visual representations, plotted in relation to the separation point, can be found in Figure 9, with the threshold to plot TKE defined as 0.01 [43, 45].

A clear distinction can be made between the sub-critical range, namely angles of attack of 12°, 13° and 14°, where moderate values of κ are exhibited, at a clear distance downstream of separation and away from the surface of the arc. Conversely, in super-critical territory, above $\alpha = 15°$, higher values of κ are found, and detected upstream of the separation point. It is therefore possible to conclude that laminar separation occurs in one case, whereas the separation is turbulent in the other, explaining the increase in lift, reduction in drag, and delayed separation previously noticed. Note that past the critical angle of attack, the fluctuations appear to weaken and move away from the surface as the incidence increases, as observed on membrane airfoils [45].

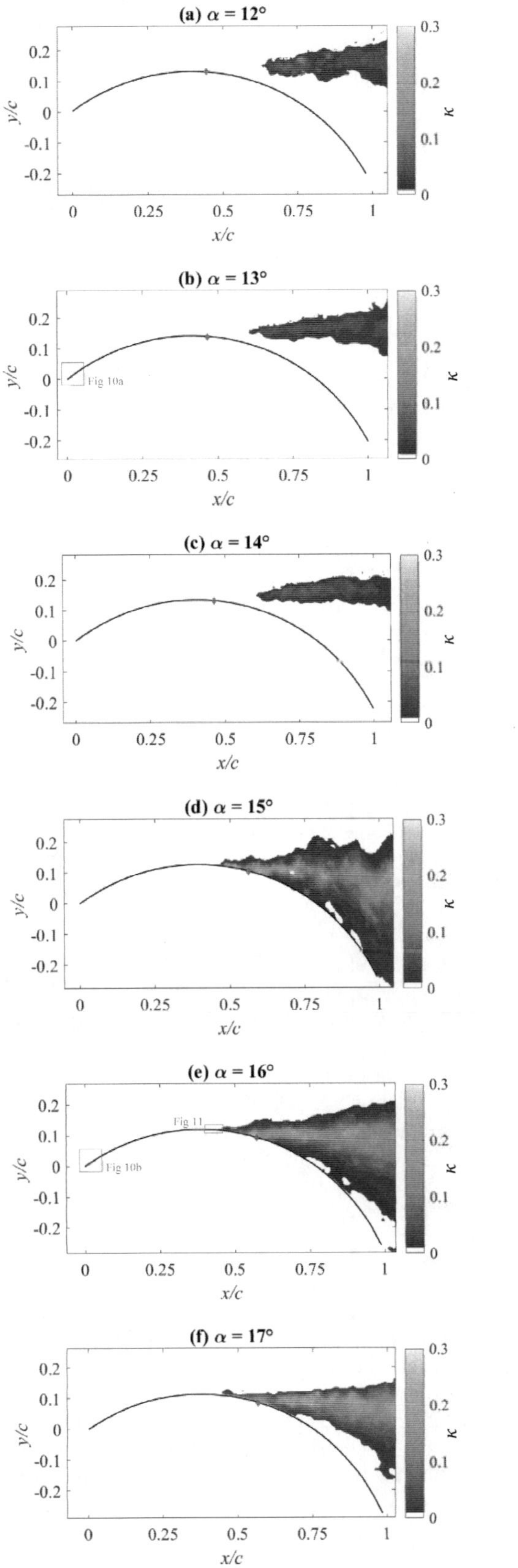

Figure 9. Nondimensional turbulent kinetic energy κ (plotted for κ > 0.01) and location of the separation point (♦) at Re = 68,200 over a range of pre (a-c) and post (d-f) critical angles of attack.

By capturing the overall flow field around the entirety of the arc, resolution is sacrificed, therefore not allowing to characterise the flow regime within the boundary layer, or at the leading edge, the latter being of particular interest on such geometries [39]. Thus, further investigation focussed on the LESB resulting from the leading-edge separation occurring past the ideal angle of attack was conducted.

6.4. Leading-edge separation bubble at Re = 68,200

A similar methodology was applied at the leading edge to quantify the TKE in both pre and post-critical regimes, at α = 13° and α = 16° respectively, for a Reynolds number of 68,200. Prior to reaching the critical angle of attack, a turbulent LESB is evident in Figure 10a. Remarkably, it is followed not by a turbulent boundary layer as might be expected, but by a laminar one, ascertained due to the absence of turbulent kinetic energy. On the other hand, at α = 16°, i.e. once the critical angle of attack has been exceeded, higher values of κ have been quantified, extending past the LESB into a turbulent boundary layer, as depicted in Figure 10b. In both cases, there is no significant difference in LESB size, or contribution to the overall lift, as later shown in section 6.5.

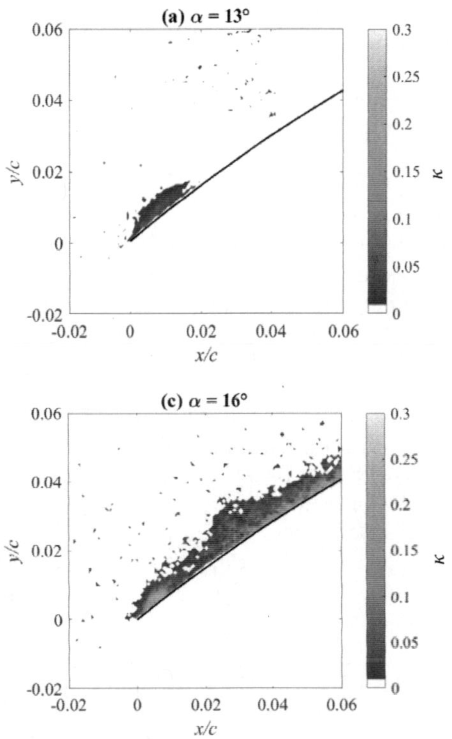

Figure 10. Nondimensional turbulent kinetic energy κ (plotted for κ > 0.01) at the leading edge pre (α = 13°) and post (α = 16°) critical angle of attack at Re = 68,200.

The nondimensional turbulent kinetic energy therefore appears to be constrained to the leading-edge separation bubble in sub-critical regime, and is followed by a laminar boundary layer. Conversely, the TKE carries on past the leading-edge separation bubble into the boundary layer when the critical angle of attack is exceeded.

In Section 6.3, the overall flow field across the whole geometry presented in Figure 9 highlighted the turbulent separation in post-critical regime, with turbulent kinetic energy visible ahead of the separation point. However, at that scale, the flow within the boundary layer was not captured. The focus on the leading edge in Figure 10 revealed that a turbulent boundary layer was indeed present past the LESB at $\alpha = 16°$. Additional PIV downstream of the leading edge but upstream of the point where turbulent kinetic energy was detected in Figure 9e yielded the results for κ presented in Figure 11.

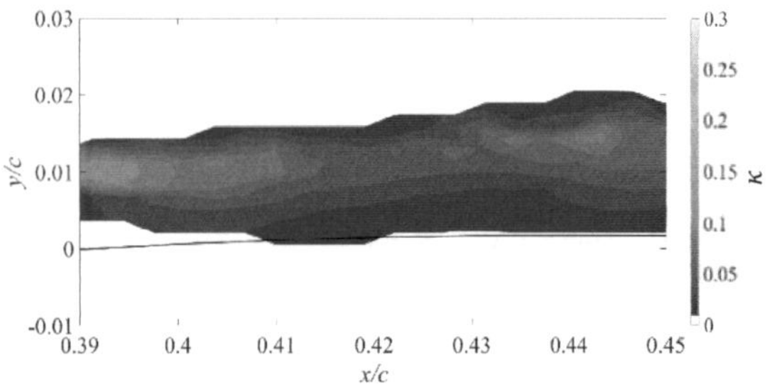

Figure 11. Nondimensional turbulent kinetic energy κ (for $\kappa > 0.01$) at $\alpha = 16°$ and $Re = 68{,}200$.

This evidence of a turbulent boundary layer between the two previously observed regions suggests that the turbulent boundary layer that follows the LESB in post-critical condition is carried along the surface of the arc all the way to the downstream trailing-edge separation.

6.5. Leading-edge separation bubble at $Re = 53{,}530$

The size of a leading-edge separation bubble increases with the angle of incidence and reduced flow speed. A case at these extremes of the presented set of experiments was therefore selected, while being in the turbulent regime. Figure 12 presents the LESB at $\alpha = 25°$ and $Re = 53{,}530$. In this condition, levels of κ consistent with previous measurements are exhibited in Figure 12a, while a primary recirculation bubble is also present in Figure 12b. However, contrarily to the work undertaken on flat plates [50] or some work on circular arcs [51] no secondary bubble could be observed at the available PIV resolution. Nevertheless, it could be speculated that the upstream inflection in the mean streamwise velocity contour ($u/U_\infty = 0$) [50] in Figure 12b may be the result of such a secondary bubble.

From the three cases focussed on the leading edge, as presented in Figures 10a, 10b and 12, the circulation Γ can be quantified as the integral of vorticity over the leading-edge separation bubble area. As the LESB does not move with respect to the arc, its circulation is part of the bounded circulation of the arc and its lift contribution can be computed with the Kutta-Joukowski lift formula:

$$L = \rho U_\infty \Gamma \tag{4}$$

Whether in pre or post-critical regime, for a similar size leading-edge separation bubble, as shown for $Re = 68{,}200$ at $\alpha = 13°$ and $\alpha = 16°$, comparable contributions to the overall lift of the circular arc are made, namely circa 2%. Conversely, for the larger LESB at $Re = 53{,}530$ and $\alpha = 25°$, the contribution to overall lift rose to 7.5%.

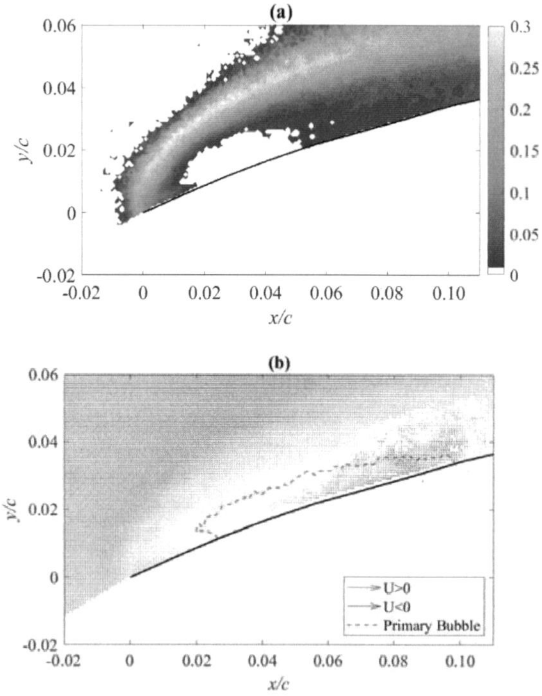

Figure 12. Leading-edge separation bubble at $\alpha = 25°$ and $Re = 53{,}530$, depicting (a) nondimensional turbulent kinetic energy κ (for $\kappa > 0.01$), and (b) velocity field with streamwise (red) and backwards (blue) flow.

It can therefore be concluded that the LESB provides a non-negligible contribution to the overall lift, and a larger LESB induces a greater contribution to lift. This is not to say that the lift is enhanced by the presence of the LESB [52], but rather that part of the lift is provided by the LESB. Furthermore, these findings reveal the critical effect of the LESB on the boundary layer regime and overall flow fields, ultimately affecting the global forces.

7. DISCUSSION

Based on the presented experimental data and literature [13, 27], Figure 13 proposes a schematic description of the lift trend with the angle of attack and the Reynolds number. This is presented without values for the lift coefficient as the onus is on its behaviour, with the actual values, which may be impacted by various factors such as blockage, not deemed relevant for this representation.

Two main trends can be distinguished, namely a laminar and turbulent one, with an abrupt transition occurring when the critical angle of attack is reached. The general behaviour of the turbulent trend for the tested arc length and leading edge shape, increasing the angle of attack from zero, can be characterised as follows:

- There is a first angle of attack range (0° < α < 8°) where the lift increases almost linearly with a slope higher than 2π/radians.

- Trailing-edge separation becomes significant as the separation point moves upstream, resulting in a reduced lift trend with incidence (8° < α < 11°).

- Ideal angle of attack (α = 11°) is reached, where the upstream stagnation point is at the leading edge.

- Subsequently, the leading-edge separation bubble grows in size with the angle of attack (11° < α < ~23°), associated with an increase in lift.

- Deep stall is eventually reached (α ≳ 23°), with the flow separating at the leading edge and no longer reattaching to the surface of the arc, with a noticeable drop in lift being generated [13].

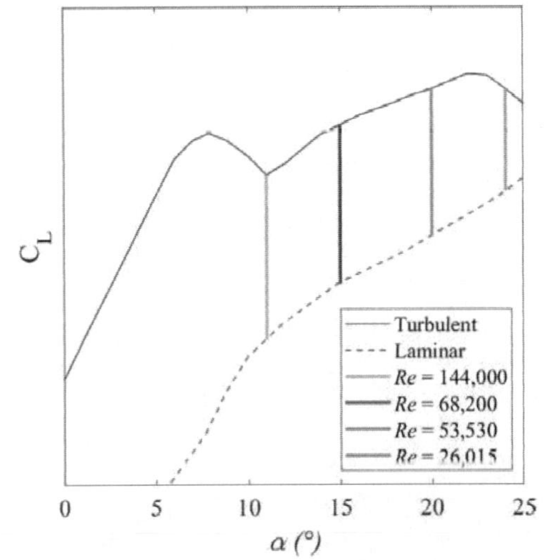

Figure 13. Idealized model for the lift generated by a highly cambered thin circular arc with sharp leading edge, including data edited from Velychko [27] and Bot [13].

With the acquired knowledge of the ideal angle of attack (11° for y_c/c = 22.32% at Re = 68,200) and corresponding critical Reynolds (144,000 +/-2,000), it is possible to recommend experimental model testing conditions so that the full-scale flow, clearly turbulent, can be replicated in a wind tunnel for instance. Indeed, as highlighted by Souppez et al. [37], evidence of laminar separation bubbles, identified thanks to a characteristic plateau in the pressure distribution, can be found in the literature, revealing inappropriate test conditions. Furthermore, the interesting case of Bot et al. [53] provides a striking example of a rigid model scale test undertaken at too low a Reynolds number in certain conditions. Indeed, pressure distribution at the shallowest incidence reveals the presence of a laminar separation bubble, suggesting the critical angle of attack has not yet been reached. However, at all subsequent and greater incidences, such flow feature is absent, and thus the critical angle of attack to trigger transition must have been reached. Given the typical Reynolds number of full scale spinnakers being of the order of 5 × 10^5 to 5 × 10^7 [39], the present findings are relevant to model scale testing.

Most remarkably, the nondimensional turbulent kinetic energy evidences the fact that, in pre-critical regime, where the downstream separation is laminar, the LESB is turbulent, but followed by a laminar boundary layer all the way to the training edge separation, before transition occurs in the wake. Conversely, in post-critical regime, the turbulent LESB is followed by a turbulent boundary layer, which is carried all the way to separation. Lastly, it was shown that, for a given leading-edge separation bubble size, the flow regime immediately downstream of the LESB does not make a significant difference to the lift contribution at the leading edge. However, a larger LESB proved to contribute to the overall lift by a greater proportion. Furthermore, it is the overall flow regime, whether laminar or turbulent, initiated at the inception of the boundary layer, that influences the global forces.

8. CONCLUSIONS

By employing a combination of force measurements and PIV flow visualisation, this study provides new insights on the flow around a circular arcs, particularly on the crucial effect of the leading-edge flow on the trailing-edge separation. More precisely, this was applied to a highly cambered (y_c/c = 22.32%), thin circular arc, with a sharp leading edge.

Evidence of a combination of critical Reynolds number and critical angle of attack necessary to trigger laminar-to-turbulent transition has been provided, with the sudden change in flow regime accounting for the sharp increase in lift and drop in drag. Additionally, the ideal angle of attack, where the stagnation point is located on the sharp leading edge has been identified, confirming previous observation of the formation of a leading-edge separation bubble once this incidence is exceeded.

Most importantly, the present research study shed a new light on the effect of the leading-edge separation bubble on the flow within the downstream boundary layer. Indeed, below the critical angle of attack, while a turbulent leading-edge separation bubble was present, the boundary layer remained laminar all the way to the trailing edge separation, with transition occurring in the detached wake. On the other hand, once the critical angle of attack is surpassed, the turbulent leading-edge separation bubble is followed by a turbulent boundary layer along the surface of the arc until turbulent separation occurs.

It is envisaged that these results will contribute to improve the design and performance of high-performance downwind yacht sails, and may further contribute to wider industrial applications, including micro aerial vehicles.

References

1. Fallow, B. J., (1996), "America's Cup sail design", *Journal of Wind Engineering and Industrial Aerodynamics*, **63**, 183-192.
2. Richards, P. J., Johnson, A. & Stanton A., (2001), "America's Cup sails – vertical wings or horizontal parachutes?", *Journal of Wind Engineering and Industrial Aerodynamics*, **89**, 1565-1577.
3. Viola, I. M. & Flay, R. G. J. (2009), "Force and pressure investigation of modern asymmetric spinnakers", *International Journal of Small Craft Technology*, **151**(B2).
4. Flay, R. G. J. & Jackson, P. S. (1992), "Flow simulation for wind tunnel Studies of sail aerodynamics", *Journal of Wind Engineering and Industrial Aerodynamics*, **41**, 2703-2714.
5. Flay, R. G. J. & Vuletich, I. J., (1995), "Development of a wind tunnel test facility for yacht aerodynamic studies", *Journal of Wind Engineering and Industrial Aerodynamics*, **58**, 231-258.
6. Hedges, K. L., (1993), *Computer modelling of downwind sails,* ME Thesis, University of Auckland, New Zealand.
7. Lasher, W. C., Sonnenmeier, J. R., Forsman, D. R. & Tomcho, J., (2005), "The aerodynamics of symmetric spinnakers", *Journal of Wind Engineering and Industrial Aerodynamics*, **93**, 311-337.
8. Hedges, K., Richards, P. & Mallison, G., (1996), "Computer modelling of downwind sails", *Journal of Wind Engineering and Industrial Aerodynamics*, **63**, 95-110.
9. Milgram, J. H., (1968), "The Aerodynamic of sails", *Proceedings of the 7th Symposium on Naval Hydrodynamic*, 1397-1434.
10. Gentry, A., (1971), *"The aerodynamics of sail interaction", The Ancient Interface III, 3rd AIAA Symposium on Sailing*, California, US.
11. Gentry, A., (1988), "The application of computational fluid dynamics to sails", *Proceedings of the Symposium of Hydrodynamic Performance Enhancement for Marine Applications.*
12. Flay, R. G. J., Piard, A. & Bot, P., (2017), "Aerodynamics of a highly cambered circular arc aerofoil: experimental investigations", *International Conference on Innovation in High Performance Sailing Yachts*, Lorient, France, 150-162.
13. Bot, P., (2020), "Force variation related to flow pattern changes around a high-camber thin wing", *AIAA Journal*, 1-7.
14. Souppez, J.-B. R. G., Arredondo-Galeana, A. & Viola, I. M., (2019), "Recent advances in numerical and experimental downwind sail aerodynamics", *Journal of Sailing Technology*, **4**(1), 45-65.
15. Wallis, R. A., (1946), *Wind tunnel tests on a series of circular arc plate aerofoils*, Melbourne: Aeronautical Research Labs.
16. Maekawa, T. & Atsumi, S., (1952), *Transition caused by the laminar flow separation,* Washington, National Advisory Committee for Aeronautics, Technical Memorandum 1352.
17. Tani, I., (1940), "On the transition caused by laminar separation of the boundary layer", *Jour. Soc. Aero. Sci. Japan,* **7**.
18. Owen, P. R. & Klanfer, L., (1955), *On the laminar boundary layer separation from the leading edge of a thin airfoil,* Ministry of Supply, Aeronautical Research Council.
19. DeLaurier, J. D. & Harris, J. M., (1974), "An experimental investigation of the aerodynamic characteristics of stepped-wedge airfoils at low speeds", *Second International Symposium on the Technology and Science of Low-Speed and Motorless Flight*, Massachusetts, US.
20. Buehring, I., (1977), *Performance characteristics of simple airfoils for windmill applications*, London, Imperial College.
21. Bruining, A., (1979), *Aerodynamic characteristics of a curved plate airfoil section at reynolds numbers 60.000 and 100.000 and angles of attack from -10 to +90 degrees*, Delft: Delft University of Technology, Department of Aerospace Engineering.
22. Pandey, M., Pandey, K. P. & Ojha, T. P., (1988). "Aerodynamic characteristics of cambered steel plates in relation to their use in wind energy conversion systems", *Wind Engineering,* **12**(2), 90-104.
23. Tse, M. C., (1991), *Overall effects of separation on thin aerofoils,* PhD Thesis, McGill University, Canada.
24. Cyr, S., (1992), *A theoretical model for flow about a circular-arc aerofoil with separation*, MSc Thesis, McGill University, Canada.
25. Sunada, S., Yasuda, T., Yasuda, K. & Kawachu, K., (2002), "Comparison of wing characteristics and an ultralow Reynolds number". *Journal of Aircraft,* **39**.
26. Okamoto, M. & Azuma, A., (2005), "Experimental study on aerodynamic characteristics of unsteady wings at low Reynolds number", *AIAA Journal,* **42**(12), 2526-2536.
27. Velychko, N., (2014), *Study of highly cambered aerofoil using JR3 sensor.* Aerodynamics laboratory report, The University of Auckland, New Zealand.

28. Brault, E., (2013), *Aerodynamics on very thin airfoil,* MSc Thesis, Newcastle University.

29. Milgram, J. H., (1971), *Section data for thin, highly cambered airfoils in incompressible flow*, NASA CR-1767, United States.

30. Cyr, S. & Newman, B. G., (1996), "Flow past two-dimension membrane aerofoils with rear separation", *Journal of Wind Engineering and Industrial Aerodynamics,* **63**, 1-16.

31. Collie, S. J., Kackson, P. & Fallow, B. J., (2004), "Two-dimensional CFD-based parametric analysis of downwind sail designs", *International Journal of Small Craft Technology,* **146**:15.

32. Lombardi, A., (2014), *Experimental analysis of highly-cambered thin profile,* MSc Thesis, IRENAV, France.

33. Lebret, C., (2013), *High cambered thin profile study,* Internship report, IRENAV, France.

34. Martin, V., (2015), *Reynolds number and angle of attack effects on a highly cambered thin profile flow topology,* MSc Thesis, IRENAV, France.

35. Thomas, G., (2015), *Flow phenomenology around a highly cambered thin profile*, MSc Thesis, IRENAV, France.

36. Bot, P., Rabaud, M., Thomas, G., Lombardi, A. & Lebret, C., (2016), "Sharp transition in the lift force of a fluid flowing past nonsymmetrical obstacles: evidence for a lift crisis in the drag crises regime", *Physical review letters,* **117**.

37. Souppez, J.-B. R. G., Arredondo-Galeana, A. & Viola, I. M., (2019), "Recent advances in downwind sail aerodynamics", *Proceedings of the 23rd Chesapeake Sailing Yacht Symposium*, Annapolis, Maryland, United States, March 2019.

38. Dewavrin, J. M. M.-A. & Souppez, J.-B. R. G., (2018), "Experimental investigation into modern hydrofoils-assisted monohulls: how hydrodynamically efficient are they?", *International Journal of Small Craft Technology,* **160**(2), 111-120.

39. Arredondo-Galeana, A. & Viola, I. M., (2017), "The leading-edge vortex of yacht sails", *Ocean Engineering,* **159**, 552-562.

40. Bird, H. J. A., Otomo, S., Ramesh, K. & Viola, I. M., (2019), "A geometrically non-linear time-domain unsteady lifting-line theory", *AIAA SciTech Forum*, California, United States.

41. Lee, D., Kawau, S., Nonomura, T., Anyoji, M., Aono, H., Oyama, A., Asai, K. & Fujia, K. (2015), "Mechanisms of surface pressure distribution within a laminar separation bubble at different Reynolds numbers", *Physics of Fluids,* **24**, 85-105.

42. Choudhry, A., Arjomandi, M. & Kelso, R., (2015), "A study of long separation bubble on thick airfoils and its consequent effects", *International Journal of Heat and Fluid Flows,* **52**, 84-96.

43. Koca, K., Genc, M. S., Acikel, H. H., Cagdas, M. & Bodur, T. M., (2018), "Identification of flow phenomena over NACA4412 wind turbine airfoil at low Reynolds numbers and role of laminar separation bubble on flow evolution", *Energy,* 750-764.

44. Istvan, M. & Yarusevych, S., (2018), "Effect of free-stream turbulence intensity on transition in a laminar separation bubble formed over an airfoil", *Experiments in Fluids,* **59**.

45. Rojratsirikul, P., Wang, Z. & Gursul, I., (2009), "Unsteady fluid-structure interactions of membrane airfoils at low Reynolds numbers", *Experiments in Fluids,* **46**, 859-872.

46. Rojratsirikul, P., Genc, M. S., Wang, Z. & Gursul, I., (2011), "Flow-induced vibrations of low aspect ratio rectangular membrane wings", *Journal of Fluids and Structures,* **27**, 1296-1309.

47. Nava, S., Carter, J. E. & Norris, S. E., (2018), "Large eddy simulation of an asymmetric spinnaker", *Ocean Engineering,* **169**, 99-109.

48. Nava, S., Bot, P., Cater, J., Norris, S., (2016), *"Modelling the Lift Crisis of a Cambered Plate at 0° Angle of Attack"*, 20th Australasian Fluid Mechanics Conference, Perth, Australia, December 2016.

49. Souppez, J.-B. R. G. & Viola, I. M., (2020), "Aerodynamics of highly cambered circular arcs with a sharp leading edge at low Reynolds numbers", *Bulletin of the American Physical Society*, 73rd Annual Meeting of the APS Division of Fluid Dynamics, Chicago, United States, November 2020.

50. Crompton, M. & Barrett, R., (2000), "Investigation of the separation bubble formed behind the sharp leading edge of a flat plate at incidence", *Proceedings of the Institution of Mechanical Engineers, Part G,* **214**, 157-176.

51. Couvrant, H., (2015), *Reynolds number and angle of attack effects on a leading edge vortex for a high cambered thin profile,* MSc Thesis, IRENAV, France.

52. Arredondo-Galeana, A., (2019), *A study of the vortex flow of downwind sails,* PhD Thesis, The University of Edinburgh, United Kindgdom.

53. Bot, P., Viola, I. M., Flay, R. G. J. & Brett, J. S., (2014), "Wind-tunnel pressure measurements on model-scale rigid downwind sails", *Ocean Engineering,* **90**, 84-92.

FORCE GENERATION MECHANISMS OF DOWNWIND SAILS

Abel Arredondo-Galeana[1], abel.arredondo-galeana@strath.ac.uk
Ignazio Maria Viola[2], I.M.Viola@ed.ac.uk

Abstract. The force generation mechanisms of a yacht sail are discussed with the aid of force and flow measurements on a model scale spinnaker. The velocity and vorticity fields are measured on five horizontal sections with particle image velocimetry. By comparing the forces measured with a balance to those computed from the vorticity field, we demonstrate how the force generation can be interpreted by the production and stretching of vortex rings. We consider vortex rings to be continuously generated and shed from the perimeter of the sail. The intersection of their vertical legs with horizontal planes are leading and trailing edge vortices. The sail force is due to the rate of change of the impulse of the vortex rings. Consequently, we show that the force can be computed from the time-averaged vorticity field using the Kutta-Joukowski lift formula, or from the strength and relative velocity of the leading and trailing edge vortices, or from the vorticity flux at the perimeter of the sail. The drag is estimated with Filon's drag formula. These results confirm experimentally the theoretical work of Viola et al. [1] and pave the way to the development of design methodologies that improve sail performance by manipulating the local vorticity field.

1. NOMENCLATURE

α	Sectional angle of attack
$\beta_{a\mathrm{OPT}}$	Optimum apparent wind angle
δ	Twist
Γ^-	Time averaged circulation of leading edge vortex
Γ^+	Time averaged circulation of trailing edge vortex
Γ_{LE}	Circulation of leading edge vortex
Γ_{TE}	Circulation of trailing edge vortex
$\dot{\Gamma}_{LE}$	Vorticity flux at leading edge
$\dot{\Gamma}_{TE}$	Vorticity flux at trailing edge
η	Rotational or trim angle
ρ	Density of fluid
ω_z	Spanwise vorticity
A	Area of sail and initial surface area of vortex ring
$\dot{A}$	Rate of change of vortex ring surface area
A_b	Area of bound vortex ring
C_{DF}	Drive force coefficient
$C_{DF,\max}$	Maximum force coefficient
c	Sectional chord length
c_0	Average chord length
I	Impulse
L	Lift force
D	Drag force
L_{LE}	Vorticity flux integration line at leading edge
L_{TE}	Vorticity flux integration line at trailing edge
V_a	Apparent wind
V_b	Boat velocity
V_t	True wind velocity
Re	Reynolds number
S	Span
t^*	Non-dimensional time
U_{LE}	Velocity of leading edge vortex
U_{TE}	Velocity of trailing edge vortex
U_∞	Freestream velocity
W	Integration line for drag computation
LEV	Leading edge vortex
TEV	Trailing edge vortex
PIV	Particle Image Velocimetry
x, y, z	Streamwise, normal, spanwise direction

2. INTRODUCTION

Downwind sails such as spinnakers are between the most powerful sails carried by a sailing yacht. However, the force generation mechanisms have often been incorrectly identified. In fact, these sails have a sharp leading edge that promotes flow separation. The bound circulation, intended as the net integral of vorticity in the thin Prandtl's boundary layer, is negligible to generate a negative lift. Hence, it cannot be the origin of the aerodynamic force [2,3]. Viola et al. [1] showed theoretically that the force generated by a sail is due to the production and stretching of vortex rings. In this paper we demonstrate this theory with physical experiments on model scale spinnakers. We show that the main force generation mechanism is associated with the convection velocity of the leading and trailing edge vortices, or by the production of vorticity that enables the associated vortex rings to grow in size. Here, we interpret the shedding vortices as the intersections of a vortex ring with a horizontal plane. We also show that the relative velocity of the leading and trailing edge vortices results, in the time average sense, in bound circulation and thus enables using the Kutta-Joukowski lift formula.

Twist of the onset flow and the sail geometry are not expected to affect the deformation and stretching of the vortex rings, but only their initial shape. We note however that non-axisymmetric vortex rings, like the ones pinched-off from a spinnaker are not in equilibrium and would probably change shape while convecting downstream [4].

3. METHODOLOGY

3.1. Sailing Conditions

In the laboratory setting, where the boat is fixed with respect to the water tunnel, the free stream represents the apparent wind $\boldsymbol{V_a}$. Hence, the sail must be set at the corresponding optimum apparent wind angle ($\beta_{a\mathrm{OPT}}$), which is the angle between the apparent wind $\boldsymbol{V_a}$ and the boat velocity $\boldsymbol{V_b}$. This angle varies with the height and thus is taken at a nominal height of 10 m. In fact, the boat

[1] Postdoctoral Research Associate, Department of Naval Architecture, Ocean & Marine Engineering, University of Strathclyde, UK
[2] Reader, School of Engineering, Institute for Energy Systems, University of Edinburgh, UK

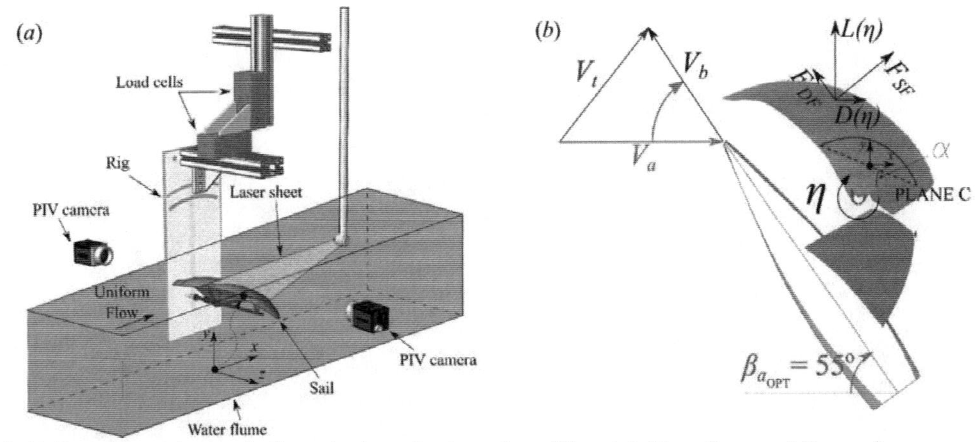

Figure 1. (a) Experimental setup in the water tunnel, where the sail is rotated by $\pi/2$ around the x-axis compared to the real sailing conditions and (b) Sail S_1 at $\beta_{a_{OPT}}$ as it would appear with respect to a boat hull from a bird's eye view

sails in the atmospheric boundary layer and thus the boat experiences an apparent wind velocity that is $\boldsymbol{V_a(z)} = \boldsymbol{V_t(z)} - \boldsymbol{V_b}$. The spinnaker considered in this paper was designed for $\beta_{aOPT} = 55^{\circ}$.

3.2. Sailing Geometry

The sail geometry was designed for the AC33 class yachts. This class, which is a set of rules for the design of the boat and the sails, was proposed for the 33rd America's Cup. The America's Cup is the world's oldest trophy and one of the most prestigious yacht races. The AC33 class was never adopted because of a legal dispute between the Defender, Alinghi, and the Challenger, Oracle BMW. Hence, the 33rd America's Cup was eventually sailed by multi-hulls under the conditions set by the Deed of Gift instead of a class rule. However, the aerodynamics of this spinnaker has been widely investigated in the last decade [5,6,7,8] and this makes it one of the best available benchmarks for downwind sails. The sail geometry together with useful notes on the spinnaker are available on https://voilab.eng.ed.ac.uk.

The original sail was designed to be 34 m in span, with a sail area of 510 m^2, while here a 1:100th scale model is considered. The particle image velocimetry (PIV) measurement planes were located at 7/8th, 5/6th, 3/4th, 1/2th and 1/4th the distance from the bottom of the sail to its tip, and labelled as planes A, B, C, D and E, respectively. We note that plane E is above the clue. A solid model was 3D printed in ABS. The sail had a span (S) of 300 mm, an area (A) of 0.045 m^2, an average chord length ($c_0 = A/S$) of 150 mm, an average thickness of 3 mm and a bevel angle at the leading edge of 20°. The twist from head to foot was $\delta = 16^{\circ}$. A 2% thickness is negligible on the sail aerodynamics, particularly because the leeward and windward sides of each sail section are parallel to the camber line. However, the leading-edge shape is critical for the direction of the separating streamline. To minimise this effect, we adopted the minimum sail thickness and bevel angle compatible with structural requirements [2].

3.3. Water tunnel setup

Figure 1a shows the experimental setup. The water tunnel is located at the University of Edinburgh. It is 9 m long and 0.4 m wide, with a flat, horizontal bed. The mean water depth was set to 0.5 m. The mean flow speed over the area occupied by the model was $U_{\infty} = 0.14\ \mathrm{ms}^{-1}$. With a span of 0.3 m, this results in a Reynolds number $Re = 2 \times 10^4$. Assuming a full-scale wind speed of 5 ms^{-1}, Re is 4×10^{-3} times smaller than at full-scale. Despite this significant difference between model and full scale Re, the differences in the flow field are expected to be moderated because the flow is fully separated at the sharp leading edge. In fact, the dynamics of leading-edge and trailing-edge vortices is only marginally affected by changes in the Reynolds number within the range from 10^4 to 10^6 [9,10].

The sail model was mounted and rotated by $\pi/2$ in the water tunnel, i.e. the water plane of the model-scale sail was parallel to a vertical sidewall of the water tunnel. A right-handed coordinate system is defined at the centre of the chord of plane C, as shown in figure 1a. The x- and y-coordinates are the streamwise and vertical directions, respectively. The sail model orientation could be adjusted by a rotation around a shaft parallel to the z-axis. The rotation is defined as the trim angle η (figure 1b). Here, $\eta = 0^{\circ}$ when the driving force is maximum.

The experiments were carried with the spinnaker in isolation. We assume that the main effect of the mainsail is to generate a uniform upwash variation of the angle of attack experienced by the spinnaker. We account for this by testing at different η. However, if we had tested with a mainsail, the mainsail circulation would deflect the spinnaker wake, resulting in potentially more attached flow in the results presented in the following sections.

3.4. Load cells

The load cells used in the experiments comprised a wind tunnel lift and drag dual-balance kit manufactured by KineOptics. The kit consisted of two Honeywell strain

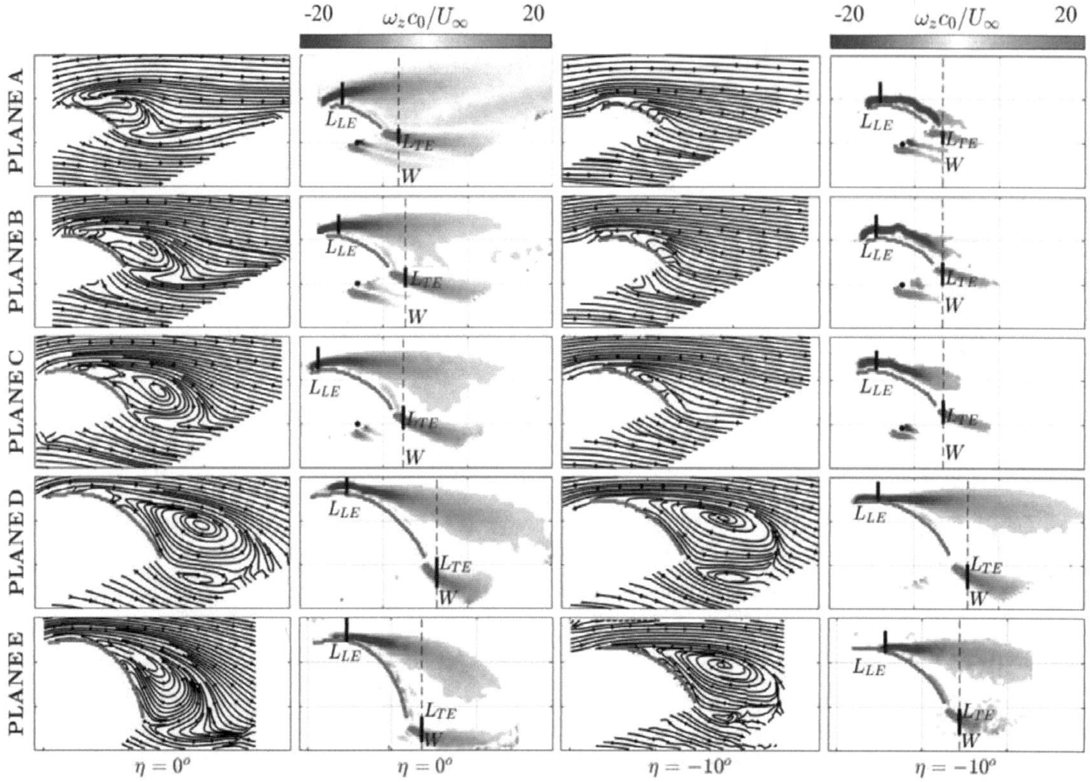

Figure 2. Time-averaged near-wake streamlines and nondimensional vorticity contours for the optimal sail trim in light wind conditions ($\eta = 0^{\circ}$, two left columns) and a depowered trim for strong wind conditions ($\eta = -10^{\circ}$, two right columns). The lines L_{LE} and L_{TE} are used in §4.6 to integrate the vorticity flux, while the dashed line W is used in §4.7 to compute the drag force

gauges connected to two SGA/A amplifiers. A low pass filter was set to 5 Hz to reduce high frequency noise coming from vibrations of the belt driving the flume propeller or electric noise. The excitation voltage for the strain gauges was 10 volts DC and 5 volts DC, for the lift and drag gauges, respectively. The amplifiers used a power voltage of 18-25 volts DC. The output analogue signals of the amplifiers were converted to digital signals, with a 16-bit National Instruments 6259 A/D board. Force signals were recorded with Wavelab.

3.5. Particle Image Velocimetry

PIV measurements were performed across *x-y* planes parallel to the free stream. The PIV system consisted of a Solo 200XT pulsed dual-head Nd:YAG laser, with an energy output of 200 mJ at a wavelength of λ = 532 nm. The laser beam was converted into a laser sheet through an array of underwater LaVision optics. The optics were fully submerged and created a laser sheet with a thickness of approximately 2 mm. The camera was a CCD Imperx 5MP with a 2448 px × 2050 px resolution and a Nikkor f/2, 50 mm lens. Seeding particles were silver coated hollow glass spheres, with an average diameter of 14 μm and a density of 1.7 g/cc. PIV image pairs were sampled at 7.5 Hz and a two-pass adaptive correlation was applied. The first pass had a 64 px × 64 px interrogation window, with a Gaussian weighting and 50% window overlap. The second pass had a 24 px × 24 px interrogation window and a 75% window overlap. Finally, a 3 × 3 Gaussian filter was used to smooth the vector fields. In order to mitigate surface reflections, a coating of matt black paint doped with rhodamine B was applied to the sail surface, allowing a notch filter on the camera to subtract the wavelength of rhodamine B and minimise the reflected light. A second coating of acrylic was applied to protect the rhodamine from dissolving in water. Additionally, background subtraction removed prevailing reflections and enabled measurements near the wall. The leading-edge region was not affected by laser reflections due to the curvature of the sail and the direction of the laser sheet. Successful background subtraction showed that no discernible deformation of the sail or strut occurred during the experimental runs. The strut was made of steel and designed with a thick root and then stepped down into a slimmer section in order to minimise deflections.

4. RESULTS

4.1. Time-averaged vorticity fields for different flow conditions

Figure 2 shows the near wake of the sail. Time- averaged streamlines and contours of nondimensional spanwise vorticity ($\omega_z\, c_0/U_\infty$) are presented for two trim angles, $\eta = 0^{\circ}$ and $\eta = -10^{\circ}$. These two angles are selected because the sail trim that allows $C_{DF,\max}$ ($\eta = 0^{\circ}$) is likely to be optimal in low wind speed conditions. Conversely,

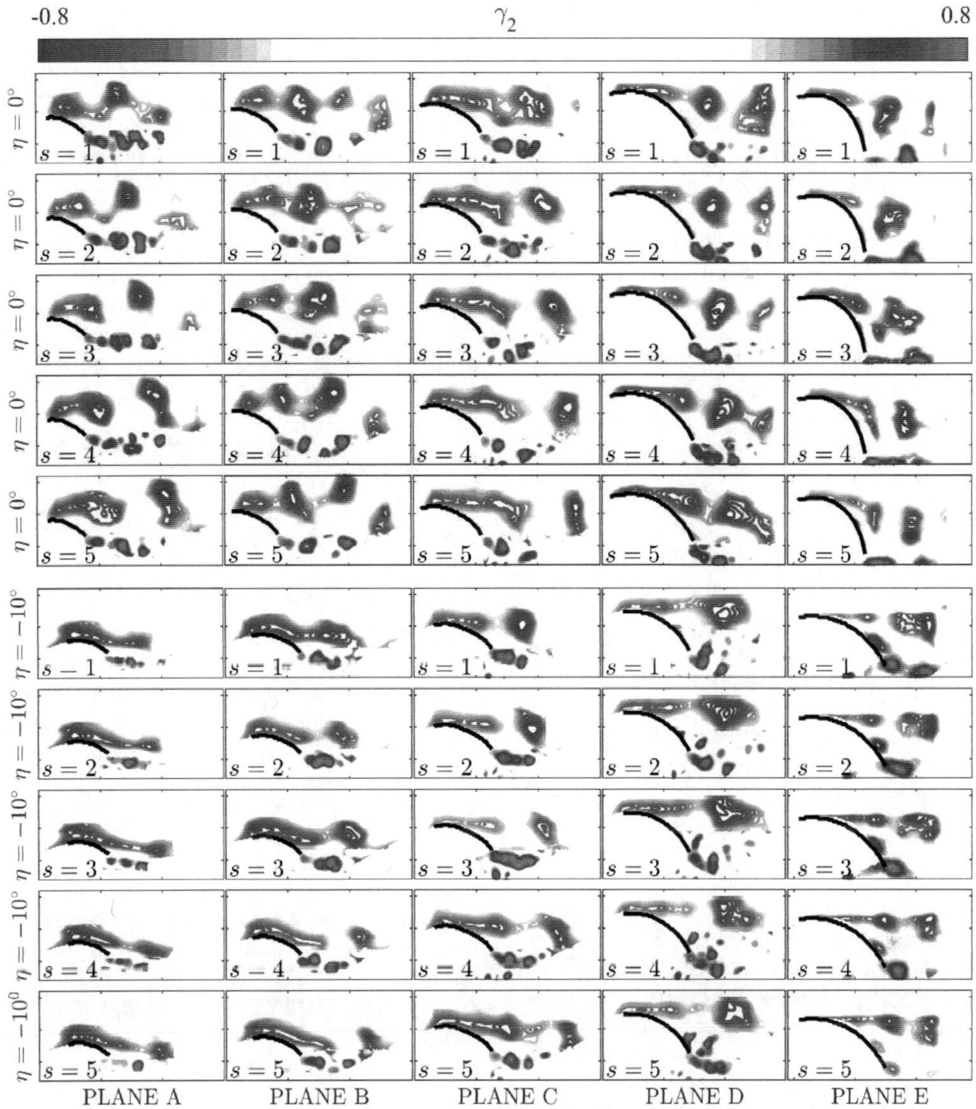

Figure 3 Sequence of γ_2 contours of sail S_1 based on vorticity measurements taken at five consecutive time steps (s = 1-5) on planes A, B, C, D and E (columns 1-5, respectively) at $\eta = 0^{\circ}$ (top array) and $\eta = -10^{\circ}$ (bottom array)

in strong wind conditions, the optimal trim is one that provides a reduced side force coefficient (C_{SF}), which increases the hydrodynamic resistance of the boat. Hence, there is a wind condition, which depends on the hydrodynamic performance of the boat, such that the optimal trim is $\eta = -10^{\circ}$.

Five flow fields are presented in figure 2, corresponding to the five PIV measurement planes (A, B, C, D, E) along the span of the sail. In the figure, plane A is the closest to the head and plane E the closest to the bottom of the sail and above the clue. A total of 500 images are averaged per plane. The lines L_{LE} and L_{TE}, in the vorticity plots of figure 2, are used in §4.6 to integrate the vorticity flux and the vorticity moment. Whilst the dotted line W is an integration line that is utilised to estimate the drag force in §4.7.

The light wind condition $\eta = 0^{\circ}$ is shown on the two left columns of figure 2. The streamlines of figure 2 reveal a large time-averaged recirculation region in most of the planes. Following the definitions of Perry & Steiner [8], stable foci are observed at the centre of the circulation regions, indicating the tridimensionality of the flow field. At the head of the sail at $\eta = 0^{\circ}$, a bifurcation line appears in plane A at the centre of the recirculation region. Contrarily, it is noted that near midspan of the sail, such as in plane D and at $\eta = 0^{\circ}$, two-dimensional nodes appear at the centre of the recirculation region.

In stronger breeze, at $\eta = -10^{\circ}$, a stable node near the surface of the sail is noted on the mid plane C. Conversely, in planes A and B, the streamline patterns close to the surface of the sail are c-shaped and indicative of vortex shedding. Perry and Steiner [8], for example, showed this same pattern behind the wake of a bluff body when the train of leading and trailing-edge vortices came in close proximity to each other.

The vorticity-contours identify two opposite sign circulation areas in all of the planes and at both $\eta = 0^{\circ}$ and $\eta = -10^{\circ}$, with negative vorticity emerging from the leading-edge and positive vorticity from the trailing-edge. On the highest sections, the positive (red) and negative (blue) vorticity wakes of the strut (indicated by a black dot) are also visible on the windward side of the sail section.

At $\eta = 0^{\circ}$, the highest sections of the sail (planes A, B and C) are stalled resulting in a large trailing-edge wake. Conversely, at $\eta = -10^{\circ}$, vorticity mostly follows the sail profile. It is to note that the bottom sections (planes D and E) are stalled at both angles η.

On any plane, there is more negative vorticity than positive vorticity. We will show in §4.5 that this is because of slower advection of negative leading-edge vorticity than of positive trailing edge vorticity. Because the time-averaged vorticity distribution is fixed with respect to the sail, the net vorticity (i.e. the sum of positive and negative vorticity) is the bound vorticity intended by Kutta [11] and Joukowski [12]. The bound circulation will be computed in §4.4 and it will be compared to that required to get the experimentally measured lift force.

4.2. Instantaneous vorticity field

The instantaneous vorticity field is investigated with a data set of 500 images, from which a subset of 5 images is presented in figure 3. To identify coherent regions of co-sign rotating flow, the γ_2-criterion [13] is used. The full data set of instantaneous flow fields is available in the supplementary material on the Edinburgh DataShare repository (datashare.is.ed.ac.uk). Each image is sampled at 7.5 Hz, corresponding to a non-dimensional convective time period $t^* = tU_\infty/c_0 = 0.13$. The instantaneous flow fields are shown in figure 3 for planes A-E at $\eta = 0^{\circ}$ and $\eta = -10^{\circ}$. Each instantaneous image is labelled with an index s, where s indicates the sequence number of the image. It should be recalled that the 5 planes were not recorded simultaneously.

The γ_2-criterion enabled the identification of vortex structures. As anticipated by the analysis of the time-averaged flow field topology in §4.1 vortex shedding occurred in all of the planes. Closed γ_2 isolines formed at the leading and trailing edges identify leading-edge vortices (LEVs) and trailing-edge vortices (TEVs), respectively. Both LEVs and TEVs are continuously generated and shed downstream.

The LEV is a more coherent vortex structure than the TEV, which instead shows a more stretched vorticity distribution in the streamwise direction. This difference is also associated with a slower streamwise velocity of the LEV. Near the sail, the LEV convects roughly at $U_\infty/2$ (see figure 3, plane D at $\eta = 0^{\circ}$), whilst away from the sail it convects at U_∞ (see figure 3, plane A at $\eta = 0^{\circ}$). The TEV convects at U_∞, both near and away from the sail, as estimated by Stevens et al. [14]. It will be shown in §4.4 that this results in about twice the integral of negative time-averaged vorticity over the entire field of view than the positive one.

4.3. Impulse theory formulae

The force production mechanisms can be described using impulse theory. For a volume of fluid V_f with constant density ρ, whose external boundaries approach infinity,

$$F = -\int_{V_f} \rho \frac{d\boldsymbol{u}}{dt} dV = -\rho \frac{d}{dt} \int_{V_f} \boldsymbol{u} dV = -\rho \frac{d\boldsymbol{I}}{dt}, \quad (1)$$

where

$$\boldsymbol{I} = \int_{V_f} \boldsymbol{u} dV \quad (2)$$

is the impulse. Wu [15] and Lighthill [16] showed that the impulse is given by the sum of the integral over the fluid volume V_f of the first moment of the vorticity ω, and the integral over the volume surface S_b with outward unit normal $\boldsymbol{n}$ of the moment of tangential velocity:

$$\boldsymbol{I} = \frac{1}{n_d - 1}\left(\int_{V_f} \boldsymbol{x} \times \boldsymbol{\omega} dV + \int_{S_b} \boldsymbol{x} \times (\boldsymbol{n} \times \boldsymbol{u})\, dS\right), \quad (3)$$

where $n = 2$ and 3 in two and three dimensions, respectively, and $\boldsymbol{x} = (x, y, z)$ is the coordinate vector. A complete derivation and discussion is available in, for instance, Eldredge [17] (p. 190). The second term of eq. 3 vanishes in a reference system fixed with the body. This, in fact, is an unsteady body force equal to the difference between the forces as observed from the reference system $O(x, y, z)$ and those observed from a reference system fixed with the body. It is proportional to the mass of the body [18,19] and thus its effect is negligible for slender bodies with small volume to surface area ratio [20] and for small body to fluid density ratio [21].

Assuming a tridimensional model, as the one considered here, and a reference system fixed with the sail in eq. 3, eq. 1 becomes

$$F = -\frac{\rho}{2} \frac{d}{dt} \int_{V_f} \boldsymbol{x} \times \boldsymbol{\omega} dV. \quad (4)$$

Let us consider the vorticity field as made of the sum of discrete vortex rings, each with a strength Γ_j, minimum surface area spanned by the vortex loop A_j, and unit vector $\boldsymbol{n}_j$ normal to the surface and pointing in the opposite direction to the surface induced velocity of the vortex ring, which acts orthogonal to its surface. The impulse of a vortex ring, which was first found by Lamb [22], is $\rho\Gamma A$. Hence, eq. 4 can be written as [14],

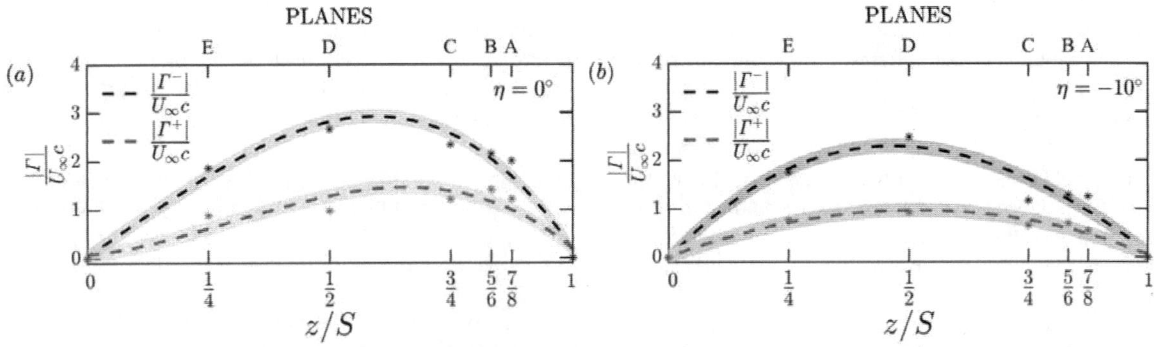

Figure 4. Nondimensional leading-edge circulation $\Gamma^-/U_\infty c$ and trailing-edge circulation $\Gamma^+/U_\infty c$ versus the spanwise nondimensional coordinate z/S for (a) $\eta = 0^\circ$ and (b) $\eta = -10^\circ$. Markers indicate measured data, whilst dotted lines are third-order polynomial fits with shaded regions showing the 95% CI of the fit.

$$F = \frac{\rho}{2} \sum_j (\dot{\Gamma}_j A_j + \Gamma_j \dot{A}_j) \boldsymbol{n}_j. \tag{5}$$

4.4. Bound vorticity

In this section, the time-averaged vorticity field is interpreted as a vortex ring bound to the sail. The ring is made by the net vorticity shown in the near wake of the sail (figure 2), the tip vortices and the starting vortex. Later on, in §4.5, discrete vortex rings are considered, whose cross-sections are the leading and trailing edge vortices from the instantaneous flow field.

Here, let us assume the vorticity as forming a single bound vortex ring with circulation $-\Gamma_b$ and a surface area A_b. Because this hypothetical vortex ring at steady state is long due to the starting vortex being at infinity, the surface of the ring is orthogonal to the y axis and thus the computed force will act in the lift direction.

The rate of change of the surface area of the vortex ring is $\dot{A}_b = SU_\infty$. We recall that S is the span, whilst $-\dot{\Gamma}_b$ is the production of the circulation needed to lengthen the tip vortex with strength $-\Gamma_b$ at a rate U_∞. Hence $-\dot{\Gamma}_b A_b = -\Gamma_b S U_\infty$. Substituting into eq. 5, we find the Kutta-Joukowski lift formula

$$L = -\rho S U_\infty \Gamma_b. \tag{6}$$

As mentioned in the previous section, in the time-averaged vorticity field, the net vorticity on each plane is the bound circulation Γ_b. An integral value of the bound circulation is given by:

$$\Gamma_b = \frac{1}{S} \int_S (\Gamma^- + \Gamma^+) dz, \tag{7}$$

where Γ^- and Γ^+ are the total negative and positive time-averaged circulations in the field of view of figure 2.

The circulation around the time-averaged leading and trailing-edge circulation zones is computed from the (discrete) area integral of vorticity. Vorticity is thresholded above the measurement uncertainty level to avoid accumulation of low-level vorticity in the results. The circulation is nondimensionalised and plotted in Figures 4a and 4b, for every plane and for the two trim angles η. The marks show the measured circulations, whilst the spanwise circulation distributions are approximated with third order polynomials showed by dotted lines. The area under the Γ^- distribution is approximately twice of that under Γ^+ for both of the trim angles η. By using the fitted curves in Fig. 4, the bound circulation is $0.9U_\infty c$ at $\eta = 0^\circ$ and $0.73U_\infty c$ at at $\eta = -10^\circ$. These values are 4% and 12% greater, respectively, than those measured with the load cells. Force comparisons with wind tunnel measurements are difficult since it is not known whether the maximum driving force trim angle ($\eta = 0^\circ$) between this experiment and wind tunnel tests of spinnakers in isolation is the same. At maximum driving force, the measured lift coefficient here was $C_L = 1.3$, whilst according to CFD studies [5], it could be $C_L = 1.6$. Differences can be attributed to the mainsail upwash which rotates the effective lift and drag axes and higher Reynolds numbers.

The comparison between Fig. 4a and 4b shows that, when the sail is depowered (from $\eta = 0^\circ$ to $\eta = -10^\circ$), the bound circulation decreases almost uniformly along the span of the sail by a proportional reduction of both positive and negative vorticity.

4.5. Vortex force

An alternative approach to use the time-averaged vorticity field as in the previous section, is to compute the vortex force based on the relative velocity of the leading and trailing edge vortices. In steady conditions, the same amount of positive and negative vorticity is shed into the wake. Hence, for every LEV, it is possible to identify a region whose integral of vorticity is opposite to the LEV circulation. Because the LEV and the TEV advect at different velocities, more leading-edge vorticity than trailing-edge vorticity is observed on average near the sail. This occurs only in the near wake though. In fact, in the far wake, positive and negative vorticity must convect at

the same velocity to ensure that the net time-averaged vorticity vanishes in accordance with Kelvin's theorem.
The instantaneous vorticity fields can be interpreted as made of vortex rings, where each LEV-TEV pair in planes A-E are the intersection of one ring with the planes. The surface A_j of the vortex ring is the distance between the LEV and the TEV, integrated over the span S. It is worth noting that, chosen the jth LEV, the choice of the corresponding jth TEV to form a vortex ring is arbitrary, as long as the sum of their circulations is null ($\Gamma_{\mathrm{LEV}_j} + \Gamma_{\mathrm{TEV}_j} = 0$).

Let us now consider the two force production mechanisms in eq. 5. The surface of the jth vortex ring with positive circulation $\Gamma_j = -\Gamma_{\mathrm{LEV}j}$ stretches in the streamwise direction at a rate $\dot{A}_j = S(U_{\mathrm{TEV}j} - U_{\mathrm{LEV}j})$. The force associated with the streamwise stretching of the vortex enclosed surface is in the lift direction. Only n vortex rings within the near wake stretch at a rate $\dot{A}_j$, because U_{LEV} gradually increases to the same velocity as U_{TEV}. Hence, only n vortex rings can be considered. To estimate the force due to the rate of production of vorticity, let us consider the circulation needed to lengthen (in the streamwise direction) the area enclosed by the vortex ring. Akin to the consideration made to derive the Kutta-Joukowski lift formula in §4.4, here we find that $\dot{\Gamma}_j A_j = \Gamma_j \left(U_{\mathrm{TEV}_j} - U_{\mathrm{LEV}_j}\right) S$. Hence the two terms between parentheses on the right hand side of eq. 5 are identical and by substitution

$$L = -\rho \int_S \sum_j^n \left(U_{\mathrm{TEV}_j} - U_{\mathrm{LEV}_j}\right) \Gamma_{\mathrm{LEV}_j} \mathrm{d}z. \quad (8)$$

An equivalent formulation for the lift of a flat plate based on the relative velocity of the LEV and TEV was derived by Stevens et al. [14].

This formulation is equivalent to that in eq. 6 and 7 as can be shown by comparing the sectional lift computed with the two approaches. Assume for example, n LEVs convecting downstream at kU_∞ while the TEVs convect with velocity U_∞, and the same circulation for both LEVs and TEVs since they both form discrete vortex rings. The time-averaged net circulation on the plane depends on the difference between the number of LEVs (n) and TEVs (nk) in the near wake, i.e.

$$\Gamma^- + \Gamma^+ = n(1-k)\Gamma_{\mathrm{LEV}} = n\frac{U_{\mathrm{TEV}} - U_{\mathrm{LEV}}}{U_\infty}\Gamma_{\mathrm{LEV}}. \quad (9)$$

We can now show that the sectional lift computed with eq. 8 is the same as that computed with eq. 6 and 7:

$$-\rho U_\infty(\Gamma^- + \Gamma^+) = -\rho n(U_{\mathrm{TEV}} - U_{\mathrm{LEV}})\Gamma_{\mathrm{LEV}}. \quad (10)$$

This equivalence was also confirmed by the measurements of the streamwise velocity of the LEV by tracking the centroids of the γ_2 isosurfaces in figure 3 and by considering $n = 3$. The convective velocity of the TEV is considered to be U_∞ [14].

The ratio of two between these two velocities matches the approximate ratio of two between Γ^- and Γ^+ showed in figure 4. Therefore, it can be concluded that when the sail is depowered (from $\eta = 0^\circ$ to $\eta = -10^\circ$), the sectional lift decreases almost uniformly along the span of the sail because of a lower circulation of the leading edge vortices, whilst near the sail, they convect downstream always at about half of the free steam velocity.

4.6. Vorticity production

Both in §4.4 and §4.5 we transformed the vortex ring production term in eq. 5 into an equivalent vortex ring growth term. Here we show that a similar force result can be achieved by considering the vortex ring production term directly. In fact, $\dot{\Gamma}$ can be estimated from the vorticity flux through the leading and trailing edge shear layers. Figure 2 shows the two segments L_{LE} and L_{TE}. The first one extends from the surface of the sail in the y direction and has a length $0.3c_0$. The latter is $0.1c_0$ downstream of the trailing edge and has a length of $0.1c_0$ The dimension of L_{TE} is such to avoid inclusion of negative vorticity from either the leading edge or the vertical strut holding the model. Vorticity is not included below the noise threshold. The vorticity fluxes are computed as

$$\dot{\Gamma}_{\mathrm{LE}} = \int_{L_{\mathrm{LE}}} u_x \omega_z dy, \quad (11)$$

$$\dot{\Gamma}_{\mathrm{TE}} = \int_{L_{\mathrm{TE}}} u_x \omega_z dy. \quad (12)$$

Results are shown in figures 5a and 5b for $\eta = 0^\circ$ and $\eta = -10^\circ$, respectively. The differences between the magnitude of the vorticity production at the leading and trailing edge are within the experimental uncertainty. When the sail is eased for strong wind conditions, the nondimensional vorticity production decreases on every spanwise section.

The lift force can be estimated from the vortex ring circulation production, which we found to be $\dot{\Gamma} \approx -\dot{\Gamma}_{\mathrm{LE}} \approx -\dot{\Gamma}_{\mathrm{TE}}$. The area of the vortex ring is the sail itself, and for the lift component of the force we consider only the area projection orthogonal to the lift direction. Hence, for each sail section along the span, the width of the vortex ring is $c\cos\alpha$, where c is the sectional chord length. Finally, assuming that most of the circulation is generated by horizontal shear layers at the leading and trailing edge, eq. 5 can be written as

$$L = \rho \int_S c\cos\alpha\, \dot{\Gamma}\, dz. \quad (13)$$

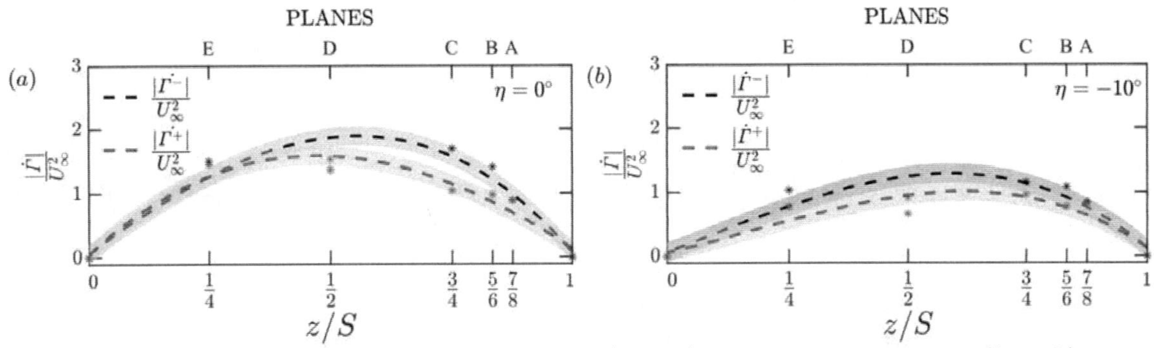

Figure 5 Nondimensional leading-edge circulation production $|\dot{\Gamma}_{\mathrm{LE}}/U_\infty^2|$ and trailing-edge circulation $|\dot{\Gamma}_{\mathrm{TE}}/U_\infty^2|$ versus the spanwise nondimensional coordinate z/S for (a) $\eta = 0^\circ$ and (b) $\eta = -10^\circ$. Markers indicate measured data, whilst dotted lines are third-order polynomial fits with shaded regions showing the 95% CI of the fit.

In Eq. 13, α is the angle between the chordline of the sail's cross-section and the horizontal axis, as shown in Figure 1b. By using the values of $\dot{\Gamma}$ in figure 5 the lift is 12% and 10% higher than that measured with the load cells for $\eta = 0^\circ$ and $\eta = -10^\circ$, respectively.

4.7. Drag

The time-averaged drag can be estimated using Taylor's formula, found in Bryant et al. [23], where it is defined as

$$D = \int_W (p - p_0)dy. \tag{14}$$

This expression states that the drag is the integral over a line W intersecting orthogonally the wake, of the difference between the pressure in the wake p and that in the far field p_0. By using the Bernoulli equation, this equation shows that the drag is equal to the momentum loss in the wake. Wu [24] recently showed that the first order approximation of Eq. 14 is

$$D = \rho U_\infty Q, \tag{15}$$

where, in 2D,

$$Q = \int_W -y\omega_z dy. \tag{16}$$

This enables the use of Taylor's formula by knowledge of only the vorticity field along W and the expression is referred to as the Filon's drag formula. A 2nd order polynomial is utilised to approximate the spanwise distribution of Q. By utilising Eq. 16 and integrating Q along the span of the sail, the steady-state drag coefficient is computed to be $C_D = 0.4$ and 0.26 for the light wind conditions ($\eta = 0^\circ$) and the strong breeze condition ($\eta = -10^\circ$), respectively.

These values lie within 15 to 20%, respectively, of the drag force coefficients measured with the load cells. The discrepancy is possibly due to the five plane data availability in the spanwise direction and to the effect of vorticity diffusion. We note that the drag of the strut and the rig were measured separately and subtracted to the force measurements, and so the strut and the rig were not the reasons for the discrepancy.

When the sail is depowered from $\eta = 0^\circ$ to $\eta = -10^\circ$, the drag decreases. Interpreting this reduction through Eq. 16, allows the use of figure 2 to identify how a reduction in vorticity enables drag reduction along the span of the sail. By decreasing the angle of attack, the width of the wake on the highest planes A and B decreases significantly, resulting in a lower drag generation. Conversely, the same angle of attack reduction on the lower sections, has a comparatively smaller effect on the wake width and thus on the drag generation.

5. FULL-SCALE SAILS CONSIDERATIONS

Firstly, we recognize that the motion of the flying shape and motion of the boat on the stability of the vortex rings are important considerations that need to be addressed in future research. Secondly, the LEVs and TEVs identified in this paper are fully turbulent and a further increase in the Reynolds number and of the turbulent intensity could enhance mixing and result in smaller vortices. However, their dynamics and the force generation mechanisms are not expected to significantly change. It is noted that that at the low Reynolds number at which we tested, the boundary layer forming from the stagnation points on the sail could be laminar while it would be most likely turbulent at full scale conditions. However, because we showed that the circulation associated with the vorticity in the boundary layer is negligible compared to the free circulation in the bulk flow field, it is unlikely that this would make a significant difference on the flow field. Lastly, at higher Reynolds number and turbulence intensity, we might expect enhanced mixing and consequent lower coherence of the vortex structures and, in turn, lower effectiveness of the different stabilising mechanisms of the leading-edge vortex. For example, the vorticity extraction through axial flow within the core of the leading-edge vortex might be lower at much higher Reynolds numbers than those tested in this work.

6. CONCLUSIONS

We presented force and flow measurements on a spinnaker in a water tunnel to analyse the force generation mechanisms. We considered the sail trim enabling the maximum driving force and a depowered trim. Using particle image velocimetry, we analysed the vorticity fields on five horizontal planes. We observed separated and tridimensional flow, with negligible bound circulation. In contrast, we showed that the vortex dynamics of leading and trailing edge vorticity is the governing force generation mechanism. This was demonstrated by comparing the force derived from the vorticity field with the force measured by the balance.

The vortex force is the rate of change (or the time derivative) of the impulse of a vortex ring in 3D, or a vortex pair in 2D. The leading and trailing edge vortices are the intersection of two of the legs of the vortex ring with the horizontal plane. We showed that considering the vortex lift from the dynamics (strength and relative velocity) of the leading and trailing edge vortices is mathematically equivalent to measuring the lift with the Kutta-Joukowski formula using the time-averaged vorticity field. Similarly, the drag force can be computed from the time-averaged vorticity field and by applying the Filon's drag formula. Furthermore, we noted that the sail force can also be computed from the vorticity flux at the leading edge and trailing edge, because this is equivalent to the production of circulation that feeds the vortex rings that emerge from the perimeter of the sail.

These results demonstrate the theoretical considerations made by Viola et al. [1]. They also pave the way to the development of design methodologies for sails that operate in separated flow conditions, where potential flow approaches are not suitable. The design can be informed by the observation of the local flow field, which can be either measured or computed with, for example, a Navier-Stokes code. Understanding the force generation mechanisms associated with the free vorticity in the field will enable tailoring the vorticity dynamics by design to achieve the desirable aerodynamic force.

Acknowledgements

This work was supported by Consejo Nacional de Ciencia y Tecnologia (CONACYT) with grant number 384490.

References

1. Viola I. M., Arredondo-Galeana, A., Pisetta, G., (2020), "How sails generate forces.", Proceedings of the 5th *Innov'sail Conference*, Gothenburgh, Sweden, June 2020, 1-16.

2. Arredondo-Galeana, A. & Viola, I. M. (2018), "The leading-edge vortex of yacht sails.", *Ocean Engineering*, 159, 552-562.

3. Viola I. M., Arredondo-Galeana, A. (2017), "The leading-edge vortex of yacht sails.", Proceedings of the 4th *Innov'sail Conference*, Lorient, France, June 2017, 115-126.

4. O'Farrell, C., & Dabiri, J. (2014), "Pinch-off of non-axisymmetric vortex rings.", *JFM,* 740, 61-96.

5. Viola, I. M. & Flay, R. G. J. (2009), "Force and pressure investigation of modern asymmetric spinnakers.", *IJSCT Technology Transaction RINA Part B2*, 151(2), 31–40.

6. Viola, I.M. & Flay, R.G.J. (2010), "Pressure distributions on modern asymmetric spinnakers.", *Int. J. Small Cr. Technol.* 152 (B1), 41-50.

7. Viola, I. M., Bartesaghi, S., Van-Renterghem, T. & Ponzini, R. (2014), "Detached Eddy simulation of a sailing yacht." *Ocean Engineering,* 90, 93-103.

8. Bot, P., Viola, I. M., Flay, R. G. J. & Brett, J.-S. (2014), "Wind-tunnel pressure measurements on model-scale rigid downwind sails.", *Ocean Engineering,* 90, 84-92.

9. Perry, AE & Steiner, TR (1987), "Large-scale vortex structures in turbulent wakes behind bluff bodies. Part 1. Vortex formation processes.", *JFM,* 174, 233–270.

10. DeVoria, A. C. & Mohseni, K. (2017), "On the mechanism of high-incidence lift generation for steadily translating low-aspect-ratio wings.", *JFM* 813, 110–126.

11. Kutta, W. M., (1902), "Auftriebskräfte in strömenden flüssigkeiten.", *Illust. Aeronaut. Mitt.*

12. Jowkowski, N. E., (1906)y, "On annexed vortices.", *Proc. Phys. Sect. Nat. Sci. Soc.,* 13 (2), 12-25.

13. Graftieaux, L., Michard, M. & Grosjean, N. (2001), "Combining PIV, POD and vortex identification algorithms for the study of unsteady turbulent swirling flows." *Meas Sci Technol* 12 (1201), 1422–1429.

14. Stevens, P. R. R. J., Babinsky, H., Manar, F., Mancini, P., Jones, A. R., Granlund, K. O., Nakata, T., Phillips, N., Bomphrey, R. J. & Gozukara, A. C. (2016), "Low Reynolds Number Acceleration of Flat Plate Wings at High Incidence (Invited).", *In 54th AIAA Aerosp. Sci. Meet. AIAA SciTech Forum,* (AIAA 2016-0286), pp. 1-15.

15. Wu, J. C. (1981), "Theory for aerodynamic force and moment in viscous flows." *AIAA Journal,* 19 (4), 432-441.

16. Lighthill, J. (1986), *An Informal Introduction to Theoretical Fluid Mechanics*, Clarendon Press, UK.

17. Eldredge, J. D. (2019), *Mathematical Modelling of Unsteady Inviscid Flows*, Springer, Switzerland.

18. Koumoutsakos, P. & Leonard, A. (1995), "High-resolution simulation of the flow around an

impulsively started cylinder using vortex methods.*", JFM,* 296, 1-38.

19. Leonard, A & Roshko, A (2001), "Aspects of flow-induced vibrations.*", J. Fluids Struct*., 15, 415-425.

20. Rival, D. E. & van Oudheusden, B. (2017), "Load-estimation techniques for unsteady incompressible flows.", *Exp. Fluids*, 58 (3), 1-11.

21. Lentink, D. (2018), "Accurate fluid force measurement based on control surface integration.", *Exp. Fluids,* 59 (1), 1-12.

22. Lamb, H. (1932), *Hydrodynamics*. Cambridge University Press.

23. Bryant, L. W., Williams, D. H., Taylor, G. I. & Bairstow, L. (1926), "An investigation of the flow of air an aerofoil of infinite span.", *Philosophical Transactions of the Royal Society of London, Series A, Containing Papers of Mathematical or Physical Character,* 225 (626-625), 199-245

24. Wu J, Ma H, Zhou M (2006), *Vorticity and vortex dynamics*, Springer, Germany.

SELECTION OF EXISTING SAIL DESIGNS FOR MULTI-FIDELITY SURROGATE MODELS

Tanya Peart[1], tanya.peart@doylesails.com
Nicolas Aubin[2], nicolas.aubin@doylesails.com
Stefano Nava[3], stefano.nava@doylesails.com
John Cater[4], j.cater@auckland.ac.nz
Stuart Norris[5], s.norris@auckland.ac.nz

Abstract. Velocity Prediction Programs (VPPs) are commonly used to help predict and compare the performance of different sail designs. A VPP requires an aerodynamic input force matrix which can be computationally expensive to calculate, limiting its application in industrial sail design projects. The use of multi-fidelity kriging surrogate models has previously been presented by the authors to reduce this cost, with the low-fidelity data provided by existing data from previously designed sails. An important consideration for the construction of these models is the choice of low-fidelity data points, which provide information about the trend of the curve between the high-fidelity data. A method is required to select the best existing sail design to use when constructing a multi-fidelity model. The suitability of an existing sail design as a low fidelity model can be evaluated based on the similarity of its geometric parameters with the new sail. It is shown in this article that for upwind jib sails, the similarity of the broadseam best indicates the ability of a design to be used as a low fidelity model for lift prediction. The predicted surrogate model error for lift proves to be close to 1% of the true error for most points. Larger discrepancies are observed for the drag coefficient.

1. NOMENCLATURE

AIC	Akaike Information Criterion
AWA	Apparent Wind Angle
AWS	Apparent Wind Speed
CFD	Computational Fluid Dynamics
FSI	Fluid-Structure Interaction
RANS	Reynolds-Averaged Navier-Stokes
RMS	Root Mean Square
VIF	Variance Inflation Factor
VLM	Vortex-Lattice Method
VPP	Velocity Prediction Program

2. GLOSSARY

Best subsets regression	A method to select between regression models that have been trained with different combinations of independent variables.
Fidelity	A measure of the relative accuracy of data points.
Independent variable	Input parameter/s for the linear regression.
Linear regression	A function which predicts the value of an output based on a linear relationship with the input parameters.
Overfitting	Training a model to fit the trend of sample data points too closely that results in trends that are not present in the additional data.
Response variable	Output parameter/s for the linear regression.

3. INTRODUCTION

In yacht sail design it is important to be able to predict the performance of a new design shape and to compare the performance of different designs [1]. A Velocity Prediction Program (VPP) can be used for this purpose. It applies the principles of equilibrium to the aero- and hydrodynamic forces of a yacht to determine the boat speed and heel angle resulting from a particular sail combination. The accuracy of a VPP is dependent on the quality of the input data [2]. When the flow is turbulent or detached, such as often occurs around sails, many Reynolds-Averaged Navier-Stokes (RANS) Computational Fluid Dynamics (CFD) simulations are required for an accurate performance prediction [3]. Consequently, the computational cost of a VPP can limit its application to industrial sail design projects since high-performance computing facilities are usually not available except for high-budget projects [4].

Surrogate models (also known as metamodels or response surface models) are a methodology to interpolate data, replacing a complex underlying function that may typically take hours or days to evaluate at a specific point, with a function that can be evaluated in a few seconds [5, 6]. This technique can be used to reduce the computational cost of obtaining a VPP prediction [7]. The functions for the aero- and hydrodynamic forces and moments can be represented by a surrogate model trained on previously obtained experimental or numerical data. This surrogate

[1] Student, University of Auckland, NZ
[2] R&D Manager, Doyle Sails, NZ
[3] Sail Designer, Doyle Sails, NZ
[4] Associate Professor, University of Auckland, NZ
[5] Senior Lecturer, University of Auckland, NZ

model can then be queried to find the required aero- and hydrodynamic forces for the VPP without the need for additional simulations or experiments.

Multi-fidelity surrogate models combine data points from two or more fidelity (accuracy) levels in a single model [5]. This results in fewer (more computationally expensive) high-fidelity points being needed, with gaps between these points interpolated using data which has a similar trend but a lower fidelity and computational cost. These differences in fidelity can be as a result of having different mesh resolutions [8], different solvers [9], a combination of experimental and numerical methods [10], or using data points from similar designs [11, 30]. Multi-fidelity methods have the potential to further reduce the computational cost of evaluating a function, or improve the accuracy for the same computational cost, when compared to conventional surrogate modelling methods [12, 11, 30].

The overall goal of this work was to construct accurate multi-fidelity surrogate models to predict the force coefficient curves of new sail designs. The multi-fidelity model methodology used is as presented in Peart et al. [11, 30], where simulations of the performance of the new sail design are treated as high-fidelity data, with previously collected data from existing sail designs used as low-fidelity data. In an industrial sail design setting, there are usually multiple existing sail designs which could be used for this purpose. In this work a method is developed using multiple linear regression that selects the best candidate from pre-existing sail designs, that when used as a low fidelity input into a multi-fidelity surrogate model is likely to result in the most accurate prediction of the force coefficient curves for a particular new sail design.

The hypothesis is that the closer two sails are in terms of their geometry (both the "design shape" which reflects the design and construction of the sail, and the "flying shape" which the sail assumes when flown [13]), the closer the shapes of their force coefficient curves are likely to be. This is because the lift and drag coefficient curves of a sail are highly dependent on the pressure distribution over the sail surface, and therefore the geometry of the flying shape of the sail [14]. By comparing the similarity of the geometric parameters of an existing sail design with the new sail being modelled, the error of a multi-fidelity surrogate model trained using those two sails should be able to be predicted.

4. BACKGROUND

4.1. Multi-fidelity Kriging

The surrogate models constructed in this work are based on the theory of multi-fidelity kriging. The model is created as a combination of high-fidelity points (the data points for the sail being modelled) and low-fidelity points (from similar existing sail designs) [11, 30]. There is no difference in the collection technique for each fidelity type. The difference in fidelity is due to the extent to which they model the sail for which the surrogate model is predicting the force coefficient curves. The surrogate models are constructed to predict the lift and drag force coefficient curves of a sail, with a separate surrogate model for each.

The choice of low-fidelity data is an important consideration [15]. The low-fidelity data points provide information about the trend of the model curve between the high-fidelity points. It is therefore important that the trends of the low-fidelity and high-fidelity curves are similar, so the final multi-fidelity curve matches the true underlying curve as closely as possible. When data fail to match, as demonstrated by Peart et al. [11, 30] for symmetric spinnaker sails, the addition of low-fidelity points can increase the error of the multi-fidelity model when compared to a single-fidelity model trained on just the new sail data points.

For details on the implementation of the multi-fidelity kriging surrogate models see Peart et al. [11, 30], and for details on the multi-fidelity kriging equations see Forrester et al. [16, 5].

4.2. Multiple Linear Regression

Multiple linear regression provides a method for predicting an output (response variable y) as a linear combination of inputs (independent variables x_n), calculated using Equation 1 [17]. In this equation, β_0 is the intercept of the model with the y-axis, β_n the gradient for the independent variable x_n, and ε the variance of the responses about the mean, or error terms. The task of fitting a linear regression model is to estimate the values from β_0 to β_n which provide the best fit to the data points [18].

$$y \approx \beta_0 + \beta_1 x_1 + \beta_2 x_2 + \cdots + \beta_n x_n + \varepsilon \qquad (1)$$

The coefficients for the linear regression in the form of Equation 1 are estimated using a least squares approach [17]. The least squares method minimises the sum of deviations of the training points from the model, so finds a model which has the best fit of these points. The deviations are calculated as the difference between the x and y values of each of the training points, and the overall mean of x and y .

In multiple linear regression the effect of each of the independent variables is considered when varied individually with all other independent variables held constant. The coefficient for each of the independent variables therefore measures the average change in the response variable for a change in only that variable.

The formulation of the multiple linear regression has several important assumptions which need to be validated to assess the suitability of the method [17, 18]:

- The relationship between each of the independent variables and the response variable

can be accurately described by a linear relationship.
- The error terms of the regression are independent.
- The error terms can be modelled by a normal distribution with a mean of zero.
- The error terms have a constant variance over the regression space.

Possible multicollinearity (where two or more independent variables are correlated with each other) also needs to be considered as its occurrence means that the effect of each independent variable on the response variable cannot be isolated, and therefore the coefficients can easily vary [17].

The process of selecting the final independent variables for use in the model is important as there are often many different independent variables which could be used as part of a multiple linear regression model, but not all of them will have a strong relationship with the response variable [18]. The "best subsets regression" method can be used to evaluate models trained using all possible combinations of independent variables, and select which combination is likely to result in the best regression predictions. As part of this process the potential for over-fitting the model needs to be considered. As the model is trained on only a small subset of the true population, small variations present in the training data may not reflect the trends in the population.

The first stage of best subsets regression is to fit a model, with a particular number of terms, for each of the possible combinations of independent variables. The model which has the minimum residual sum of squares is selected. This process is repeated for regression models having numbers of terms ranging from one to the total number of independent variables available. From these, the single best model is chosen based on a criterion. In this work the Akaike Information Criterion (AIC) [19] is used.

The AIC balances the goodness-of-fit of the model with the number of independent variables used to reduce overfitting [19]. A lower AIC value is better. It is calculated from Equation 2 where L is the likelihood and k the number of parameters. Having a higher likelihood (used as the goodness-of-fit measure) will reduce the AIC value. The value will increase if there are more parameters in the model.

$$AIC = 2k - 2\ln(L) \quad (2)$$

5. METHODOLOGY

5.1. Sail Modelling

The focus of this work is on upwind sails, with 10 jibs from 5 different yachts tested. The yachts ranged in length from 38 ft (11.6 m) to 52 ft (15.9 m). The SailPack Fluid Structure Interaction (FSI) tool was used for calculating the lift and drag coefficients for each of the sails, and each was tested with a mainsail present and trimmed in the domain.

The SailPack FSI tool uses the Vortex Lattice Method (VLM) solver AVANTI [20] to solve the flow equations. Using VLM reduces the run time of the FSI solver, allowing more sails to be tested in a specified duration. However, it has a reduced accuracy when compared to RANS modelling for flows where flow separation is important. The structural equations are solved using the structural solver ARA [21].

The trimming of the jib and mainsail is important as it impacts the flying shape and angle of attack of the sails, and therefore the value of the force coefficients [1]. A consistent trimming method was used to increase the consistency of force coefficients between different wind conditions and sails. The sails were trimmed manually in a similar way to real-world sailing by maximising the drive force of the sails while keeping the heeling moment less than a target value [22]. Maximising drive force is important as this provides the forward-motion of the yacht and therefore speed [23]. However, most yachts have an optimal heeling moment when sailing upwind, and exceeding this will increase hydrodynamic drag, reducing the performance of the yacht.

Each of the sails was tested at seven Apparent Wind Angles (AWAs) (18°, 20°, 22°, 25°, 30°, 35°, 40°). The range of AWAs was concentrated at the lower values as these are usually of most interest when designing a jib sail. A sail "code" (e.g. "J2") is used to denote the conditions, particularly wind speeds, for which different jib sails are designed, and it would be expected that sails with the same sail code would have a similar performance [25]. The Apparent Wind Speed (AWS) was therefore changed depending on the sail code. The AWS was 15 knots for a J1, 20 knots for a J2 and 25 knots for a J3 at a reference height of 10m. An atmospheric boundary layer was used for the modelling in the AVANTI solver.

5.2. Multi-fidelity Kriging Surrogate Models

The aim of the regression formulations is to predict the RMS error of a multi-fidelity surrogate model trained with a particular combination of sails. Models were constructed using the methodology previously developed by the authors [11, 30] based on the Surrogate Modelling Toolbox [24] Python package using ordinary multi-fidelity kriging.

The models were all trained using three high-fidelity points at AWAs of 18°, 25° and 40°. The placement of these high-fidelity points was chosen to reduce the error of the models while remaining consistent between the combinations. From the authors' previous experience, having high-fidelity points at the lowest and highest extremes of the surrogate model AWA range helps to correct the gradients of the force coefficient curves in these areas. This improves the predictive capability of the

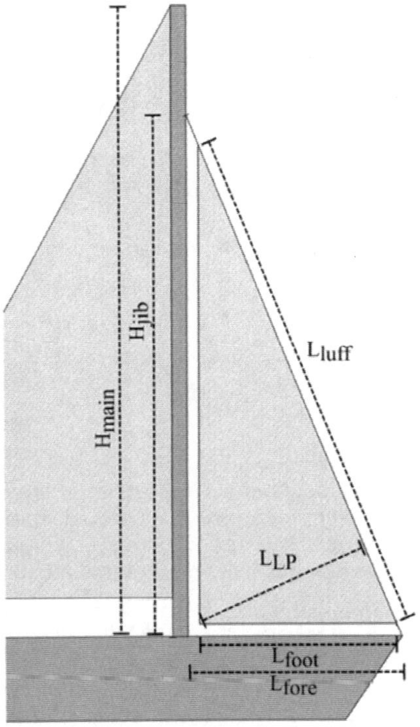

Figure 1: Overall sail plan dimensions for calculating the geometric design shape parameters.

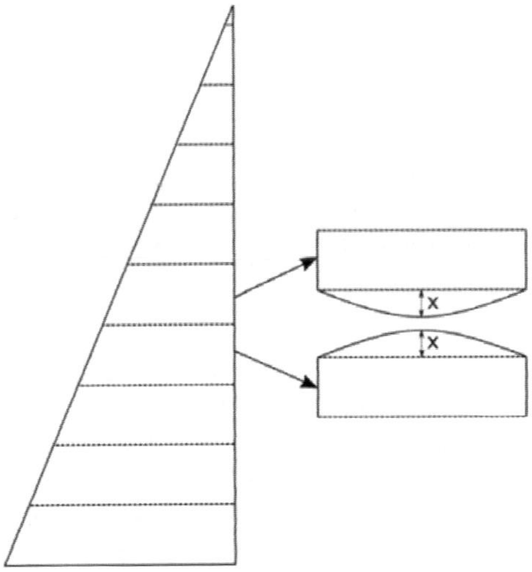

Figure 2: Definition of the broadseam for each sail panel.

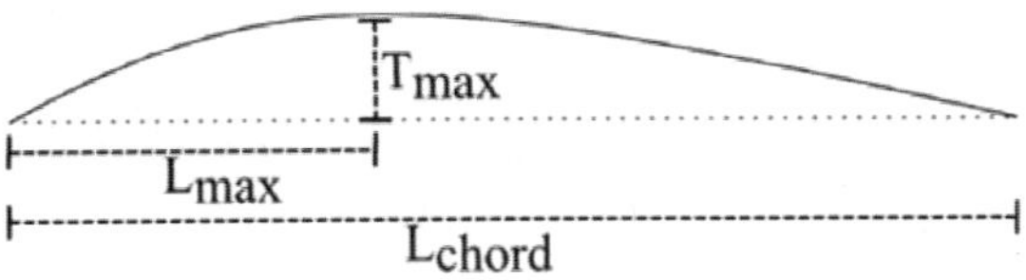

Figure 3: Definition of the maximum camber and maximum camber position for a particular sail section.

Table 1: Geometric Parameters Considered for the Regression Model, Using Parameters from Figures 1-3

Parameter	Type	Description
Aspect ratio	Design shape	The ratio of the sail's height to width. $AR = \frac{L_{luff}}{0.5 \times L_{foot}}$
Overlap	Design shape	The proportion the jib overlaps with the mainsail. $Overlap = \frac{L_{LP}}{L_{fore}}$
Jib height	Design shape	The relative height of the jib compared to the mainsail. $Jib\ height = \frac{H_{jib}}{H_{main}}$
Broadseam	Design shape	The shaping in the sail added through curvature on the top and bottom of each sail section. $Broadseam = \frac{\sum x}{L_{luff}}$
Camber	Flying shape	The maximum depth of the sail at the specified height - 25%, 50%, and 75% of the luff length. $Camber = \frac{T_{max}}{L_{chord}}$
Camber position	Flying shape	The position of the maximum depth of the sail, measured from the luff at the specified height - 25%, 50%, and 75% of the luff length. $Camber\ position = \frac{L_{max}}{L_{chord}}$
Mitre depth position	Flying shape	The position of the maximum camber of the mitre curve (depth in the vertical direction) measured from the head of the sail. $Mitre\ depth\ position = \frac{L_{max}}{L_{chord}}$
Twist	Flying shape	The twist of the sail at 50% of the luff height compared to the position of the foot of the sail.

models. All seven AWAs were used for the low-fidelity sail.

The selected measure of the surrogate model accuracy was the RMS error between the model predictions and the gathered FSI data point. This is a suitable approach when the focus of the surrogate models is exploration and all the points in the model are therefore weighted equally [5].

5.3. Multiple Linear Regression

Literature on geometric parameters which affect lift and drag forces [1, 25], as well as the authors' sail design experience, informed the initial selection of fourteen different geometric parameters. These measured different aspects of the sail's shape, with some being from the design shape of the sail, and others from the flying shape.

The flying shape of the sail depends on the trim and the AWA, and therefore varies for each data point on the force coefficient curve. To not restrict comparisons at specific AWAs for new sail designs, an average measure of the flying shape geometry was used. This was computed as the mean value of each parameter calculated across the AWAs of 20°, 25°, 30°, and 35°. These were chosen to ensure that the average was evenly spaced and not weighted towards the geometries at lower AWAs. The magnitude of the difference between the parameter values of the two sails was used rather than the value for each sail.

Each data point used to train the regression represents a multi-fidelity surrogate model trained on a combination of two sails. Ninety different surrogate models were constructed from the simulation of 10 sails, and therefore 90 data points were used to train the regression. It is important to note that the same two sails can be used to train two different multi-fidelity surrogate models, thereby creating two different data points. For example, a model can be trained to predict the force coefficient curve of sail A using the "existing" data of sail B, and another model trained to predict the force coefficient curve of sail B using the "existing" data of sail A. These two models have been trained with the same two sails and therefore the magnitude of difference between the two sail geometries is identical. As the final regression formulation predicts the surrogate model error based solely on the difference in geometric parameters of the two sails, it will predict the same error for the two models.

The authors previously observed that the error of a multi-fidelity surrogate model changes when sail A is used to predict sail B and vice versa for symmetric spinnaker sails [11, 30]. The relative complexity of the force coefficient curves was identified as a potential cause. The two models will therefore likely have a different prediction error. If the difference is significant, this will be a source of variation not accounted for by the regression formulation.

The geometric parameters were analysed to assess the potential for multicollinearity, as some parameters may capture the same geometric characteristics as others. This was assessed using the Variance Inflation Factor (VIF), where $VIF < 5$ is desirable [18]. Pairs of parameters which did not meet this criterion were assessed, with the parameter from the pair which had the highest average VIF with the other parameters removed from the model. Two parameters were removed in this way. Descriptions of the remaining parameters are found in Table 1, with the variables used in the calculations illustrated in Figures 1-3. These parameters were entered into the best subsets regression, as described in Section 3.3. As there were twelve geometric parameters remaining after the multicollinearity assessment, best subsets regression was a feasible option in terms of computational cost.

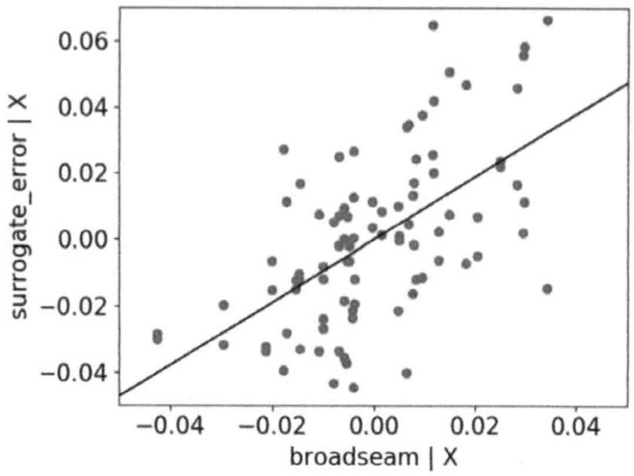

Figure 4: Partial regression plot for the lift coefficient surrogate model error. The y-axis represents the error of the surrogate model regressed against the independent variables excluding broadseam. The x-axis represents the broadseam regressed against the other independent variables.

After the best subsets regression method identified the best combination of independent variables and the associated regression coefficients, the coefficients were analysed for statistical significance using p-values and standard error. Those which were not considered to be statistically significant (with $p > 0.005$) were removed from the model.

The multiple linear regression was coded using the *statsmodel* [26] Python package, with the best subsets regression code based on James et al. [18].

5.4. Regression Testing

The accuracy of the proposed regression formulations were tested in a similar manner to that which would be used in an industrial study. The force coefficients of a new test sail were modelled using multi-fidelity surrogate models, with an existing sail design used for the low-fidelity data. For each combination of the test sail and one of the 10 existing sail designs (which were used to train the regression), the error of the corresponding multi-fidelity model was predicted using the regression formulation, referred to as the "predicted error". The existing sail with the lowest predicted surrogate model error was chosen as the best sail to use. To assess the accuracy of this selection, the "true error", the measured error between the surrogate model predictions and the known data points, of each of the surrogate models was also measured. This enabled a comparison of the true surrogate model error and the predicted error, as well as assessing whether the selection of the best existing sail was correct.

The test sail was simulated using the SailPack FSI tool. The sail chosen was a J2 from a 52 ft (15.9 m) yacht. The results for this sail were not used to train the regression, but only to assess the regression accuracy.

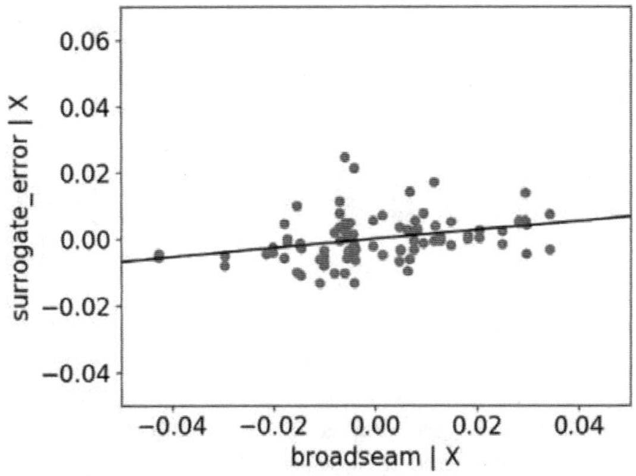

Figure 5: Partial regression plot for the drag coefficient surrogate model error. The y-axis represents the error of the surrogate model regressed against the independent variables excluding broadseam. The x-axis represents the broadseam regressed against the other independent variables.

6. RESULTS

6.1. Regression Assumptions

It is important to consider whether the use of a linear model is suitable to capture the relationships between each independent variable and the response variable [17]. This was assessed by analysing the partial regression plots which illustrate the relationship between each of the independent variables (geometric parameters) and the response variable (error of the multi-fidelity surrogate model) when all other independent variables are held constant [27].

For both the regressions of the lift and drag coefficient surrogate model errors, the broadseam is the only independent variable to show a trend. The partial regression plot for the broadseam in the lift coefficient surrogate model regression is shown in Figure 4. Each point represents one of the 90 multi-fidelity surrogate models used to train the regression. The y-axis represents the residuals of the error of the lift coefficient surrogate models when it is regressed against all the independent variables other than the broadseam, and the x-axis represents the residuals of the broadseam when it is regressed against all the other independent variables.

In Figure 4 the broadseam appears to have a positive linear trend. This is expected, as a smaller difference in broadseam between two sails represents a closer geometric shape, and a lower error in the resulting lift coefficient surrogate model.

There is scatter present in the broadseam values in Figure 4, with the linear model having $r^2 = 0.35$. The linear model for the other independent variables had a higher variation not explained by a linear model, with $r^2 <= 0.05$ for all. None of the models showed a polynomial trend.

The partial regression plots for the drag coefficient surrogate model error show a similar pattern to that of the lift coefficients. Again, the broadseam partial regression in Figure 5 has the only distinguishable trend, with a positive linear relationship. However, this relationship has a flatter gradient of 0.14 compared to a gradient of 0.95 for the lift. The linear fit of the broadseam for the drag coefficient surrogate model error does not explain as much variation as it does for the lift, with $r^2 = 0.10$. The other independent variables show a similar level of scatter as the lift, with $r^2 <= 0.06$, with none having a polynomial trend.

Similar r^2 values were observed when the geometric parameters at each AWA were used, rather than the average values of the flying shape parameters which is used for Figure 5 and the final regression. This suggests that the averaging of the flying shape geometric parameters is not a significant cause of the partial regression variation.

Other assumptions used for the multiple linear regression were also assessed. The presence of constant variance of the regressions for the lift and drag coefficient surrogate model errors was checked by looking at the residuals plot, with approximately uniform variance found in both cases [18]. Neither regression was found to have outliers, assessed using the residuals plots and Cook's distance [17]. Finally, the normality of both regressions was assessed using a Q-Q plot and the Shapiro-Wilk test [28]. From this, the error terms from the regression for the lift coefficient surrogate model error were likely to follow a normal distribution, while the random errors for the regression for the drag coefficient surrogate model error were not. This means that confidence intervals for this regression were likely to be inaccurate.

6.2. Regression Formulations

The broadseam was found to be a statistically significant variable for each model. The resulting regression formula for both lift and drag is shown in Equation 3, with the coefficients for each model set out in Table 2.

$$\text{model\_error} = \beta_0 + \beta_1 \times \text{broadseam} \quad (3)$$

Table 2: Lift and Drag Regression Parameters

	β_0	β_1
Lift	0.021	0.82
Drag	0.0076	0.11

The broadseam of the sail is a measure of the shaping given to a sail at the design stage, and therefore influences the depth which the sail has while flying, see Figure 2. The depth of a sail influences the fluid flow around it and therefore the forces which are produced, with a deeper sail producing higher lift and drag forces [1].

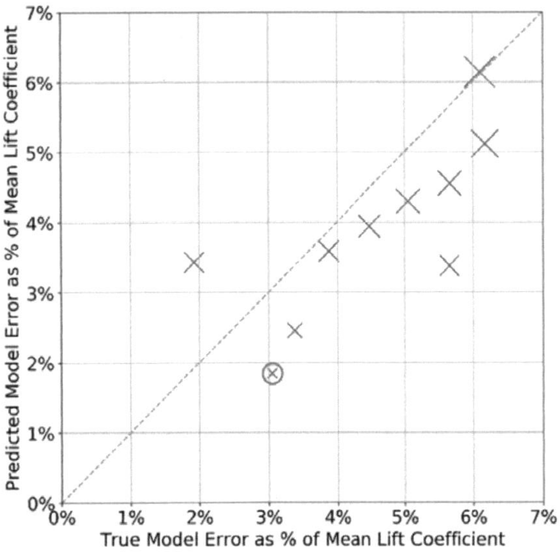

Figure 6: Predicted error of the lift surrogate models compared to the true error, both as a percentage of the test sail mean lift coefficient value. Each symbol represents a sail combination of the test sail with one of the existing sails. The size of the symbols is based on the difference in broadseam between the two sails, with a larger difference having a larger symbol size. The dashed line represents the error being predicted perfectly. The circled symbol is the sail combination with the lowest predicted error.

6.3. Prediction Accuracy

A test sail, which was not used in the training of the regression, was used to assess the regression accuracy. This is similar to the way the regression method would be applied in an industrial context.

In Figure 6, each symbol represents a multi-fidelity surrogate model trained to model the lift coefficient curve of the test sail, using one of the existing sails for the low-fidelity data. The predictions of the regression for the lift coefficient surrogate model error are plotted on the y-axis as a percentage of the mean lift coefficient value for the test sail. The true error of the equivalent surrogate model, also as a percentage of the test sail lift coefficient, is plotted on the x-axis.

The dashed line in Figure 6 represents the case where the predicted error of the surrogate model exactly matched the true error of the surrogate model. Most of the points are within 1% of this line, suggesting that the regression predictions have an acceptable accuracy in general. However, this assessment was undertaken for only one test sail, and more testing would be needed to have more confidence in the model's predictions and robustness.

There are two notable exceptions to this pattern which demonstrate that, as expected, the difference in the broadseam of two sails is not the only parameter which influences the error of the surrogate models. The positive linear trend implies that the sail with the lowest predicted error will also have the lowest true error. This is important

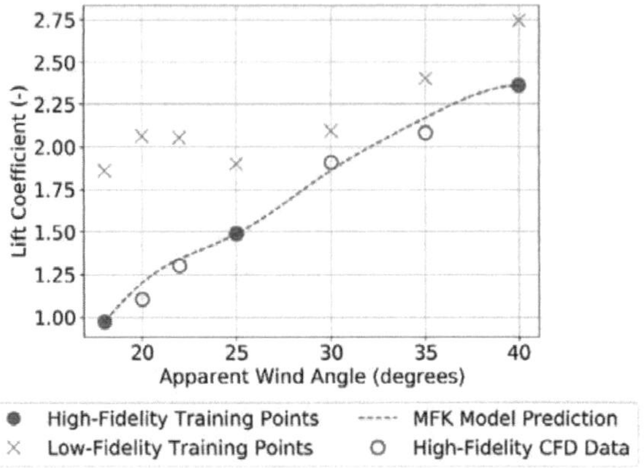

Figure 7: Multi-fidelity surrogate model predicted lift coefficient curve of the test sail with low-fidelity points from the predicted "best" existing sail from Figure 6.

when using the regression to choose the best existing sail design for use in the multi-fidelity surrogate model. As the true error of the sail combinations is not known, the sail with the lowest predicted surrogate model error, circled in Figure 6, would be chosen. This has a predicted surrogate model error of ~2%, with a true model error of ~3%. As not all the points follow the linear trend, this sail combination with the lowest predicted error has only the second-lowest true error. However, this represents only a 1% difference in true error which is unlikely to be an issue for industrial design cases.

In contrast to what would be expected, the point with the lowest predicted error in Figure 6 represents the combination using a J1 sail as the existing sail, while the test sail was a J2. Having a J1 sail with the second-lowest true error shows that it is the geometric parameters, or other characteristics, of the sail which is important, not the code of the sail. Sails which are classified as the same code do not necessarily share the same geometric characteristics or have similar force coefficient curves.

The multi-fidelity surrogate model for the test sail trained with low-fidelity data from the sail with the lowest predicted model accuracy from Figure 6 is presented in Figure 7. While there are differences between the low- and high-fidelity trends at low AWAs, the two curves have a similar gradient at higher AWAs. The resulting surrogate model is smooth and passes close to the known data points for the test sail. It therefore appears to be a suitable estimation of the lift coefficient curve of the test sail and could be used to predict input data for a VPP.

The results for the accuracy of the drag coefficient regression are less conclusive, as shown in Figure 8. The data points do not follow the linear trend of the predicted error matching the true error, with the points instead showing a primarily horizontal pattern. The range of predicted errors is significantly smaller than the true errors, suggesting that there are one or more significant factors influencing the model error which are not captured by the broadseam alone.

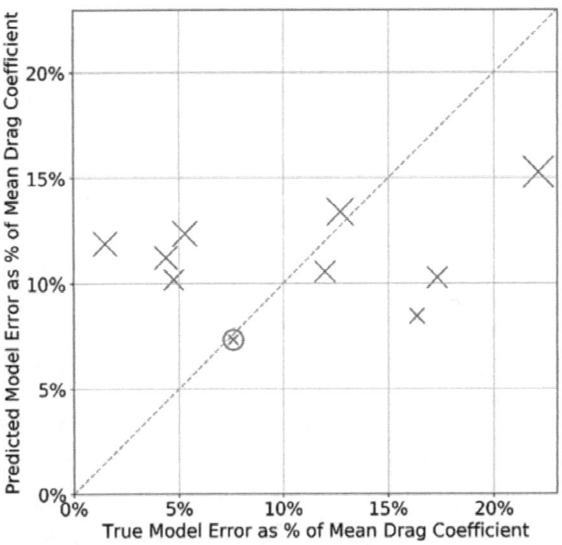

Figure 8: Predicted error of the drag surrogate models compared to the true error, both as a percentage of the test sail mean drag coefficient value. Each symbol represents a sail combination of the test sail with one of the existing sails. The size of the symbols is based on the difference in broadseam between the two sails, with a larger difference having a larger symbol size. The dashed line represents the error being predicted perfectly. The circled symbol is the sail combination with the lowest predicted error.

The error for the drag coefficient surrogate models in Figure 8 is overall significantly higher than the lift coefficient models in Figure 6, with errors as high as 25%. This is likely a result of the drag coefficient curves being more complex than those of the lift coefficients. While the lift coefficient curves are predominantly straight and smooth, the drag coefficient curves have significant changes in gradient between lower and higher AWAs.

The sail combination with the lowest predicted error is again circled in Figure 8 for the drag coefficients. While having the lowest predicted error, it has the fifth lowest true error, with a difference of 6%. Due to the horizontal nature of the data points, the second lowest-predicted error has a predicted error increase of only 1%, but an increase in the true model error of 9%. This will likely have significant implications for the resulting predicted drag force coefficients of the new sail design. The prediction of the error of the drag coefficient models is therefore unstable.

The drag coefficient curves themselves contain significant inaccuracies as the VLM only considers induced drag, not the friction or separation drags. This results in inaccuracies in the measured drag coefficient forces, as analysed in previous papers such as in Sacher et al. [29]. It is therefore likely infeasible to resolve the relative difference in drag between different sail designs when the absolute error in the drag is likely to be large. To improve the prediction of the drag coefficient surrogate model error RANS simulations should be considered which will model the drag force coefficients with a higher accuracy.

7. DISCUSSION

For both regression formulations there is a significant variation in the error of the surrogate models for each sail combination which is not explained by the regression formulation. There are several factors which may contribute to this variance.

An optimisation method was not used to trim the sails at each AWA. The trim of the sails may therefore not be at the true global optimum for the drive force for the target heeling moment due to the large number of potential trimming combinations. Differences in the trim of the sail will lead to differences in the measured flying shape as well as the force coefficients of the sail and may lead to different regression coefficients being estimated.

Another cause of the variation may be the two surrogate models constructed for each pair of sails, which share the same magnitude of difference of geometric parameters between them, but which may have a variation in model error. As the regression only considers this geometric difference to explain the error of a particular surrogate model, this cause of variation in the surrogate model error is not considered in the regression.

The number of data points used may have also contributed to this variance. In general, the more data points that a regression model is trained on, the more statistically reliable the results of the regression are and the less the model coefficients vary when data points are added or removed [17]. Due to the time required to trim the sails for each AWA, the number of sails which could be included in this analysis was limited to 10. While these sails are from different sized boats and cover three jib sail codes, they cannot be representative of all potential jib sails which may be designed. The combinations of each of these sails led to 90 different surrogate models being trained. Having more data points may have identified trends with some of the geometric parameters. There may be variance associated with uncertainty in the model caused by the limited number of sails and therefore sail combinations which could be used in the regression.

Another source of variance was parameters or factors of the analysis which were not considered as part of the regression. These may have impacted the shape of the force coefficient curves and the resulting error of the surrogate models, without being considered by the regression. This is likely to be the case, as only the broadseam was considered for both regression formulations. It is unlikely that the broadseam is the only factor causing the surrogate models of each sail combination to have a certain error. While the other geometric parameters analysed were not found to have statistically significant relationships, there may be others which were not considered which have an influence, such as the angle of attack of the sail.

Factors specific to the construction of the surrogate model may also have an influence, including the placement of high-fidelity training points, which while consistent for all of the models will influence the shape of the resulting force coefficient curve prediction. This influence can differ depending on the existing sail force coefficient curve as the placement of the high-fidelity points determines the locations where the difference model between the high- and low-fidelity models is evaluated, and therefore the influence of the low-fidelity model on the rest of the force coefficient curve prediction.

The multiple linear regression models discussed in this work have been restricted to upwind sails, in part due to the computational cost required to compute the large number of RANS simulations which would be required for the regression training. While upwind sails are worth considering with the multi-fidelity surrogate modelling method, due to the ability to analyse these sails using more computationally efficient techniques such as VLM, the cost savings which can be achieved are more limited. Reaching and downwind sails with higher separation, where more expensive RANS simulations are required, is a more likely use of the multi-fidelity surrogate method. Developing a similar regression to choose the best existing sail to use for reaching and downwind sails is therefore an important area of future work. The use of RANS, and therefore more accurate drag coefficient modelling, may help to improve the drag regression accuracy. This would be necessary for the regression method to be of use for sails where accurately modelling the drag is essential.

8. CONCLUSIONS

- For the regressions of both the lift and drag coefficient surrogate model errors, the broadseam was the only geometric parameter found to have a distinguishable relationship with the surrogate model error. The broadseam of the sail therefore influences the shape of the force coefficient curves and the resulting accuracy of a surrogate model.
- For the regression of the lift coefficient surrogate model error, the similarity in broadseam between two sails was used to predict the error of a surrogate model, with an error of less than 1% achieved for most test sail combinations.
- For the regression of the drag coefficient surrogate model error, reliable predictions were not able to be made with the broadseam. Further analysis with more accurate drag coefficient values is required.
- The regression of the lift coefficient surrogate model error was used to choose between existing sail designs for use in a surrogate model, by selecting the sail which had the lowest predicted model error. The selected sail had a true error 1% higher than the sail with the lowest true error, and resulted in a smooth surrogate model prediction of the test sail lift coefficient curve.
- There was variation in the true surrogate model errors which was not explained by the regression model, this needs to be further investigated.

Acknowledgements

This work is supported by Doyle Sails. The authors would like to thank Callaghan Innovation for their funding of this work through the R&D Fellowship Grant number DSYX1801.

References

1. F. Fossati, (2009), *Aero-Hydrodynamics and the Performance of Sailing Yachts: The Science Behind Sailing Yachts and Their Design.* A&C Black.
2. R. Korpus, (2007), "Performance prediction without empiricism: A RANS - based VPP and design optimization capability", in *18th Chesapeake Sailing Yacht Symposium*, Annapolis, USA.
3. P. Alza, (2012), "Numerical and experimental studies of sail aerodynamics", Ph.D. dissertation, Polytechnic University of Madrid, Spain.
4. T. Doyle, B. Knight, and D. Swain, (2016), "A comparison of a RANS based VPP to on the water sailing performance", in *22nd Chesapeake Sailing Yacht Symposium*, Annapolis, USA.
5. A. Forrester, A. Sobester, and A. Keane, (2008), *Engineering Design via Surrogate Modelling: A Practical Guide.* Wiley.
6. S. Harries and C. Abt, (2019), "CAESES - the HOLISHIP platform for process integration and design optimization", in *A Holistic Approach to Ship Design: volume 1: Optimisation of Ship Design and Operation for Life Cycle.* Springer International Publishing, pp. 247–293.
7. M. Ruiz de Elvira, (2015), "Design of a generalized tool for the performance assessment under sail based on analytical, numerical and empirical results", Ph.D. dissertation, The Polytechnic University of Madrid, Spain.
8. A. Serani, R. Pellegrini, J. Wackers, C.-E. Jeanson, P. Queutey, M. Visonneau, and M. Diez, (2019), "Adaptive multi-fidelity sampling for CFD-based optimisation via radial basis function metamodels", *International Journal of Computational Fluid Dynamics*, vol. 33.
9. R. Pellegrini, C. Leotardi, S. Zaghi, R. Broglia, E. Campana, U. Iemma, and M. Diez, (2016), "Multi-fidelity adaptive metamodel for ship hull performance via CFD", in *19th Numerical Towing Tank Symposium*, France.
10. Y. Kuya, K. Takeda, X. Zhang, and A. I. J. Forrester, (2011), "Multifidelity surrogate modeling of experimental and computational aerodynamic data sets", *AIAA Journal*, vol. 49, no. 2, pp. 289–298.

11. T. Peart, N. Aubin, S. Nava, J. Cater, and S. Norris, (2020), "Multi-fidelity surrogate models for VPP aerodynamic input data", in *5th International Conference on Innovation in High Performance Sailing Yachts*, Sweden, pp. 37–48.

12. J. de Baar and S. Roberts, (2015), "Multi-fidelity surrogate-based parameter estimation for a sailing yacht hull", in *21st International Congress on Modelling and Simulation*, Gold Coast, Australia.

13. K. Graf and O. Muller, (2009), "Photogrammetric investigation of the flying shape of spinnakers in a twisted flow wind tunnel", in *19th Chesapeake Sailing Yacht Symposium*, Annapolis, USA.

14. F. Fossati, S. Muggiasca, and F. Martina, (2008), "Experimental database of sails performance and flying shapes in upwind conditions", in *International Conference on Innovation in High Performance Sailing Yachts*, Lorient, France.

15. D. J. Toal, (2015), "Some considerations regarding the use of multi-fidelity kriging in the construction of surrogate models", *Structural and Multidisciplinary Optimization*, vol. 51, no. 6, pp. 1223–1245.

16. A. Forrester, A. Sobester, and A. Keane, (2007), "Multi-fidelity optimization via surrogate modelling", *Proceedings of the Royal Society A*, vol. 463, pp. 3251–3269.

17. R. Freund, Q. Wilson, and P. Sa, (2006), *Regression Analysis*. Academic Press.

18. G. James, D. Witten, D. Hastie, and R. Tibshirani, (2013), *An Introduction to Statistical Learning*. Springer.

19. H. Akaike, (1974), "A new look at the statistical model identification," *IEEE Transactions on Automatic Control*, vol. 19, no. 6, pp. 716–723.

20. M. Durand, (2012), "Interaction fluide-structure souple et légère, application aux voiliers", Ph.D. dissertation, Ecole Centrale de Nantes, France.

21. B. Augier, (2012), "Etudes expérimentales de l'interaction fluide-structure sur surface souple: application aux voiles de bateaux", Ph.D. dissertation, The University of Western Brittany, France.

22. H. Hansen, P. Jackson, and K. Hochkirch, (2002), "Comparison of wind tunnel and full-scale aerodynamic sail force measurements", in *High Performance Yacht Design Conference*, Auckland, New Zealand.

23. I. M. Viola and R. G. J. Flay, (2011), "Sail erodynamics: Understanding pressure distributions on upwind sails", *Experimental Thermal and Fluid Science*, vol. 35, no. 8, pp. 1497–1504.

24. M. A. Bouhlel, J. T. Hwang, N. Bartoli, R. Lafage, J. Morlier, and J. R. R. A. Martins, (2019), "A python surrogate modeling framework with derivatives", *Advances in Engineering Software*.

25. T. Whidden and M. Levitt, (2017), *The Art and Science of Sails*. Seapoint Books and Media.

26. S. Seabold and J. Perktold, (2010), "Statsmodels: Econometric and statistical modeling with python", in *9th Python in Science Conference*.

27. R. De Veaux, P. Velleman, and D. Bock, (2013), *Intro Stats*. Pearson.

28. M. Williams, C. Grajales, and D. Kurkiewicz, (2013), "Assumptions of multiple regression: Correcting two misconceptions", *Practical Assessment, Research & Evaluation*, vol. 18, no. 11.

29. M. Sacher, F. Hauville, P. Bot, and M. Durand, (2015), "Sail trimming FSI simulation - comparison of viscous and inviscid flow models to optimise upwind sails trim", in *5th High Performance Yacht Design Conference*, Auckland, New Zealand.

30. T. Peart, N. Aubin, S. Nava, J. Cater, and S. Norris, (2021), "Multi-fidelity surrogate models for VPP aerodynamic input data", *Journal of Sailing Technology*, vol. 6, no. 1, pp. 21-43.

AERODYNAMIC PERFORMANCE - TRANSFERRING LOAD BETWEEN CABLE AND SAIL

JB Braun[1], jb.braun@northsails.com

Abstract. Explore headsail design opportunity as a function of transferring load from the headsails luff cable/stay to the sail membrane through a range of apparent wind angles. Performance is quantified by aerodynamic ratio of drive force divided by roll moment and using the North Sails velocity prediction program to quantify predicted performance. Fluid structural interaction was used to create the flying sail shapes utilizing North Sails Design Suite and evaluated with computational fluid dynamics. Several key outcomes include larger amounts of head stay sag benefit more from sharing load on the membrane of the sail, while less initial head stay sag decrease the effect. With lower head stay sags, performance optimums become a more complicated coupled solution for optimum performance.

1. NOMENCLATURE

AWA	Apparent Wind Angle
AWS	Apparent Wind Speed
PLT	Tecplot unstructured (FE) ascii data
Cl	Lift Coefficient
Cd	Drag Coefficient
Cp	Pressure Coefficient
CoE	Center of Effort
CFD	Computational Fluid Dynamics
DOF	Degree Of Freedom
Delta Cp	Pressure Coefficient Difference
FEA	Finite-Element Analysis
FSI	Fluid-Structure Interactions
Lu(Fan)	Luff Fan-Tangent to sail surface
Lu(Nor)	Luff Normal- Normal to sail surface
Mi(Nor)	Miter Normal- Sail design volume
Miter	Vertical camber distribution
NDS	North Design Suite
RANS	Reynolds-Averaged Navier-Stokes
TWA	True Wind Angle
TWD	True Wind Direction
TWS	True Wind Speed
VPP	Velocity Prediction Program

2. INTRODUCTION

Sail designers strive for better aerodynamic efficiencies, which is a key factor in boat performance, the engine above the deck. Boat performance advantages can in part be related to sail design concepts, materials and manufacturing techniques and/or constraints. Evaluation of headsail design aerodynamics can be quantified by evaluating flying sail shape changes due to sharing load between a head stay or luff cable with the sails membrane. The following performance predictions are evaluated as a function of the Luff Cable load divided by sail membrane load (load sharing ratio) through a range of apparent wind angle (AWA).

Looking back through the last century of head sail design there is a notable trend in sail structure applied to headsail luff sections. Headsail design trends have progressed from soft woven cross cut polyester 65+ years ago to Miter cut softer finished woven polyester 55+ years ago, to tri-radial designs 45 + years ago, to tri-radial designs incorporating lower stretch materials in sails luff section 35+ years ago, string sails with catenary luff support 25+ years ago to the current state of the art of laminate structure sails 10+ years ago. This trend has seen attention to luff section support by several factors. These include fiber alignment and material modulus, progressing from rotated cross cut panels to radial panels to lower stretch materials to catenaries and currently to laminate structures, these material provide the foundation for our design space opportunity.

Headsail design space is broken into three sections, each incorporating different analysis techniques. These highlight design space opportunity into useful sail design controls by transferring part of a headsail loading from a head stay to the sails membrane.

The flying sail shapes used to generate the data and analysis are based off fully coupled FSI simulations where FEA solutions are coupled with CFD. Both panel codes and RANS codes were used for the pressure loadings and continually updated in both directions until converged. This process could be a paper in itself. This technique is North Sails design engine, North Sails widely utilizes this process for all type of research and development and used by top North Sails designers. The CFD used incorporated North Sails Flow (a panel code) and RANS programs including the Wolfson Unit's extended version of the standard OpenFOAM with marine specific modules called North Sails VWT. Also a check in with Star-CCM+ was used for VPP aero response surfaces.

For simplicity experiments 3.2 & 3.3 use an aerodynamic ratio of Drive Force divided by Roll Moment as a means to quantify the performance of sail design parameters. Drive to Roll is a useful ratio once boats reach their heel limit and are constrained by righting moment. Almost all boats aerodynamic power is governed by righting moment, thus increasing the drive force for a given roll moment is a useful ratio to predict potential performance deltas. One might consider this a 1 + Degree Of Freedom approach (DOF), and provides a method to quantify an aspect of performance without considering other forces

[1] Director Design and Engineering, North Sails, Marblehead, MA, USA

and moments. While 3.4 utilizes North Sails VPP to quantify performance which can provide more accurate performance prediction.

3. EXPERIMENT DESCRIPTION

Section 3 introduces the different performance prediction methods and section 6 quantifies the findings.

3.1. Concept of Load Sharing and Terminology:

Load sharing is presented as the load on a stay divided by the load on the sails membrane, with the combination of the two loads equal to the total tack load in general terms. Dividing ninety by ten, represents a load share ratio of 90 / 10 based on the total tack load. Ninety percent of the total load on the head stay/cable and ten percent on the sails membrane. Thus cable load / sail load – this will be used as our general terminology for headsail load sharing.

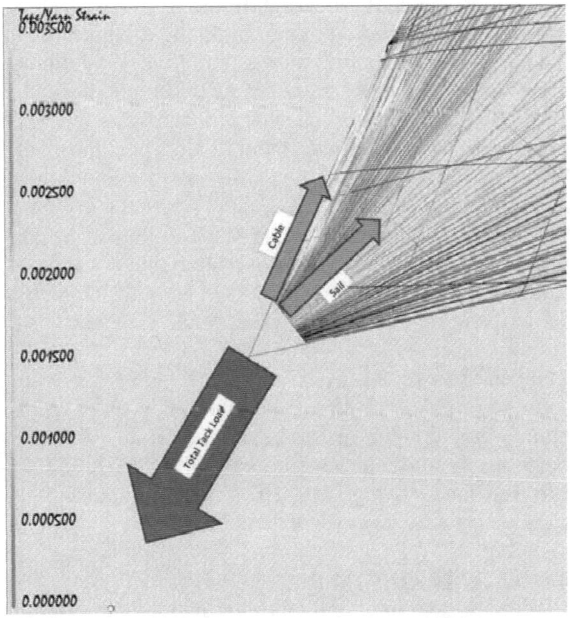

Figure 1: Total Tack Load is a combination of the head stay or luff cable plus the sail or sail membrane load: Sail load + Cable load = Total Tack Load

Tack load and direction change as function of several variables including total load, AWA, TWS, Sag percent, sail trim, sail design, sail structure etc…., this ratio simplifies these variations into one ratio by using the total loads without considering the changes in load directions. Changes in this ratio help define global head stay movement/position relative to boats centreline for each load sharing ratio through a range of AWA.

Detailed description of headsails luff position defined by Luff Fan and Luff Normal is in section 4.1. In general sail design terminology defines a sails luff position and is defined relative to the sails cross sectional surface at the luff edge normal to the surface and tangential to the sail surface at the luff.

The tangential direction is tangent to the sail surface at the luff and is referred to as Lu(Fan), and in a similar fashion Lu(Nor) is at the luff point, normal to the sails surface.

Luff fan is generally referenced as hollowing the luff in the direction of the sails surface and the normal is allowing the luff to fly or project or perpendicular to the sail surface.

3.2. Segregation of Design Parameters

Segregation of sail design parameters helps to quantify their effect on aero efficiency (Drive Force / Roll Moment) specific design variables related to the headsails luff position are Lu(Fan), Lu(Nor) and incorporating span wise camber distribution described as Miter position Mi(Nor). North's Design Suite (NDS) is used to generate a baseline flying sail shape converged with FSI. This highlights the position of the headsails luff for a given load sharing ratio. The baseline sail set has a load sharing ratio of 90/10 (Head stay load / Sail load).

The force and moment data from CFD runs on the FSI baseline flying sail shape is used to normalize the performance data. Each of the three sail design controls are modified individually while holding the other flying sail shape controls constant. They are compared against the baseline CFD data.

By holding all other design parameters constant, the FSI flying sail shapes were morphed to a new luff position. The initial FSI flying sail shape is the baseline. Then modifying each shape by applying sail designer's controls of headsail luff fan, luff normal. The sails edge length, girths, and total sail area were fixed. Holding sail shape constant, moves or morphs the sail sections to the new luff position creating a new flying sail shape. This process maintains a constant sail section camber, twist amount and vertical profile. The sails sectional shapes, draft position, front and back percentages, this is sail making terminology, are the same as the base line. This process isolates the luff position change as a function of sail design parameter Lu(Fan), Lu(Nor) highlighting the effectiveness of changing the sails luff position.

The aerodynamic efficiency is quantified with the ratio of Drive force divided by Roll moment and plotted for a sweep of Luff Normal, Luff Fan & Vertical Miter Normal position (span wise camber distribution). This analysis is intended to identify individual design parameter effects on efficiency as a function of load sharing and AWA's in Section 6.1.

3.3. Luff Position

Changes in load sharing ratio effect luff position and flying shape of a headsail change as a function of load sharing percentages and thus have an effect on aerodynamic efficiency. The North Design Suite of tools was utilized to produce FSI shapes focused on the position of headsails luff as a function of load sharing ratio and head stay sag. From these luff positions a baseline cross sectional shape and twist profile are applied to the headsail luff edge for different Load sharing ratios. This experiment maintains the same flying sail shapes by superimposing the same sectional shapes on the different load sharing ratios FSI flying shape head stay positions. This eliminates differences in sails sectional shape created by load sharing ratio changes. Different sail loads (sharing ratios) change flying sail shapes. Section 6.2 experiments maintain the sectional shapes and twists constant for each load sharing ratio change. These changes in load sharing ratio produce differences in flying sail shapes and head stay position. Each load sharing ratio change the target (baseline) flying sail shape and could be achieved with changes in sail design/structure. However a new sail design and or structure for each load share ratio would be needed. The same flying sail shapes could not be achieved with a single sail design for different load sharing ratios. To ensure the same flying sail shapes were analyzed, the baseline cross sections were used for each load sharing ratio analysis and placed on the FSI head stay position. This eliminated the effect of sail shape changes with load sharing ratio change. Head stay position or sag amount were quantified by calculating the effect on aero efficiency of Drive Force / Roll Moment.

3.4. VPP Simulation

Two unique aero response surfaces powered VPP simulations by creating fully coupled FSI sail shapes for two load sharing ratios. A baseline of 90/10 and 50/50 load sharing ratio sail shape created the aero response surfaces for section 6.3.

VPP performance of the baseline jib design 90/10 was compared against a load sharing jib design of 50/50 load sharing ratio. The sail shapes and luff sag are normalized by changes to sail design and sail structural of the 50/50. Optimal trim utilizing FSI and coupled to a converged sail balanced with expected surge force and roll moment. North Flow was used as the CFD tool for FSI pressure and aero response surface data. Trimming sequence and coupled FSI solution achieved a well-trimmed sail at each AWA.

The baseline hydro model was based off modern MM72 utilizing SYRF's Wide-Light project. Mast and rigging properties were based off the current racing fleet as well as the sail designs used in the simulations.

A sweep of Apparent Wind Angles was analyzed from a tight angle of 15 to a broader angle of 45 degrees for the Jibs, with mainsail Travel, Jib track angle and Heel sweeps were run. The FSI coupling pressure files were generated with North Sails Flow (a panel code) set up with the same mythology as the final data set.

The force and moment equilibrium, commonly referred to as degrees of freedom (DOF), are treated as constraints in the optimization. Hence the user can specify which DOFs to enable or disable in the VPP. The number of enabled DOFs depends on the problem at hand. Typically all 6 DOF's are enabled for a foiling multihull, whereas a displacement monohull is often run with only 4 DOF enabled. Additional constraints, such as lower/upper bounds can be placed on the user defined parameters. Likewise certain relationships can be enforced on selected sets of parameters.

The advantage of the reduced 4 DOFs compared to the full 6 DOFs is that when generating the data for the VPP components you need not include variations in Sink and Trim. This reduces the data set required to run the VPP significantly. In the most extreme case, one could run a single DOF VPP, if for instance only a righting moment curve was available. However, reducing the DOFs simultaneously reduces the accuracy of the VPP.

The Wide-Light tank model was tested with free Sink and Trim, which is common practice for such experiments. As the purpose of the SYRF Wide-Light Project experiment was to provide a bench mark for CFD computations, the 120 test point matrix covered two configurations: 45 points with the un-appended model and 75 points with the fully appended model. Furthermore, both these configurations included a 12 point Trim variation. As it was originally designed for the Wide-Light project, the matrix was geared towards covering a wider range of sailing conditions for the CFD bench mark than a matrix for providing a VPP hydro data set would have been.

The data fits of the above components were achieved by splitting the VPP hydro component into three parts, namely an upright resistance curve (HydUpRt) plus components for the change in drag (HydHeelD) and the lift, roll and yaw (HydHeelL), both components in response to heel, leeway and rudder angles.

4. SAILMAKING TERMONOLGY

4.1. Head Stay Sag - Lu(Fan) & Lu(Nor)

As load transfers from the cable to sails membrane the luff fan decreases (straightens) and the luff normal increases, pushing it away from the boat. The magnitude of these changes are a function of AWA, tack load and sag percentages or amount of sag, sail design and structure.

Sail design terminology defines the sail luff position relative to the sails cross section in a normal and tangential direction to the sail surface at the luff.

The tangential direction is tangent to the sail surface at the luff and is referred to as Lu(Fan), and in a similar fashion Lu(Nor) is at the luff point normal to the sails surface. The magnitude of these values is a percentage of the total luff length generally presented as a percentage, or luff sag percent where 1% sag indicate the luff edge is sagging 1% of the luff length, from a reference straight line between the head and tack of the headsail.

Luff fan is generally referenced as hollowing the luff in the direction of the sails surface and the normal is allowing the luff to fly or project more perpendicular to the sail surface.

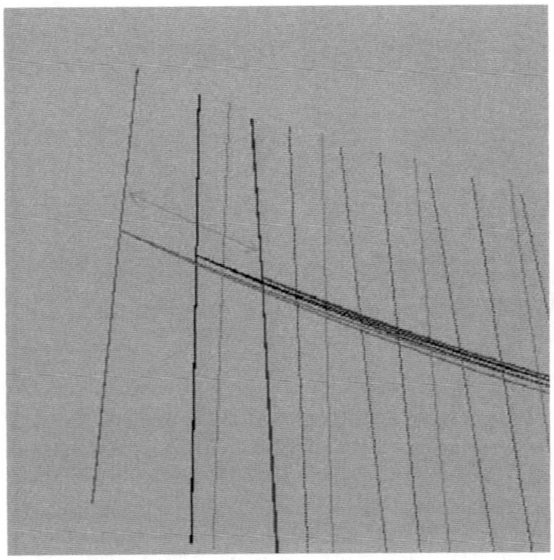

Figure 2: Lu(Fan) - Image of Lu(Fan) with the black a baseline Luff fan, left side sails luff in red decreasing fan and the right sails luff in blue increasing luff fan.

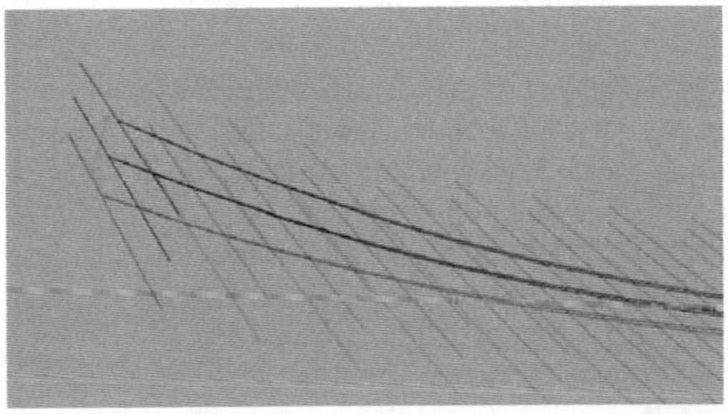

Figure 3: Lu(Nor) - Image of Luff Normal, middle sail in black a base Normal with bottom sail in red increasing Normal and the top sail in blue decreasing luff Normal.

Plots of luff position change as a function of load sharing ratio from 90/10 to 40/60 load sharing ratio in figures 4 and 5, which illustrates load transferring from the cable to sail effect the head stay position defined as Lu(Fan) & Lu(nor) for a Mini Max 72 in TWS 12 knots. In general as load sharing ratio increases loading in the sails luff fan decreases and luff normal increases. Dimension relative to MM72 jib with a nominal luff length equal to 28.600 meters.

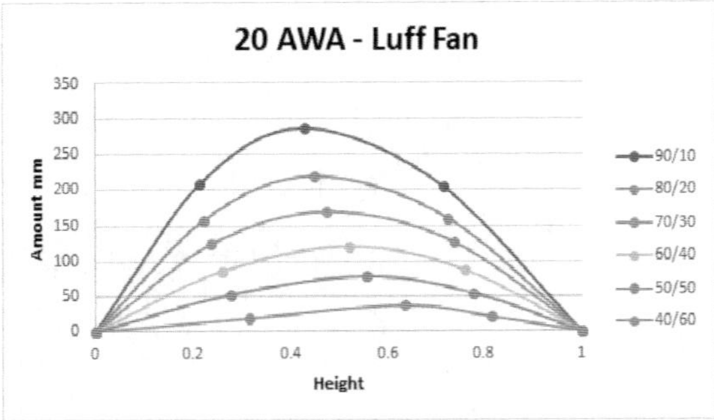

Figure 4: Lu(Fan) Tangential movement for variations in load sharing ratio for the same total tack load.

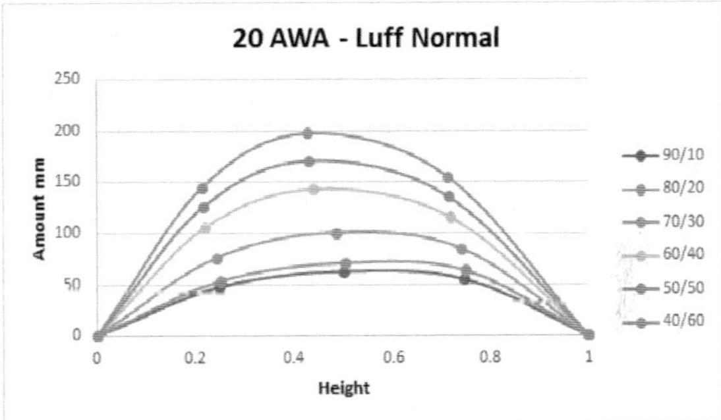

Figure 5: Lu(Nor) Normal movement for variations in load sharing ratio for the same total tack load.

Figure 6 and 7, display an overlay of the FSI flying sail shapes presented in Figure 5 and 6. The Lu(Fan) and Lu(Nor) graphs are presented in a 3D view as FSI overlay of flying sail shapes as a function of load sharing ratio for two AWA. The standard trim of 90/10 load sharing ratio in Red, through a load sharing ratio of 40/60 in grey. This highlights the changes in luff position in 3D for AWA 20 and 40 degrees for the specified load sharing ratio, also displaying how jib cambers flatten, entry angle decreases and the sails membrane load increases.

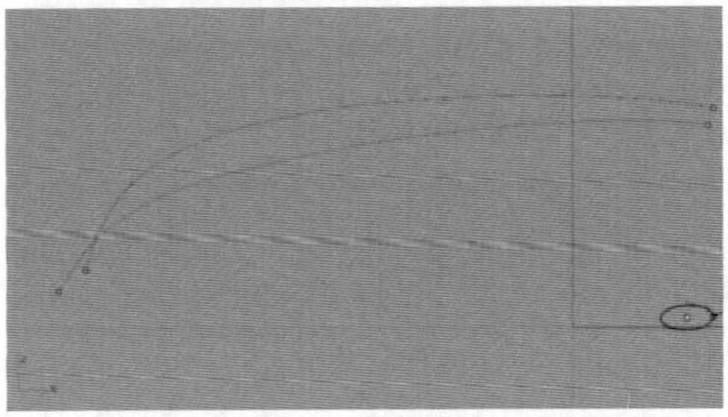

Figure 6: 20 AWA Luff position overlay in section as a function of load sharing ratio for the same sail design and total tack load. This presents the effect of load sharing ratio change in luff position and sail shape with Red being the highest cable load and fullest sail (90/10) and Grey being the lowest cable load and flattest sail (40/60).

Figure 7: 40 AWA Luff position overlay in 3D as a function of load sharing ratio for the same sail design and total tack load. This presents the effect of load sharing ratio change in luff position and sail shape with Red being the highest cable load and fullest sail and Grey being the lowest cable load and flattest sail. Note the flat sail is visible illustrating the change in sail entry and sectional shape.

4.2. Head Stay Sag Reference Value

Figure 6 & 7 display how head stay sag changes is three dimension and for our purposes of describing head stay sag we use a simplified term to define sag from a straight line between the sails head and tack. Simplifying sag into one number works for the purpose of reporting in this paper. However head stay position or sag for load sharing is more complicated than a single number and important to understand the differences in position. More refined reporting of head stay location will help quantify performance drivers, for example referencing the boats centreline with a fore & aft and sideways reference frame may be more meaningful. The simplified sag term used in this data has the follow terminology to define sag.

Head stay Sag = Absolute Value (Luff Fan%) + Absolute Value (Luff Normal%), which produces a luff sag as a percent of the total luff length of the sail.

The absolute value are used due to cases where the Fan and Normal change sign. You could use the SQRT of the two but it's not used here. Looking at sag as the SQRT of the fore and aft and sideways position may have some merit, however it looks like defining the actual position would be more meaningful for future references.

4.3. Fiber Strain Differences:

Displayed in figure 8 sail fiber strain display the sail structure strain e, simulated in TWS 12 at Max RM of a representative boat in upwind maximum VMG sail trim and onset flow angles. The intention of these images is to highlight the fiber strain changes of the headsail as loading is transferred form head stay to sail membrane. Load sharing ratio of 70/30 left to more standard load sharing ratio of 90/10 right image which is a more typical upwind load share ratio. The fiber strain rate is displayed with color range defined from 0.000 e (strain) in black and up to fiber strain of 0.0011e (strain) in white.

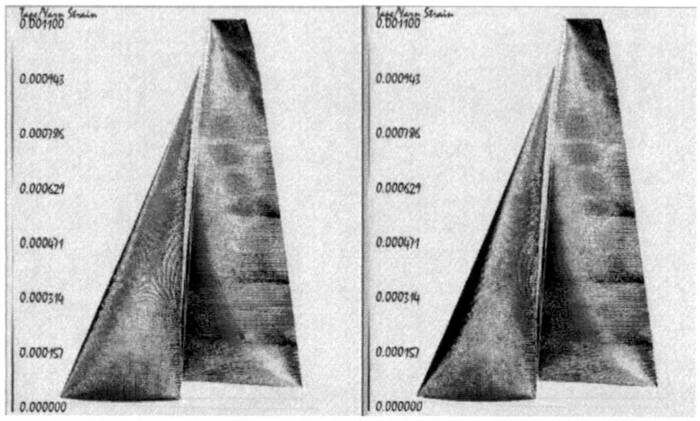

Figure 8: Fiber strain e (the amount of deformation in the direction of the applied force divided by the initial length of the material). The color range from 0.000 e (strain) in black to 0.0011 e (strain) in white. The two images illustrating the strain transfer of different load sharing ratios, displayed are load sharing ratio 70/30 left compared to more standard ratio of 90/10 load sharing ratio right.

4.4. Sail Shape Changes as Function of Load Share Ratio:

Applying more load to the sails membrane affects flying sail shapes, in general, flattening the sails volume resulting in flatter flying camber as illustrated in Figure 6 and 7. With more sail membrane load special attention on the distribution of flying twist profile should be noted, as these can have negative aerodynamics and is directly coupled to sail design and structural considerations.

Customized sail structure for load sharing under normal sailing loads can results in a flying sail shape that closely replicates current design (same target flying shape).
In a general load sharing sense a standard trim resultant tack load has a range of loading at ~ 10% to 15% of the total tack load applied to the sail. This load will vary due several variables, like sails aspect ratio, head stay sag amount, sail design etc.....

Figure 9 is an example of the flying sails shape change as more load is applies to the sails membrane. The load on the sails membrane is ~ doubled, reducing the sails flying mid camber by ~ 10% -15%. For reference this amount of load change can be handled by most hardware on current boats and could be applied without major modification to existing yachts. Camber reduction is noted with the foot camber and mid leech twist held to similar values. Figure 9 shape changes are achieved with North's 3Di sail structure and NDS designs, other constriction methods and designs may have different flying sail shapes as a function of load sharing ratio change.

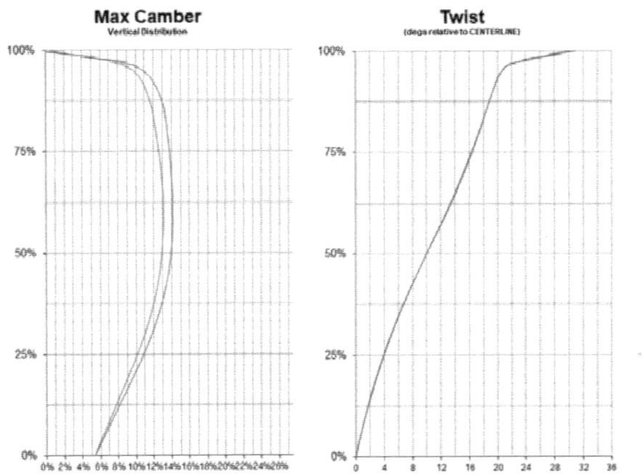

Figure 9: Red plot standard trim load sharing ratio of 90/10 and blue plot higher sail loading with load share ratio ~ 70/30 stay/sail. Both graphs plot the camber and twist from foot to head with the vertical axis 0% being the foot and 100% being the head. The max camber percentage is plotted horizontally as a fraction of the local chord (as is the standard for all sail shape analysis). The twist is plotted horizontally in degrees plotting the angle change of the chord from the 0% height which is the tack/clew chord.

5. SAILMAKING SUMMARY

The effect of load sharing has a large effect on the flying sail shapes including camber, entry/exit angle, and front and back %'s. Most of these changes can be addressed with the initial sail design and sail structure for a given load share ratio. Changes in the load share ratio have an effect on flying sail shapes. As a result load sharing ratio could be targeted to optimize aerodynamic efficiencies.

Practical challenges with the range of mast bend, mainsail shape and limits to the sails range will have affects in shape control tendencies which might have larger effect on aerodynamic efficiencies than the head sail shape changes, they may or may not be desirable changes in performance. These variables and other considerations are part of overall design space are touched on but not investigated/researched here.

Identifying performance drivers for load sharing highlighting the effect of individual design control feature, i.e. Lu(Fan), Lu(Nor), miter shape as sail design control is helpful in understanding the design space. Segregation of these individual controls is not realistic in a flying or FSI sense, but the trends should be similar for a coupled FSI solution, thus these experiments to target the trends of the design space and not the actual design targets can be helpful exploring design space and setting sail design targets.

6. EXPERIMENT DETAIL AND RESULTS

To evaluate performance advantages/disadvantages of load sharing, section 6 experiment is designed to quantify design parameters effect on aerodynamic efficiencies. Isolating effects of design parameters that change as the load sharing percentages change. As illustrated the position of head stay change as a function of load sharing, inquisitive designers may question what a target should be. Sections 6.1, 6.2 & 6.3 help simplify the design space and provide some guidelines as to the performance drivers associated with load sharing ratios.

6.1. Effects of Design Parameters Luff Fan and Normal

Isolating the effects of design parameters, North's FSI suite of tools NDS were used to generate baseline flying sail shape as a standard starting point/baseline to morph design parameters around (These fully coupled FSI flying sail shape are baseline design). At each AWA the shape geometry of each sail combination was generated by means of Fluid Structure Interaction (an iterative process) targeting the expected surge force and roll moment in order to achieve a sequence of equally well-trimmed sails; for the sake of predicting optimal performance, this study assumed optimal sail trim for each configuration. Running simulations with FSI, rather than solid shapes, limited the number of points analyzed in each base sails shape. However, the FSI approach was selected for its higher accuracy. The mast and more importantly the hull were present in all CFD simulations. The vertical profile of the wind was modeled with the power law with a .1 exponent. Each simulation onset flow was based off accepted performance tables for a MM72.

The baseline FSI flying sail shape was used to create new shapes by morphing a sweep of Luff Fan, Luff Normal and miter position. The baseline sail shape started with a 90/10 load share ratio setting, then baseline sails sections / twist held constant we superimposed on the morphed luffs for changes in the design parameters of Lu(Nor), miter distribution, and Lu(Fan). This is unrealistic in physical terms without changing the sail design/structure for each load share ratio. But does provide some trends in how the load sharing and resulting luff position affects efficiency.

These new shapes were analysed in CFD simulations running North's Flow CFD to evaluate the aerodynamic gain/loss between the different design parameters. Table 1 displays Drive Force to Roll Moment with percentages above 100% producing greater efficiencies than the base line flying sail shape.

The design parameter percentages listed indicate the amount of change as a function of the luff length from the base shape. A sweep of Lu(Nor), miter distribution, Lu(Fan) was simulated. The luff normal changed from -1% to +1% show that a negative luff normal provide greater efficiencies. More positive Luff normal starts to gain efficiencies and advantage at wider AWA's. This may have to do with the added sail twist at wider AWA's.

The Normal performance plot was interesting and opposite to the anticipated result, the negative Normal or the sail sagging back closer to the main was more efficient and it was not until the wider AWA's the performance started to equalize. From the Lu(Nor) results a "Normalized Leech" position was added to the data set to help understand these results (Efficiencies of Normalized leech position highlighted in Table 1 at AWA20 & 40 and listed as normalized leech). For the Normalized Leech simulate the leech was fixed in space, as if a sail trimmer adjusted the leech to a mark on the spreader defined with the leech position of the base line flying shape. The results of normalizing the leech position had a similar trend on performance as Luff Normal and Luff Fan, thus there are other interactions effecting performance to consider. This could be an area of further research to understand why performance changes with the interaction of the relative sail positions. There are changes in pressure on the main and jib through the range of AWA's this interaction is displayed in Figures 12, 13, 14.

In Figure 10 the image displays an overlay of Lu(Nor) minimum and maximum amount explored, from -1% to +1% Lu(Nor). This helps visualize the design space explored and the change in geometry and position of the sail at AWA 20.

Figure 10: AWA 20 -1% to +2% Luff Normal shape overlay. The grey displays the -1% luff normal and in red displays the +2% luff normal. It's also important to note the sail edge length, girth and area are constant and the sail sections are morphed into the new luff position.

6.2. Span Wise loading

Figure 11 plots at the aerodynamic effect of Lu(Nor) changes at AWA 20 degrees. The spanwise loading is plotted with the solid lines representing Negative -1% Normal (Grey sail in Figure 10) and the dashed line the positive +1% Normal (red sail in Figure 10) with loadings on the horizontal axis in N/M for X1 (fore & aft) and X3 (sideways) loadings. Note the increased loadings of the negative Normal. The red plot is the mainsail the green is the headsails and the blue is the combination of the two.

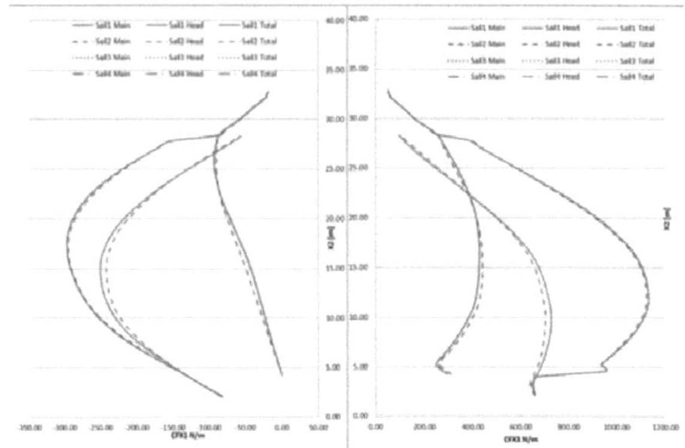

Figure 11: AWA 20 +1% to -1% Luff Normal plotting span wise loading. The horizontal axis plots in N/m and the vertical axis is the span plotted in meters. The left plots the sideways force component and the right plots fore and aft force component.

6.3. Chord Wise Loading

Plotting sail section (from Figure 11) chord wise pressure distribution Delta Cp at the sails height at ~ 42.5% span display pressure distribution change with the sails interaction. The solid lines plot the negative, -1% Normal and the dashed lines plot the +1% Normal. The sail shape section position in meters from an origin (front face of mast at waterline) the horizontal and right vertical axis, with the Delta Cp on the left vertical axis. Figure 13 and 14 normalize the sails position and highlight the delta Cp pressure difference, note how the main & jib pressures flip with the change in the jib's normal Lu(Nor).

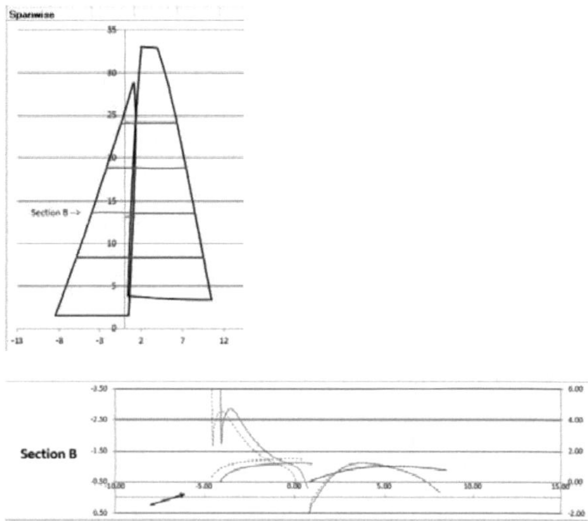

Figure 12: Illustrates the span wise position of section B at 13.6 m above the waterline. Section B plots at AWA 20, two Luff Normals, +1% (dashed lines) to -1% (solid lines). Displaying the jib Luff Normal change in sail shape with Main Section and delta Cp on both sails. The horizontal and right vertical axis plot 2-D sail shape section position in meters from an origin (front face of mast at waterline), with the Delta Cp on the left vertical axis.

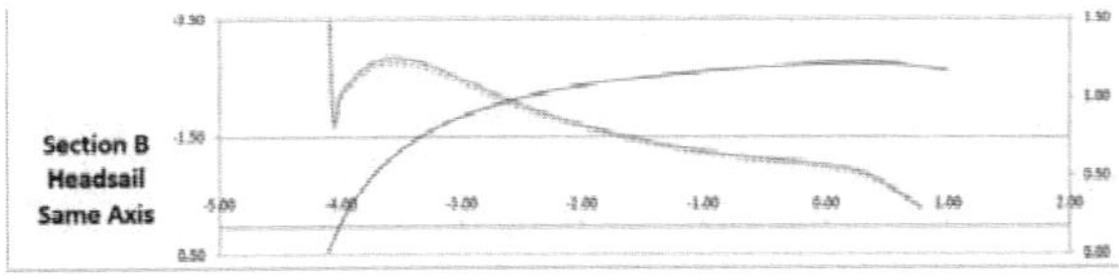

Figure 13: Normalizing Jib sails position at AWA 20, the +1% (dashed line) to -1% (solid line) Normal Jib Section and delta Cp. The horizontal and right vertical axis plot sail shape section position in meters from an origin (front face of mast at waterline), with the Delta Cp on the left vertical axis. Note the globally increase in sail pressure for the -1 Normal.

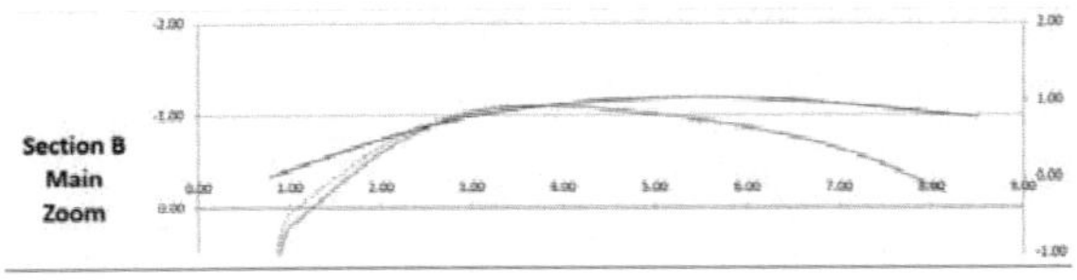

Figure 14: Normalizing Main sail position, the delta Cp plots at AWA 20 +1% (dashed line) to -1% (solid line) illustrates how changing the jib's luff normal has an effect on the Mains delta Cp. The horizontal and right vertical axis plot sail shape section position in meters from an origin (front face of mast at waterline), with the Delta Cp on the left vertical axis.

Table 1: Isolating the effects of design parameters efficiencies with Drive Force / Roll Moment deltas from a Base Line design, Normalized leech position noted as (normalized leech). It's noted that the Drive Force divided Roll Moment is not dimensionless N/N-m. However for illustration purpose it's presented as a percent.

DriveRoll

Lu(nor)

AWA	-1.0%	-0.5%	0	+0.5%	+1.0%
15	101.3%	100.5%	100%	99.4%	99.2%
20 (normalised Leech)	100.5% (100.6%)	100.2%	100%	99.7%	99.5% (99.0%)
30		100.2%	100%	99.7%	100%
40 (normalised Leech)		100% (100.2%)	100%	99.9%	99.8% (99.6%)

Mitre Distribution

AWA	-8.0%	0	+8.0%	+16.0%
20	99.9%	100%	100.1%	100.2%
30	99.9%	100%	100.1%	100.1%
40	99.9%	100%	100.2%	100.3%

Lu(fan)

AWA	0	+0.5%	+1.0%
20 (normalised Leech)	100%	100.2% (100.2%)	100.2% (100.1%)
40 (normalised Leech)	100%	101.3% (101.1%)	101.7% (98.7%)

6.4. Miter Distribution

Miter distribution is the spanwise camber distribution. Simulations indicated higher miter predict performance advantage. Specified miter change percentages reflecting the change in camber at the 1/4 height and 3/4 heights, negative change is less camber is positive change in greater camber. An 8% change would change the ¼ or ¾ height camber depth by 8% from example from 10% deep to 10.8% camber.

Lu(Fan) is tangent to the sails surface, Positive values reflect a change in less hollow, a straighter luff and is reflected with increasing efficiencies.

6.5. Effect of Head Stay Sag

The second method of evaluating the performance drivers was to analyze the effect of sag on the efficiency. In general terms closer or upwind AWA sails tend to have or be trimmed with less sag due to several variables like sea state, rig stability/movement and mainsail shape. At wider more reaching AWA sails tend to set with or accept more sag in general.

Investigations centered around three different sags with simulations conducted on a reaching sag amount of 2.5%, a closer angle at 1.0%, and a racing upwind angle at .5%. As the load was transferred from the stay to the sail the total load was held constant. The luff sag and sail shape baseline started at 90/10 load share ratio setting to 20/80 load share ratio. From these FSI defined head stay positions the baseline sail sections / twist were superimposed onto the new (head stay) luff position defined by FSI simulations for the load share ratios. Each baseline sag percentage defined the sails cross section and vertical camber distribution and applied to the different load share ratio luff position; this maintained the sail shape with the intention of isolating the load sharing percentage luff position. This is unrealistic in physical terms, flying sail shape could be achieved by changing the sail design/structure for each load share ratio. This experimental technique was designed to provide trends in how the load sharing affects efficiency for different sag amounts.

The results indicate that higher sag percentages the larger the performance gains available and the lower the sag or tighter the head stay the less effect load sharing has. This can be seen by the slope of the green (2.5% sag) vs. the slope of the Red (.5% sag) in Figure 16.

The different colors represent different sag %'s with Green 2.5% sag, Gold 1% sag and Red .5% sag. The solid line plot the efficiency (Drive to roll) on the left vertical axis with higher negative numbers having more performance. The dashed lines plotting against the right vertical axis plot sag% and a function of luff length. The horizontal axis represents the load on the luff cable as a percent of the total tack load, with 90 being 90 percent of the total load is carried by the head stay or cable. More AWA's and sail design options should be explored to add to global understanding and quantify these effects in more detail.

6.6. Result Plots

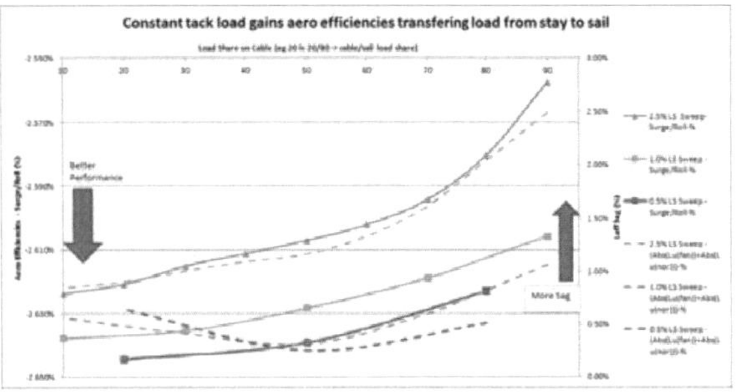

Figure 16: Plots load sharing ratio against performance and sag for a constant tack load. The load sharing ratio on the horizontal axis plots three different total luff sag, Green = 2.5% sag, Gold = 1% sag and Red = .5% sag. The luff Sag percentages on the right… aerodynamic efficiencies on the left vertical axis. Indicates increases in load sharing reduce sag % for the same tack load, which also increase performance. Note higher sag percentages have greater performance deltas.

The plot in Figure 17 is useful to highlight a constant sag percentage, the total tack load can be reduced with increased load on the sail, as indicated with dashed red line (total tack load reduction can vary greatly depending on variables like aspect ratio, pressure loading, sag percentages, etc…..). The red dashed line plots a reduction in load as a percent of the standard trim with a load sharing ratio of 90/10, the dashed blue line plots the increased sail loads. As the total tack load decreases the sail corner loads are increasing. The solid grey line plots the efficiency and a function of load sharing. Again the efficiency line (solid grey) is misleading if you consider one sail design and structure the fly shape changes with change in load share %'s, as will efficiency. Higher load share changes flying sail shape by reducing camber, flatter or less entry angle and rounder back %'s, which generally decrease aerodynamic efficiencies. However this provides some insight into the general trend or effect.

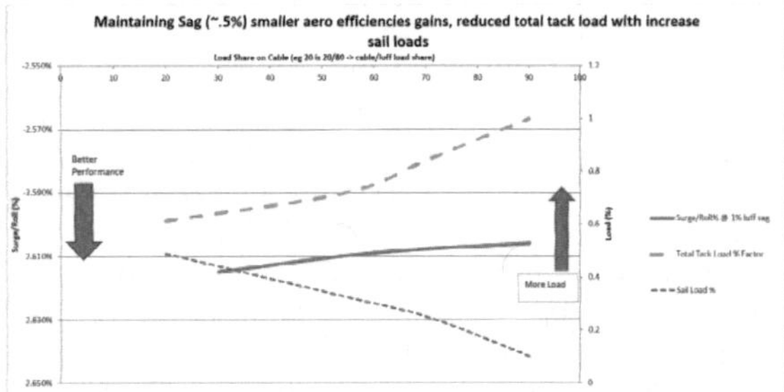

Figure 17: Plots the effect of head stay sag percentage maintained as load sharing varies on the horizontal axis. Red dashed line plots total tack load, smaller dashed blue line plots the increase in corner loads normalized value on the right vertical axis. Grey solid line plots aero efficiencies on the left vertical axis.

6.7. Experiment VPP Results

A Maxi 72 (MM72) rig and sail plan was chosen to baseline this work, primarily due to SYRF's tank testing and hydro CFD data available under the Wide-Light Project 1 and report for SYRF on Performance Prediction Program, CFD Modelling Options for Hull & Sails, Phase 1, which was Prepared for: SYRF, September 2017.

A MM72 jib design was selected to analyse in the VPP acknowledging that different headsail designs perform differently with AWA ranges. VPP runs were conducted with a Standard Jib design (90/10) and load sharing Jib (50/50) design targeting the same headsail sag %. This resulted in different rigging loads, the main sail design was adjusted for the 50/50 jib's VPP response surface, creating the same main flying shape for the two VPP response surfaces. AWA sweeps from a tight angle of 15 to a broader angle of 45 degrees in increments of 5 degrees covering the effective AWA range of these sails. North Sails VPP is used to quantify the performance difference associated with different force components. Figure 18 overlays the base line FSI sail shapes corresponding to the targeted AWA. Base line sail shapes were created for both load sharing ratios in the same manor of utilizing the North Design Suite of fully coupled FSI solutions used to generate the VPP aero response surfaces.

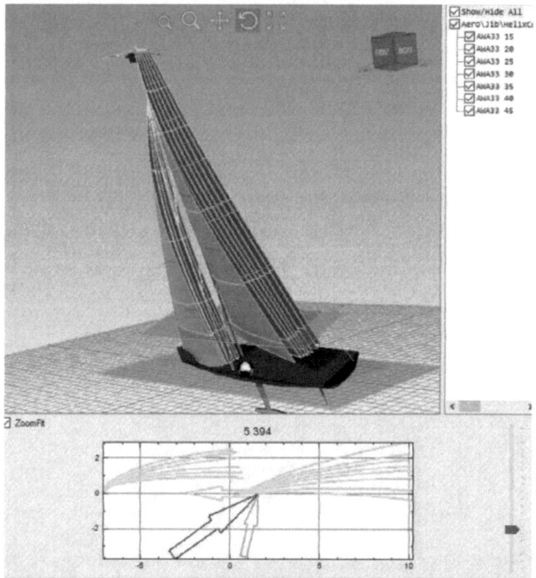

Figure 18: A range of FSI sail shape overlaid used to create the aero response surfaces for VPP simulations. In the 3-D view there is a section cut through the sails just at 5.394 m above the origin (front face of mast at waterline). In the lower box the sail sections are displayed with the sail shape section position in meters from an origin (front face of mast at waterline) the horizontal and left vertical axis.

The 50/50 load sharing jib design was changed to achieve the flying sail shape targeting the same cross sectional flying sail shape as the standard (90/10) load share sail. This was done to isolate the effect of luff position due to headsail load sharing and not a sail shape experiment. The target sectional shape can be achieved with a combination of sail design shape and structural considerations. Thus the focus simulations have been on different load sharing effect on performance for a given sectional sail shape.

In order to get to the same flying shape (close to the same forces) there needs to be large adjustments in sail designed luff normal and luff fan (less hollow).
Adding luff NOR and adjusting the fan with the new luff position to achieve similar flying sail shapes as the baseline FSI shapes. The shapes were achieved with FSI design simulations utilizing North Sails NDS, Flow and MemBrain.

These different performance deltas are displayed by Figure 19 polar plots of the predicted boat speeds based on the two aero models (standard sail design and 50/50 load share design) created with NDS, Flow CFD code.

The polar plot curves connect the predicted boat speeds at each of the AWAs in the Aero Matrix. The Jibs coupled solutions were computed in 12 TWS, to cover a range of TWS to either side of these base lines, the VPP was run at 8, 10, 12, 14 and 16 TWS for the Jibs. Since the Aero Matrix did not cover these TWS variations when running at a lower or higher TWS than the base line, the VPP scales the base line aero force and moment components by the lower or higher AWS^2 relative to baseline AWS^2. This scaling ignores any change in sail camber or twist across the TWS range, which a sail trimmer might normally do to the sails. The VPP can only adjust the traveler and the boat heading when up- and de-powering the sails, Jib track angles were locked.

The base line polar curves are in red for the lower TWS, and the curves in black are for the upper TWS. The solid lines represent a 50/50 load share Jib, the dotted lines are standard jib. The numbers at each point are the AWAs, the horizontal and radial lines represent boat speed in knots and the TWA's are listed at the radial intersections with the outer perimeter.

The VPP predicted performance is close however the 50/50 load sharing jib is slower than the standard jib at upwind AWA. At the deeper angles it gets closer with a performance cross over around 30 -35 AWA. One might expect this as the Load share design is deeper than the standard jib design and may set differently at wider AWA's. This sail design change is needed to achieve the same FSI flying sail shapes at the targeted upwind AWA's.

6.8. VPP - Polar Plot Results

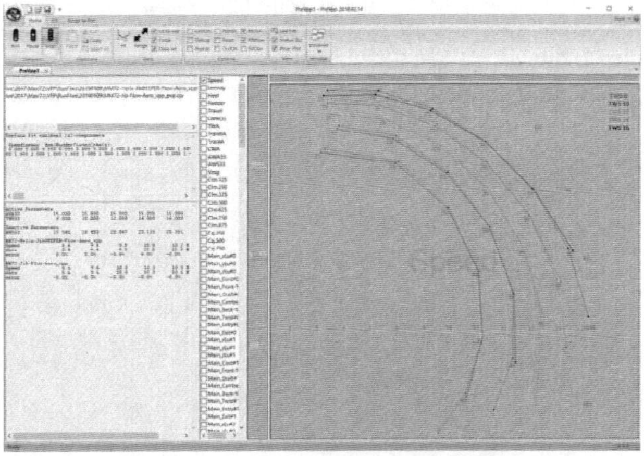

Figure 19: VPP polar plot of predicted performance. A polar plot is an easier way to view performance differences of two jib designs as a function of TWS and TWA. In Fig. 19 a display of boats performance as a function of TWA, where the horizontal axis represents a TWA equal to 90 degrees, and the vertical axis would represent a TWA equal to 0 degrees. The distance away from the origin represents boat speed in knots, the further away from the origin the faster the predicted boat speed. The different colored lines represent different TWS from 8 to 16 knots. Varying TWS, TWA for a standard jib with a load share ratio of 90/10 plotted as dashed lines and comparing against a load share ratio of 50/50 as solid lines. Results in a performance crossover around 30 to 35 AWA, where the 90/10 ratio looks better at the tighter angles and the 50/50 ratio at the wider angles.

Looking into the performance results and achieved flying sail shapes, the results may be slightly skewed for the 50/50 load sharing data by the effect of the runner offset from centreline. The baseline data normalized the headsail sag; to normalize the 50/50 sag less runner tension was used. The typical Maxi72 sailed @ 14.5 ton head stay load upwind. The load shard 50/50 sail setup with 8.6 ton head stay load to achieve the same sag upwind in TWS12. This change seems significant, for a boat whose primary objective is not 100% VMG upwind there could be potential gains to be made in both mast and sail design for reaching application. However the sail tack load may have little or no effect on mast and hull structure design point. There may be other load cases that are more critical for structural design. For true VMG upwind performance the gain might be balanced by reducing sag as opposed to load.

The 50/50 load shared design required lower runner loads to normalize sag percentage and this change increased vertical rigging loads, resulting in mast fall off to leeward much more than the standard trim. This may have produced aerodynamic disadvantages that affect the tighter AWA VPP results for the 50/50 jib design. This highlights that these types of design changes are coupled and considering one alone will not provide the entire performance answer.

Figure 20 overlays of the baseline FSI flying sail shape baseline and the 50/50 load sharing FSI shape used in the VPP aero response surfaces. The Gold image is a standard trim with a load sharing setting ~ 90/10 while the Blue image is a 50/50 load share targeted to achieve similar sag amount. (Note the leeward deflection of the 50/50 setting).

Figure 20: Gold image is a standard trim with a load sharing ratio setting ~ 90/10 while the Blue image is a 50/50 load share ratio.

7. CONCLUSIONS

Segregation of sail design basic controls highlights how effective these design parameter are in terms of aerodynamic efficiencies. These parameters can help designers understand flying sail shape targets. Achieved flying shapes will be different than the shapes tested in this section but provided useful insight into the cause and effect of load sharing. Some of the tested shapes are unrealistic or unachievable in an FSI sense, however Luff Fan is a more powerful driver in efficiencies, especially at wider AWA's, than either Lu(Nor) or Mi(Nor) the vertical miter position.

The second analysis method of defining the luff position for a given load sharing ratio indicated load sharing's aerodynamic efficiencies have a greater effect on sails with higher amount of initial head stay sag than sails with lower initial sags. In these simulations the position of the sails luff is realistic (the head stay position) which was defined as a result of FSI simulations. All sectional shapes were normalized to the baseline. It was demonstrated that the lower the initial sag the less efficiency change with load sharing. However changes in load sharing percentages change sail sectional shape which may have larger effect in performance than the change in head stay position. This needs further investigation to quantify.

The third analysis method utilizing more complex VPP simulations, results indicating aerodynamic efficiencies and boat performance are coupled solutions that cannot fully be evaluated without looking at the entire FSI aero package. Utilizing FSI in combination with VPP's helps evaluate this trade-off, which better represent what happens on the water. This process helps define the design opportunity for sailors and designers alike. VPP results indicate at AWA wider than 35 AWA the 50/50 load sharing started to outperform the standard sail set at 90/10 load sharing ratio.

Acknowledgements

Acknowledgement to North Sails Designers for their contributions:

Michael Richelsen
Steve Calder
Mickey Ickert
Phillip Williamson
Graeme Willcox
Juan Meseguer

References

1. Claughton, A.R., SYRF Wide-Light Project, 2015, Sailing Yacht Research Foundation, Electronic Retrieval Location: http://sailyachtresearch.org/projects/wide-light-project

2. JB Braun & Michael Richelsen, Performance Prediction Program, CFD Modelling Options for Hull & Sails, Phase 1, which was Prepared for: SYRF, September 2017.

QUANTITATIVE DYNAMIC ENERGY ABSORPTION OF SANDWICH CORE MATERIAL

Thomas Basset[1], thomas.basset@gurit.com
Mark Battley[2], m.battley@auckland.ac.nz
Raphael Gerard[3], raphael.gerard@gurit.com
Tom Allen[4], tom.allen@auckland.ac.nz

Abstract. When considering structural failure during an impact, the energy that can be absorbed by the material can be more important than its ultimate static strength. Often the dynamic energy absorption is assumed to be approximated by the strain energy accumulated during a quasi-static test before failure, which depends on both the elongation at break and the maximum load sustained. As a result, for core materials subjected to impact loads, the shear elongation at break has thus become considered to be as important a property as ultimate shear strength [10].

Nonetheless, it has also been found that the transient nature of impact loads can cause stress and strain rates that are high enough to affect the behavior of the core material, including the resulting strength and elongation, particularly for polymeric foams [17]. As a result, the static energy absorption is only a qualitative indication of the likely dynamic behavior of a core material but cannot be taken as a quantitative measure of the dynamic property by itself. The aim of this paper is to undertake characterization of the dynamic shear energy absorption of different core types and manufacturing treatments at various loading rates, in order to build a quantitative database of material performance to inform structural design and better understand the relative performance of these materials.

1. INTRODUCTION

Current engineering practice for the structural design of a yacht's hull laminate is generally reliant on international scantling rules such as DNV-GL [19], ISO[21] and previously ABS[22]. These rules standardise and simplify the approach to a yacht's structural design in three broad steps:

1. Deriving likely motions of a yacht based on its overall geometry and operating condition (length, beam, displacement, speed, deadrise, area restriction). These motions are simplified to a unique design vertical acceleration.
2. Deriving local panel pressures based on the yacht's vertical acceleration, panel location and size, and orientation to oncoming flow (deadrise and waterline entry angles). These pressures are assumed uniformly distributed and static.
3. Deriving panel response to the design pressure, based on panel geometry (aspect ratio, curvature), boundary conditions, and somewhat laminate (anisotropy).

There are simplifications and limitations in each of these three steps, with the associated opportunity for progress and relevant research.

The first steps revolve around the area of seakeeping, or "flight simulation" for more recent yachts. Progress in this area have been made in the form of full size instrumented research vessels, and data acquisition systems embarked on race yachts such as during the Volvo Ocean Race 2014-2015 and presented as a keynote at HPYD 5. Nonetheless structural design cannot cover all possible impact scenarii in terms of a yacht motion and on-coming wave surface, as such there will always need to be a simplification into a small subset of critical design cases agreed upon, which are aimed at covering worst probable likely impact in operation, but which inherently leave a small probability of being exceeded.

The second step is currently based on analytical models of slamming that were begun by von Karman on seaplane floats in 1929 [1]. Much research has been carried out since then both experimentally and numerically [3], [5], [9], [12], [13], [14], [16]. One area of focus is hydro-elasticity, whereby the structural response of the plating influences the hydrodynamic loading. There are two main aspects to hydro elasticity: As the panel deflects, it moves in the same direction as the fluid, i.e. that the relative motion between panel and fluid reduces. This leads to a reduction in pressure, particularly at the position of maximum deformation, such as mid panel. Unfortunately the reverse is true at the longitudinal edge of the panels: as the panel deflects, its local deadrise decreases, thus increasing the local pressure as the peak pressure evolves with the inverse of tan(deadrise) [15]. This is especially critical as peak pressure close to an edge leads to high through thickness shear forces and stresses. In general, scantling rules are still based on a rigid-wedge approach, and include a "panel area" factor that reduces the pressure the bigger the panel is. This is in-line with the first aspect of hydro-elasticity, as a bigger panel, for the same stiffness, would deflect more, but unfortunately is unconservative with regards the second aspect of hydro-elasticity mentioned above.

Composite materials, especially in a sandwich construction, have a much lower through-thickness shear to in-plane stiffness ratio, hence they deflect differently from the thin metallic plates that traditional scantling

[1] Senior design engineer, Gurit Composite Engineering APAC, Auckland, NZ
[2] Associate Professor, Department of Engineering Science, University of Auckland, NZ
[3] Design engineer, Gurit Composite Engineering EMEA, Antibes, France
[4] Senior Lecturer, Centre for Advanced Composite Materials, University of Auckland, NZ

design is based on. Even neglecting hydro-elastic effects, slamming is better approximated by a rolling point load than by a uniformly distributed pressure. In common with an actual slamming load, a rolling point load tends to lead to a through-thickness shear critical stress state, whereas uniformly distributed pressure tends to lead to a bending critical stress state. At the time of the early slamming studies, typical marine going structures were thin-shelled, typical marine going structures were thin-shelled, so bending critical, whereas nowadays yachts hull are mostly sandwich, hence through thickness shear critical. Furthermore the bending to in-plane stiffness ratio is also very different for a sandwich panel or a solid plate, influencing the membrane to bending load sharing. None of these aspects are accounted for explicitly in traditional scantling design but could be in a fairly inexpensive linear static Finite Element Analysis (FEA) at a panel level. In practice however, true material behaviour during a real slamming impact is both non linear and dynamic.

Experimental studies have highlighted the importance of the non-linear behaviour of various materials in simulated slamming events [4],[6],[7]. In particular, materials with low elongation at failure such as Balsa and Nomex Honeycomb fail "earlier" (at a lower impact speed) than most polymeric foam cores. Within the foam core family a distinction was found between standard "cross linked" or "rigid" PolyVinyl Chloride (PVC), and modified "linear" PVC specially formulated for impact toughness or high elongation, as well as Styrene-AcryloNitrile (SAN) foam [17],[18]. Impact toughness is often measured in engineering terms as energy absorption. The first law of thermodynamics (conservation of energy) means that the energy that is stored within a material as it undergoes deformation is equal to the work of the force applied to it:

$$U = \int_0^{\Delta} F d\Delta \qquad (1)$$

Where U is the strain energy, F the force and Δ the displacement.
This is thus equal to the area under the force-displacement curve. For a similar shear strength, a material having a linear behaviour up to maximum load with a brittle fracture thereafter would thus store much less energy than a material with a similar strength but having a non-linear behaviour with a long plastic phase after the linear portion of the load displacement curve, as shown in Figure 1.

This behaviour has been accounted for by classification societies in the form of core selection: for a core to be suitable for use in slamming area, it needs to achieve a minimum elongation at failure, and in some scantling codes the material safety factor depends on the shear elongation [19],[20]. Based on their work on the Visby class corvette [2], DNV[19] did so, and developed a core shear test at a higher strain rate than typically used in core shear test standards [23], with the test coupon having to incorporate a bond line between two pieces of core [20].

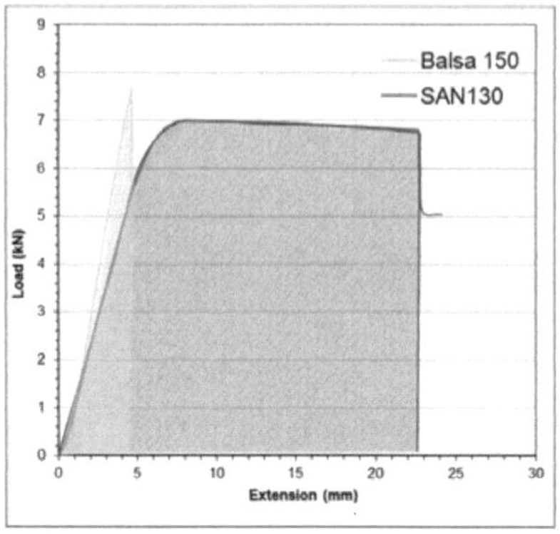

Figure 1. Comparison of Balsa150 vs SAN130 load displacement curves.

This test is mostly aimed at identifying whether the core, or the presence of the bond line, would react adversely to elevated strain rate loading. Should the dynamic shear strength match or exceed the static results, the core is deemed suitable for use in slamming area, and engineers can use the measured dynamic strength instead of the static value. This is useful progress towards incorporating experience-gained knowledge in rule-based design practice, but it suffers from some limitations. Some material, as shown in this paper, can have an increasing strength with increasing strain rate, but decreasing elongation at failure, thus decreasing energy absorption: The dynamic energy absorption is often different and can be less than the static strain energy measured in a standard test. It can even be less. Even if a material is able to sustain more stress in a dynamic situation, it may actually loose its ability to absorb energy, hence prematurely fail in an impact. As scantling societies look at structural design in terms of stress (calculations) or static elongation (selection), but not as energy, and even less dynamic energy, this approach can lead to structural designs that aren't fit for purpose. Structural engineers designing a yacht's hull cannot solely rely on scantling rules, but also need to have experience, or better, experimental results to inform their design practice. Previous research having investigated the effect of strain rate in core shear tests simulating slamming events [7] , [17] are of significant value, but, to date, their results was mostly expressed in terms of stress or equivalent pressure, not yet in terms of energy.

Furthermore, a structural engineer designing a yacht's hull cannot solely rely on pure material data, but instead needs to account for in-situ material performance. As recognised by the presence of the bond line in the DNV slamming grade test, the process of assembling core and skin plies into a large-scale structure results in features that are different from pure material test coupons:

- The sheets of cores need to be bonded together (also called core splicing), and the bond line between sheets forms a resin vertical membrane which behaves differently from the core.

- The core needs to be perforated in order to allow for the air entrapped into the lower skin to escape out during vacuum consolidation.
- In order to conform into curved shapes such as a yacht's hull, the core needs to be treated in one of the following ways:
  -either cut partially as in various "core treatment" as offered by most core manufacturers (e.g. Double Cut or Triple Cut from Gurit as shown on Figure 2 below),

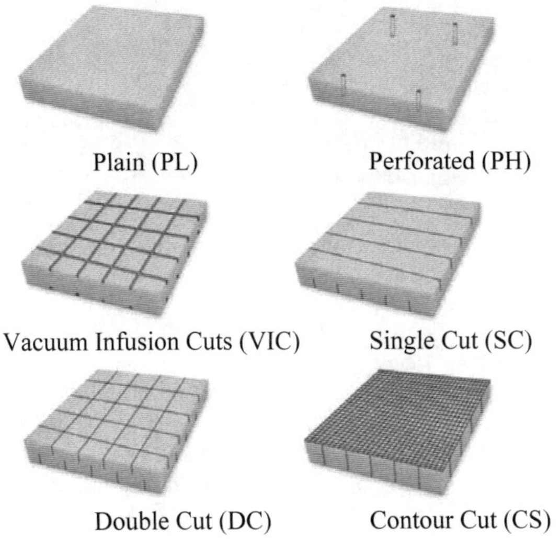

Figure 2. Typical core treatments.

  -either cut fully such as for strip planking,
  -either thermoformed.
- Finally, the completed structure undergoes a cure cycle, at a temperature necessary to gel the resin used to laminate the skins. This can be a lower temperature of typically 50°C for lamination processes such as infusion, typically 70-85°C for large sandwich structures with pre-preg skins, and sometimes in excess of 100°C for special components.

Each of these processing aspects can have an impact on the dynamic energy absorption of a real structure, compared to a pure material test sample. For a builder to produce a yacht that performs as expected by the engineer that designed it, both the builder and engineer need to have an understanding of the implication of the manufacturing process, either for the builder to select processes that aren't detrimental to the material's ability to absorb impact energy, or for the engineer to use design data that is relevant for the process used. The present study aims at providing both builders and engineers with such data.

Ideally, designers and scantling authorities would develop a new design approach, changing from stress based to (dynamic) energy based. In that sense, engineers would then need to be provided with material property data that includes (dynamic) strain energy absorption. The present study lays the basis of such a quantitative dynamic shear energy absorption database.

For special projects that can afford to run slamming simulation (dynamic, non-linear simulation similar to crash analysis), over and above the default scantling requirement, engineers will need to be provided with material property data that includes full (dynamic) load-displacement curves to calibrate non-linear, strain-rate dependant, material models.
Already without a design shift by scantling rules or full dynamic simulation, the current paper can be used, as an experimental reference to compare the impact performance of various core materials, but at least by rating them based on the material property that is most suited to impact analysis: dynamic energy absorption.

2. EXPERIMENTAL METHODOLOGY

2.1. Summary

A drop weight dynamic 4pt bending test was used to characterise core materials across a range of chemistry, densities, processing, and strain rates. The different chemistries included in this study are cross-linked polyvinyl chloride (PVC, 80, 100 and 130 kg/m3), polyethylene terephthalate (PET, 90, 110 and 135 kg/m3) and styrene acrylonitrile foams (SAN, 80, 100 and 130 kg/m3), representing a range of different levels of ductility and maximum elongation. The different core processes include plain, perforated, cut, bonded and thermoformed. Quasi-static 4pt bending was also undertaken of the same specimen types to provide benchmark static data.

2.2. Specimens

The nominal specimen geometry was 60 x 500 x 33mm (W x L x T). The beam skins were made of 2 layers of 1200g/sqm of quasi-isotropic E-glass fibres (@0°,+45°, -45°, 90°), infused with Epoxy resin. The material was doubled over a 35mm wide region where the loading bars contact the specimen to minimize the likelihood of localized skin failure. Specimens were manufactured by Gurit with 12 specimens provided for each core material and density.

2.3. Dynamic Beam Testing System

The high-rate beam testing was undertaken in an IMATEK IM10 Drop Weight Impact Testing system as shown in Figure 3. Four point load and support spans of 133 and 330mm respectively were used to test the specimens. Impacts of the SAN and PVC cored beams were performed at 3.5 m/s or higher with an impacting mass of 34.7 kg. Impacts of the PET cored beams were performed at 2.0 m/s with an impacting mass of 14.7 kg due to the low failure energy of the PET. Rubber and steel pads were placed on top of the fiberglass pads to reduce dynamic signal noise during the impact event. A high speed camera (PhantomVEO410L) was used to record the failure process and assist in identifying and comparing failure modes.

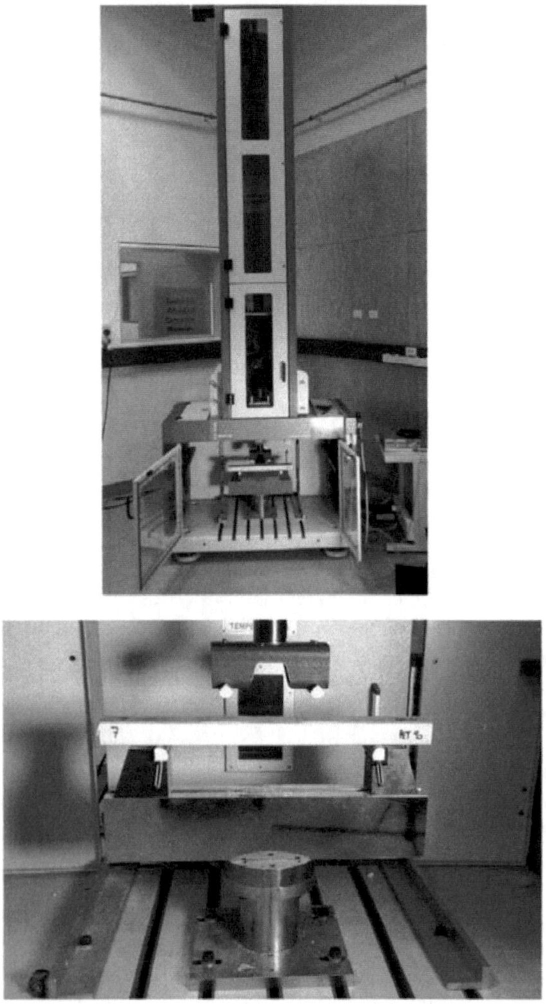

Figure 3. Drop weight impact testing system.

3. RESULTS ACROSS CHEMISTRIES AND DENSITIES

3.1. Failure Modes

Under dynamic loading all of the core types failed predominantly by transverse shear as expected, however the shear failure differed between materials. The SAN foam was the most ductile, undergoing significant plastic deformation before failure, which is discussed further below. The PVC core had some plastic deformation before also failing in transverse shear fracture. The PET cores failed as brittle transverse shear fractures. Figure 4 compares failure modes for 100J impacts, with only plastic deformation for the SAN 100 core (top), a single crack transverse shear fracture for PVC 100 (middle) and complex multiple shear fractures for the PET (bottom).

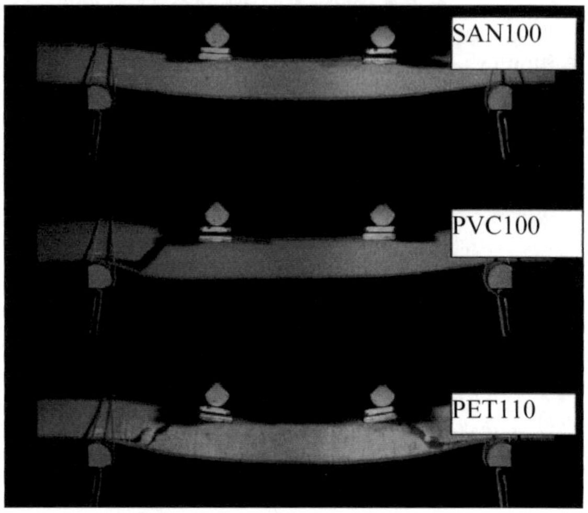

Figure 4. Failure modes for 100J impacts.

For SAN foam, the plastic deformation leads to an ultimate failure that is not necessarily shear fracture through the foam. The plasticity leads to high displacements, which allows the skin to bend all the way to its bending failure, as shown on Figure 5. This is a failure mode that is different from the traditional sandwich bend-testing failure modes, such as indenting, micro buckling, or skin wrinkling. All these failure modes are core modulus (shear and compressive) driven failure mode. It was verified that increasing the skin thickness did not influence the ultimate failure. As such the ultimate failure mode, even if skin in nature, is still core-driven, hence representative of the core's ability to absorb energy. Similar skins were used throughout the comparative testing, accepting the skin failure as a core yield failure mode.

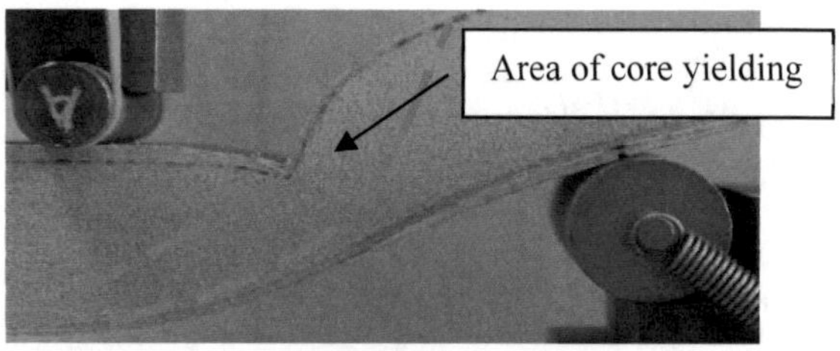

Figure 5. SAN ultimate failure

3.2. Shear strength and energy to failure

Figure 6 presents typical load displacement curves for static and dynamic tests, taking as an example the 100 J impacts shown above for the SAN and PET samples. The high-speed camera images were used to identify the time at which fracture occurred for the energy calculations.

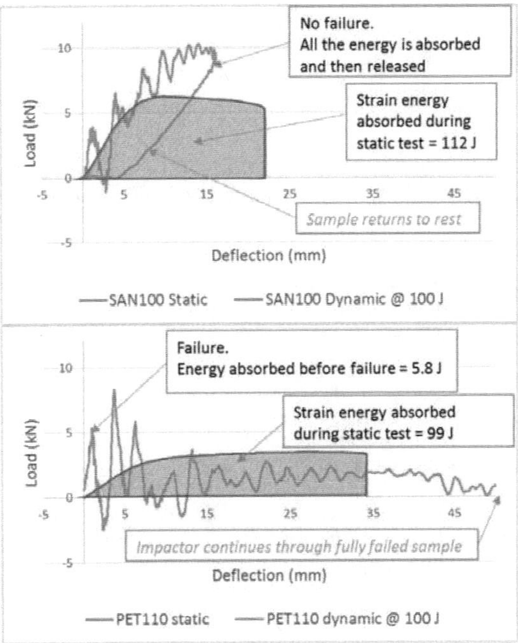

Figure 6. Load Displacement curves for 100J impacts.

Figure 7 compares the static and dynamic transverse shear strength and energy to failure for each of the materials.

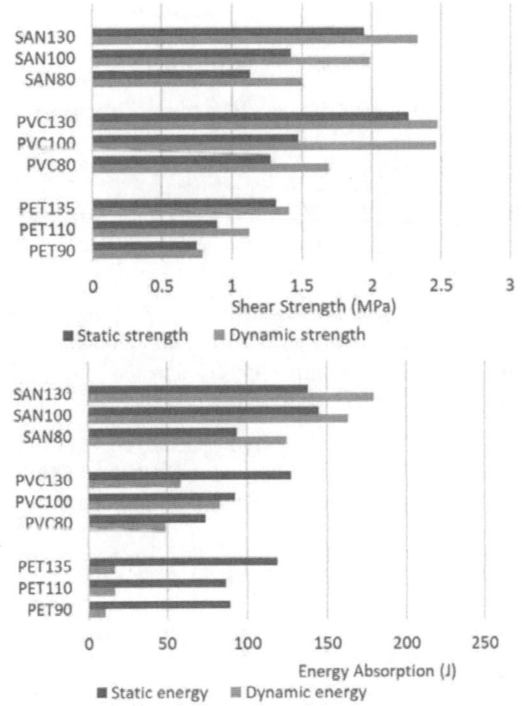

Figure 7. Static and dynamic transverse shear strength (top) and failure energy (bottom) for various core chemistry and densities.

As explained in the introduction, all materials do indeed display an increase in shear strength with an elevated strain rate. But only SAN foam displays an increase in energy absorption at elevated strain rate. Both PVC and PET have lower dynamic energy absorption than static, PET dramatically so. The range of static energy absorption is fairly similar across the three chemistry: it varies only by a factor of 2 between PVC 80 at 74 J to SAN 100 at 145 J, the dynamic energy absorption varies greatly across material types, increasing from PET to PVC to SAN, with a ratio of 17 between the material able to absorb the least energy, PET 90 at 11J, and the material able to absorb the most, SAN 130 at 180J. There is also only SAN for which the dynamic energy absorption increases with density. For PET, the dynamic energy absorption ability for the 135 density is the same as for the 110 density, and for PVC the 130 density is actually lower than for the 100 density. This is an important consideration as it differs from static core shear strength, which is proportional to density. In a static core shear stress-based design approach, should a panel be found not to match the requirement, a natural design practice would be to use a denser core instead. The present study shows that this practice could actually yield a weaker scantling with PVC foam.

4. RESULTS ACROSS CORE PROCESSES

In order to study the effect the processes mentioned in Section 1 have on the dynamic energy absorption ability of core materials, a series of SAN100 foam samples were tested in the following configurations:

- Plain
- Perforated
- Perforated, thermoformed (at 110°C[5])
- Perforated, bonded with standard laminating resin[6]
- Perforated, bonded with specially formulated very high elongation resin[7]
- Double and Triple cut

The effect of cure temperature has not yet been investigated.

4.1. Failure Modes

The perforations, thermoforming and high elongation adhesive didn't affect the failure mechanism, leading to skin failure by core plasticity, whereas the core cutting, and the standard adhesive core splicing changed the SAN failure mode to transverse shear fracture.

4.2. Shear strength and energy to failure

Figure 8 compares the static and dynamic transverse shear strength and energy to failure for each of the various processes. The perforation did not degrade the core's ability to absorb dynamic energy, actually increasing slightly both the static and dynamic shear strength and energy absorption. Core splicing increases both the static and dynamic strength, but decreases the static energy absorption, with standard laminating resin affecting the core's ability to absorb dynamic energy, whereas high

[5] Refer to [25]
[6] Gurit Prime 27 with a tensile elongation of 2.2%
[7] Resoltech DS1600 with a tensile elongation of 72%

elongation resin enables the core join to achieve as high a dynamic energy absorption as virgin foam. Even if increasing the shear strength, core treatments aimed at rendering the core conformable to curved surfaces greatly affect the ability of the core to absorb energy, both static and dynamic, basically "embrittling" the core, whereas thermoforming preserves the core ability to absorb dynamic energy. As a result, when designing and building a structure aimed at sustaining high impact loads, composite engineers and manufacturer should plan to use thermoforming rather than core treatment to conform into curved shapes. Alternatively, strip planking could be used, but with a dedicated high elongation adhesive.

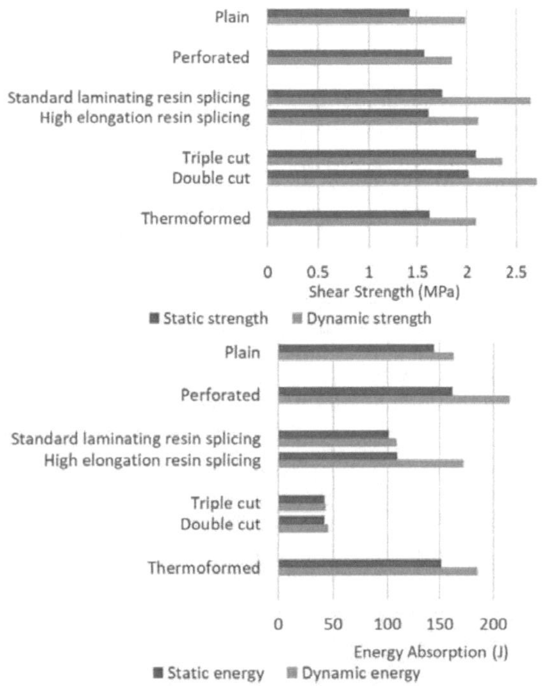

Figure 8. Static and dynamic transverse shear strength (top) and failure energy (bottom) for various core processes.

5. CONCLUSIONS

There are significant differences in the dynamic transverse shear strength and failure energy of cores depending on the type of material, and processes. Very ductile cores such as SAN have significantly higher dynamic strengths and failure energies than static, and the dynamic strength increases with density. For moderately ductile materials such as PVC the energy absorption becomes lower dynamically than statically. More brittle cores such as PET can have very low energy to failure during impact loading.

The results also demonstrate that the static shear strength and strain energies to failure are not good indicators of the core material's performance during impact loading, and highlight the need to better understand the behavior of polymeric core materials during dynamic loading. Consideration also needs to be given to how best to incorporate this behaviour into material selection and design processes.

Furthermore, the core processing steps necessary to produce a large sandwich structure also affect the plating's ability to absorb impact energy. Perforations needed to remove the air from the underlying layers don't affect the core's performance, but core splicing needs to be done with specialist high elongation resin, while conforming the core into curved surfaces needs to be done via thermoforming (or strip planking) rather than using "core treatment" such as cuts.

References

1. (1929) "The Impact on Seaplane Floats During Landing," NACA Technical Note NACA-TN-32,
2. A.Rosen, K.Garme, J.Kuttenkeuler, (2007) "Full-Scale Design Evaluation of the Visby Class Corvette", *Proceedings of 9th International Conference on Fast Sea Transportation (FAST07), Shanghai,*
3. Battley, Stenius, Breder, Edinger, (2005),"Dynamic Characterisation of Marine Sandwich Structures", *Proceedings of the 7th International Conference on Sandwich Structure (ICSS7),*
4. Downs-Honey, Edinger, Battley, (2005), "Slam Testing of Sandwich Panels", *Proceedings of Society for the Advancement of Material and Process Engineering conference (SAMPE),*
5. Paolo Manganelli, (2006), "Experimental Investigation of Dynamic Loads", *PhD Thesis, University of Southampton*
6. Battley, Lake, (2006) "Designing composite structures for slamming loads" - *Proceedings of 2nd High Performance Yacht Design (HPYD2)*, Auckland
7. Battley & Lake, (2007), "Dynamic performance of sandwich core materials", *Proceedings of the 16th International Committee On Composite Materials (ICCM16)*, Kyoto,
8. Battley, et al. (2007), "Hydroelastic behaviour of slam loaded composite hull panels" *Proceedings of 3rd High Performance Yacht Design (HPYD3)*, Auckland
9. Battley, et al. (2009) "Effects of panel stiffness on slamming responses of composite hull panels", *Proceedings of the 17th International Committee On Composite Materials (ICCM17)*, Edinburg
10. M. Burman et al. (2010), "Characterisation of ductile core materials", *J. Sandwich Structures and Materials*, 12-2: 237-252.
11. Allen, Battley, Stenius, (2010), "Experimental Methods for determining shear loads", *Proceedings of the 9th International Conference on Sandwich Structures (ICSS 9),* Passadena

12. M.Battley, T.Allen, (2011), "Servo-hydraulic System for Controlled Velocity Water Impact of Marine Sandwich Panels", *J. Society for Experimental Mechanics*

13. Allen, Battley, Casari, Kerling, Stenius, Westlund, (2011) "Structural Responses of High Performance Sailing Yachts to Slamming Loads", *Proceedings of 11th International Conference on Fast Sea Transportation (FAST11),* Honolulu

14. Goutard, Das, Battley, Allen, Norris, (2012) "Evaluation of the influence of hydro-elasticity in slamming of elastic panels using a coupled fe-fv method", *Proceedings of the 4th High Performance Yacht Design (HPYD4)*, Auckland

15. Allen, (2013), "Mechanics of flexible composite hull panels subjected to water impacts" Mechanics of Flexible Composite, *PhD Thesis*, *University of Auckland*

16. Razola, Rosen, Garme, (2014), "Experimental evaluation of slamming pressure models used in structural design of high-speed craft", J. International Shipbuildin Progress 61(1-2):17-39

17. M. Battley, et al., (2014), "Shear Strength of Sandwich Core Materials Subjected to Loading Rates Relevant to Water Slamming", *J. Reinforced Plastic and Composites*, 33-6: 506-513.

18. M.Battley, T.Allen, (2015), "Failure Mechanics of Sandwich Panels Subjected to Water Slamming", *Proceedings of the 20th International Committee On Composite Materials (ICCM20)*, Copenhagen

19. DNV GL rules for classification: Yachts (RU-YACHT), Edition December 2015 Amended July 2017.

20. CLASS PROGRAMME Type approval DNVGL-CP-0084 Sandwich core materials, Edition March 2016

21. ISO 12215-5:2019, Small craft — Hull construction and scantlings — Part 5: Design pressures for monohulls, design stresses, scantlings determination

22. American Bureau of Shipping, Guide for Building and Classing Offshore Racing Yachts, 1994

23. ASTM C393 / C393M – 20, Standard Test Method for Core Shear Properties of Sandwich Constructions by Beam Flexure

24. Core material brochure (v19): https://www.gurit.com/-/media/Gurit/Datasheets/Core_brochure.pdf

25. Core material processing guide (v5): https://www.gurit.com/-/media/Gurit/Datasheets/structuralcoreprocessingguidere.pdf

LOCALISED CORE SHEAR STRESS IN SANDWICHES WITH FACESHEET THICKNESS VARIATIONS

Tom Allen[1], tom.allen@auckland.ac.nz
Elizabeth Grant[2], egra109@aucklanduni.ac.nz

Abstract. Local loads and complex geometries in yacht hulls require additional structural reinforcement. For sandwich structures these structural reinforcements often take the form of local thickening of the facesheets. Existing studies in literature have identified the benefits of having gradually tapered ply drop-offs in order to reduce laminate stresses. This paper investigates the effect of the thickness and taper angle of local strengthening on the core shear stress for sandwich structures. A series of static experimental tests have been undertaken on beams with varying strengthening configurations and the transverse core shear strains captured through digital image correlation. A finite element numerical model of the same testing scenarios has been developed and validated against the experimental results. The validated model has been used to undertake a parametric study, investigating the effects of varying the thickness and taper angle of the local facesheet strengthening on the transverse core shear stress. The shear stress in the region of the doubler has been shown to vary significantly from the analytical predictions, deviating as the local strengthening thickness increases. The distribution of the core shear stress within the strengthened region has also been shown to be affected by the taper angle at the end of the strengthening. In all cases, increasing the local facesheet thickness or taper angle results in a lower-than-expected core shear stress in the numerical model within the doubler region, some small increases were seen just beyond the region.

1. NOMENCLATURE

τ_c	Transverse shear stress in sandwich core	[Pa]
τ_{xz}	Shear stress	[Pa]
D	Flexural rigidity	[N/m]
d	Distance been facesheet neutral axes	[m]
E_c	Youngs modulus of core	[Pa]
E_f	Youngs modulus of facesheet	[Pa]
t_c	Core thickness	[m]
t_f	Facesheet thickness	[m]
T_x	Transverse shear force	[N]
z	Distance from neutral axis	[m]

2. INTRODUCTION

Yacht hulls typically have a number of locally applied loads and complex geometric features which require the application of local structural reinforcement [1]. The approach employed for strengthening composite sandwich constructions locally is the addition of thickness to the facesheets, replacement of the core or both, depending on the specific scenario.

A composite laminate can easily be thickened through the application of additional plies. This is how a facesheet would typically be strengthened for localised loads or complex geometric features. The number and length of the local plies laminated into or onto the facesheet not only effects the thickness but will also create a tapered transition where the thickness of the facesheet returns to its base laminate. The tapering of the local laminate is often undertaken to reduce the potential for local bridging and voids at the ply drop-offs as well as to decrease the possibility of a stress raiser at the transition.

This work focuses on the effects of locally thickening the facesheets, particularly establishing how varying the thickness and taper of these plies effects the transverse shear stress in the core. This information can be used to determine an optimal taper angle and thickness.

Transverse shear stress must balance the change in direct stress field. This leads to Equation 1 for the shear stress, τ_{xz}, in a sandwich structure with symmetric facesheets with a thickness of t_f, separated by a distance d [2]. This can be solved for calculation of the transverse shear stress, τ_c, in the core (Equation 2).

$$\tau_{xz}(z) = \int_z^{(d+t_f)/2} \frac{d\sigma_x}{dx} dz \tag{1}$$

$$\tau_c(z) = \frac{T_x}{D}\left[\frac{E_f t_f d}{2} + \frac{E_c}{2}\left(\frac{t_c^2}{4} - z^2\right)\right] \tag{2}$$

Based on Equation 2, the core shear stress in a region with additional facesheet thickness should only vary relative to the additional thickness. The addition of a taper should result in a gradual change in shear stress in line with the taper angle.

Significant work has previously been undertaken looking at the effect of tapering ply drop-offs on the stress distributions within monolithic laminates [3]. These have shown gradual tapers reduce stress concentrations within the laminate. Less work has been done on the effect of local facesheet ply drop-offs in sandwich composites. That which has been done, has focused on the effects on the stress within the facesheet and local strengthening, rather than the core [4].

The additional plies can be thought of as analogously to bonding a doubler onto the facesheet, which are also commonly used to locally strengthen composite structures [5]. To ensure consistency of the thickness, doublers have been used in this study to represent the local

[1] Senior lecturer, Centre for Advanced Composite Materials, Department of Mechanical Engineering, University of Auckland, NZ
[2] Research Assistant, Centre for Advanced Composite Materials, Department of Mechanical Engineering, University of Auckland, NZ

reinforcement. Aluminium doublers have been used as they can be accurately machined to different tapers and thickness allowing different cases to be accurately tested.

Testing has been carried out by investigating a range of different doubler thicknesses and tapers on the shear stress in the core. This has been completed both experimentally and numerically. The experimental specimens were manufactured with a foam core and aluminium facesheets and were tested statically under an out-of-place flexural load using an Instron 1185 universal testing machine. A numerical model of the same configuration was constructed and validated using the experimental results. The validated numerical model has been used to undertake a parametric study to further establish the relationships between doubler thickness and taper and the transverse shear stress state in the core.

3. EXPERIMENTAL SPECIMENS

Experimental specimens were designed to investigate varying levels of local facesheet reinforcement on the core shear stress distribution. The experimental specimens were tested in a universal 1185 Instron and its results compared to numerical and analytical results. This comparison was used to validate the numerical model ensuring it gave accurate and reliable results for a variety of doubler types.

The specimens are a sandwich structure with aluminium facesheets and polymeric foam core. Aluminium facesheets have been used to; minimise orthogonality, manufacturing induced local thickness and mechanical property variations seen in composite facesheets. Aluminium doublers are used because they can be accurately machined, meaning different tapers and thickness could be tested. Different specimens of the same doubler type were also manufactured to ensure consistency between results. Foam core was used as it is a common core in yacht hull structures. It also provides a more homogeneous surface than honeycomb for extraction of strain data through Digital Image Correlation (DIC). Although aluminium is not a commonly used facesheet and core replacement material in yacht structures, its stiffness makes it analogous to multi-axial glass or carbon-fibre reinforced laminates. Not substantial variation in the outcomes is expected between using aluminium and a traditional composite material system for the purposes of this work.

The experimental specimens are manufactured from 25 mm thick Gurit Corecell M80 foam with 0.45 mm thick aluminium 5052 facesheets. In order to induce a combination of bending and shear, the specimens have been fixed at either end and loaded in the centre of the span. Embedded aluminium blocks have been used in the regions of the fixing and loading bolts. Inclusion of embedded blocks gives consistency to locations of the loading and represents the inclusion of hardpoints or core-replacements in the region of high point loads. Furthermore, the embedded blocks will restrict compressive displacement of the two facesheets relative to each other at the loading locations. The combination of embedded blocks and bolted supports restricts all six degrees of freedom at the supports. As a result, the specimens effectively have clamped ends. The geometry of the specimens is shown in Figure 1.

The facesheets were bonded onto the foam core and embedded blocks using Gurit SA-70 150 gsm epoxy adhesive film. A cure cycle of 65 minutes at 85° Celsius was carried out with the specimens consolidated under vacuum pressure of approximately 10 mbar. The doublers are then bonded onto the facesheets with the same adhesive film, cure cycle and vacuum consolidation.

A range of doubler thicknesses and end tapering ratios have been tested. The details of the experimental specimens are given in Table 1. Three of each specimen type were tested. For doublers with a taper angle these were post-machined after specimens were fully bonded.

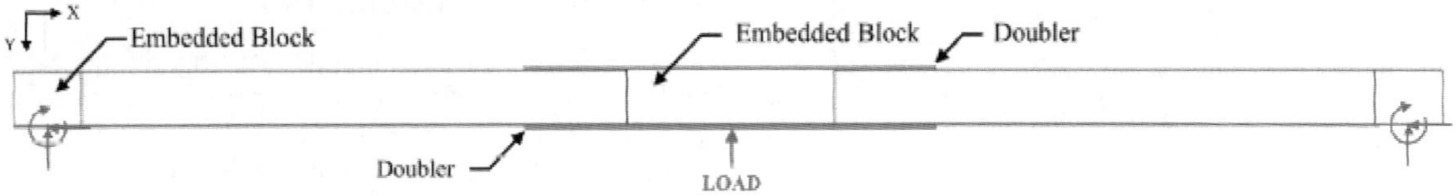

Figure 1 Schematic of test specimen with doubler

Table 1 Experimental specimen summary

	Core thickness [mm]	Skin thickness [mm]	Doubler thickness [mm]	Doubler taper
No Taper	25	0.45	0	0
Thin	25	0.45	1	0
Thick	25	0.45	3	0
Long Taper	25	0.45	3	1:10
Short Taper	25	0.45	3	1:4

Digital Image Correlation was used to obtain full-field strain fields during the experimental test. Therefore, one face of each of the samples was spray painted matte black with a white speckle pattern overtop. The quality of the speckle pattern was assessed initially within the DIC software to ensure sufficient features were found and could be track. For all cases the number of features exceeded 7500.

4. EXPERIMENTAL TESTING

The five specimen types listed in Table 1 were tested on an Instron 1185 universal testing machine. The specimens were bolted to the Instron using a mounting plate. A three-point bend test was carried out through applying a force through the central bolt. The ends of the panel were attached to the mounting plate and all torqued to 12.5 Nm.

The deformation of the specimens throughout the test was captured with a UI-1490SE-M-GL DS uEyeSE camera using a FUJINON 1:1.4/35 mm HF35SA-1 fixed focal length lens. The camera took one photo per second. The specimen was lit with polarised light from a dedicated light source. A match polarised filter was attached to the camera lens, reducing the effects of any variations in ambient lighting on the grayscale values captured by the camera. These images were then processed with in-house DIC software to calculate the shear strain on the surface of the specimens.

The specimen was loaded in three steps. First the specimen was loaded to 500 N at a rate of 2 mm/min. A 500 N load was then held for 30 seconds. The specimen was then loaded at the same rate until 1000 N was reached. The specimen was then held for 30 seconds before being loaded at 2 mm/min until 1500 N. The specimen was held for 30 seconds at 1500 N before being unloaded. A loading rate of 2 mm/min was selected as it was in accordance with ASTM-C393 [6], which is an industrially accepted standard for testing of foam core materials in a beam configuration. A loading schematic is given in Figure 2.

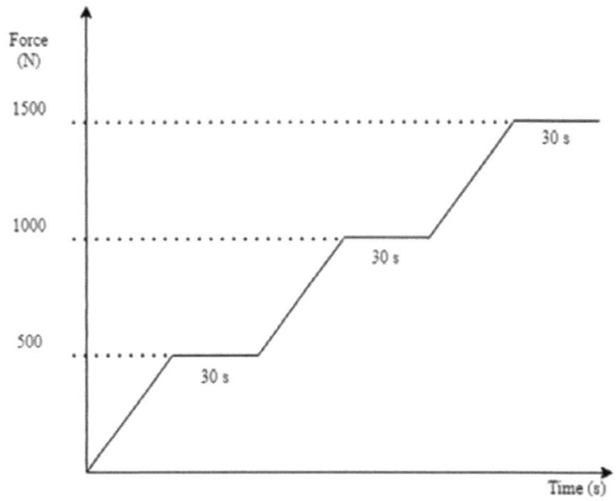

Figure 2 Loading schematic

Figure 4 plots the spatial strain distribution obtained for each of the three load levels for a specimen with a 3 mm doubler and a taper of 1:10, normalised to the applied load. The normalised position follows the path outlined in Figure 3. This figure indicates, for the loads considered, the foam core is in a linear-elastic region of its material behaviour.

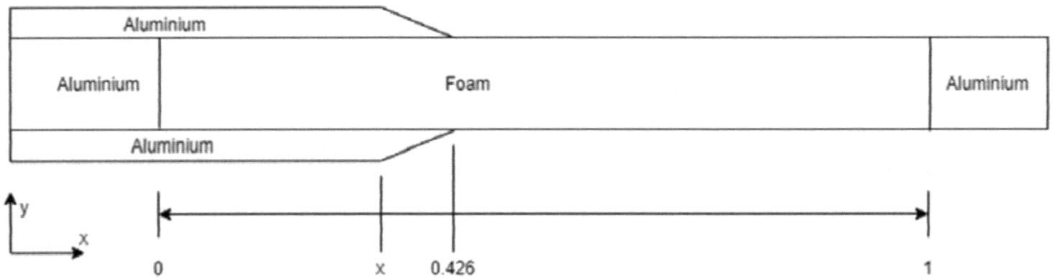

Figure 3 Diagram of specimen

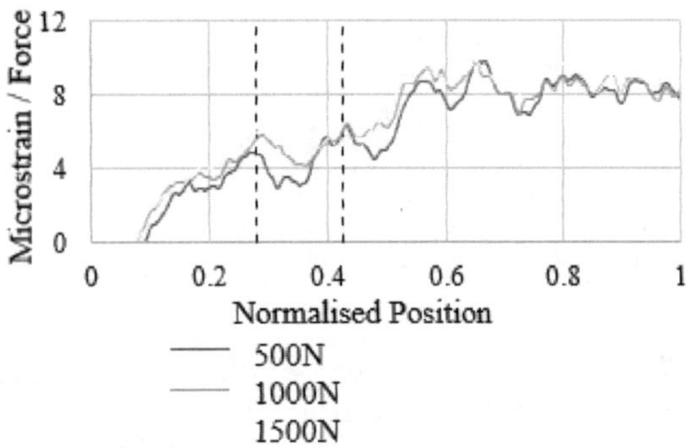

Figure 4 Microstrain normalised by applied force for a 3 mm doubler with a taper of 1:10

5. NUMERICAL MODELLING

A finite element numerical model has been developed to replicate the specimens and experimental testing configuration. The model has been developed in ABAQUS and is solved using an implicit solver. A half-model of the specimen has been created, due to symmetry. The geometry is shown in Figure 5. The geometry has been meshed using three dimensional solid C3D8R and C3D4 elements and verified through a mesh convergence study.

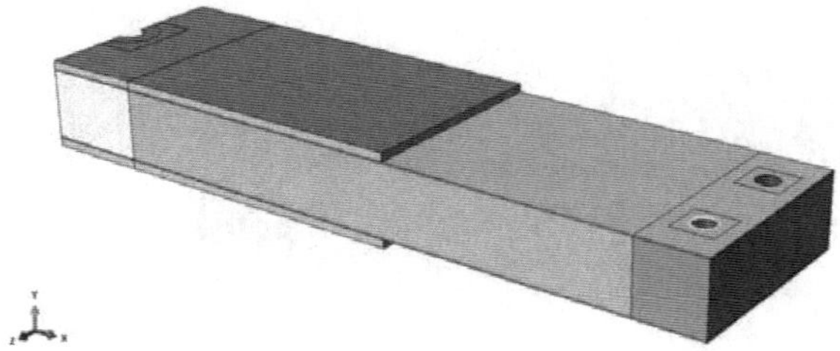

Figure 5 Finite element model geometry

A pressure load of 3.51 MPa was applied on the middle insert to simulate a force of 1000 N applied through a bolt. "Encastre" boundary conditions were applied to the end insert bolt holes and a symmetry constraint through the symmetry axis. The constraints are shown in Figure 6.

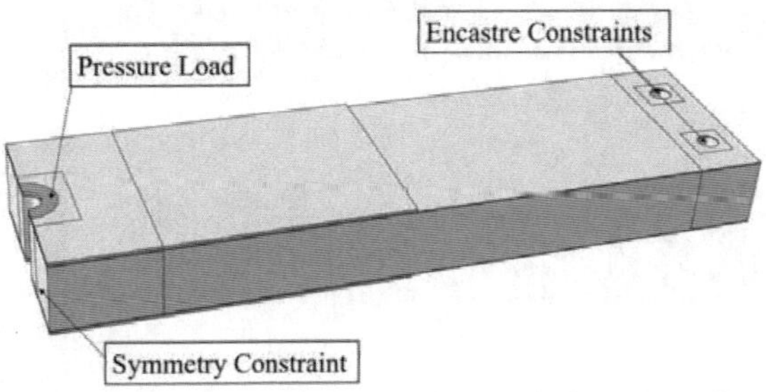

Figure 6 Finite element forces and constraints

5.1. Material Models

Both the aluminium and Corecell M80 foam materials have been modelled in the finite element model as linear-elastic and isotropic. The load levels are such that both materials should be within their linear-elastic limits, as demonstrated by Figure 4. Isotropy is a simplification for the anisotropic foam, however previous researchers have shown this to be sufficiently accurate, especially considering the computation infeasibility of modelling a large structure at micro-scale numerically [7]. The adhesive film has not been modelled as it is assumed to have negligible effect of the structural response of the beam. Both materials were specified based on properties obtained from relevant manufacturer datasheets. These properties are given in Table 2.

Table 2 Material models

Material	Young's Modulus	Poisson's Ratio	Density (gcm^{-3})
Aluminium	68 GPa	0.36	2.700
Corecell M80 Foam	72 MPa	0.24	0.085

5.2. Mesh Convergence Study

A mesh convergence study was carried out to ensure the mesh selected gave reliable results with a minimal number of elements. For the mesh convergence study, the maximum principal shear stress was plotted against the natural logarithm of one over the number of elements. This was done for different mesh configurations and is shown in Figure 7. Mesh type 7 was selected which consists of an overall mess size of 3 mm with 2 mm used around the edges of the taper. A hexahedral dominated mesh type was used in the taper of the doubler to allow for a better fit of the mesh to the acute angles. The mesh is shown in Figure 8.

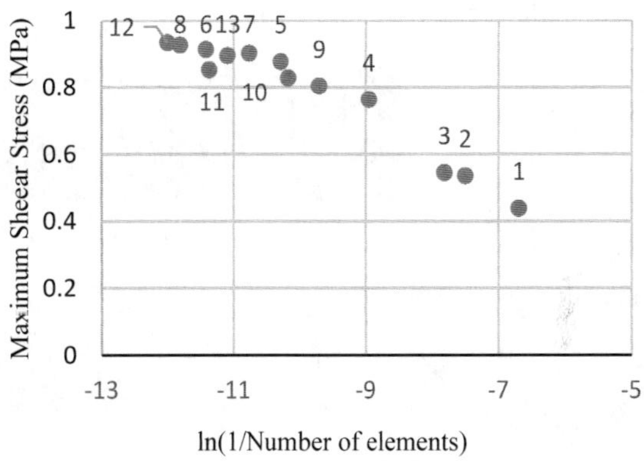

Figure 7 Mesh convergence study

Figure 8 Finite element mesh for model with a 3 mm doubler and a 10:1 taper

6. RESULTS

Figure 9 and Figure 10 show typical transverse shear strain (ε_{12}) contours predicted by the numerical model and measured during the experimental testing, both for a 1000 N load. Figure 9 shows contours for a specimen with a 3 mm doubler without any taper while Figure 10 shows the contours for a 3 mm doubler with a 1:10 taper. In the numerical model the load is applied as a force at the centre (left side of the image). In the experimental testing the load is applied by displacing the edges downwards (right side of the image) keeping the centre fixed. The inversion of the sign of the strain, positive in the numerical model and negative in the experimental results, can be attributed to differences in coordinate system orientation.

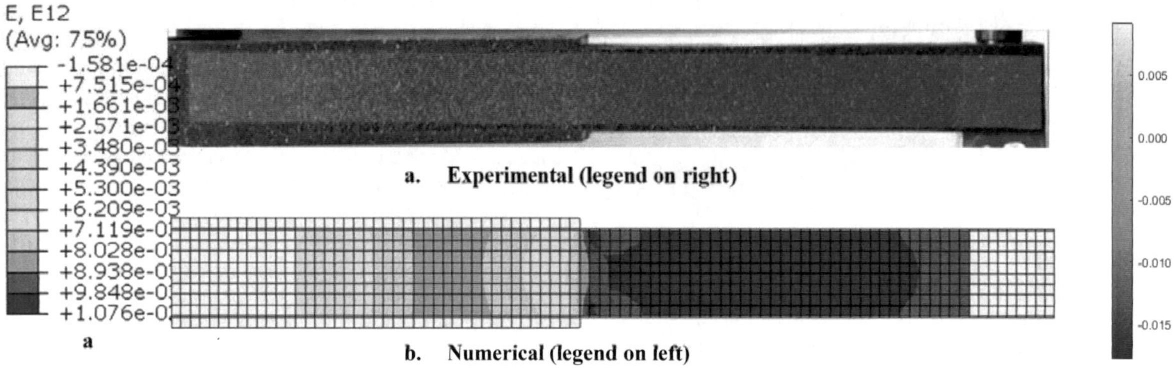

Figure 9 Contour of shear strain in a 3mm doubler with no taper

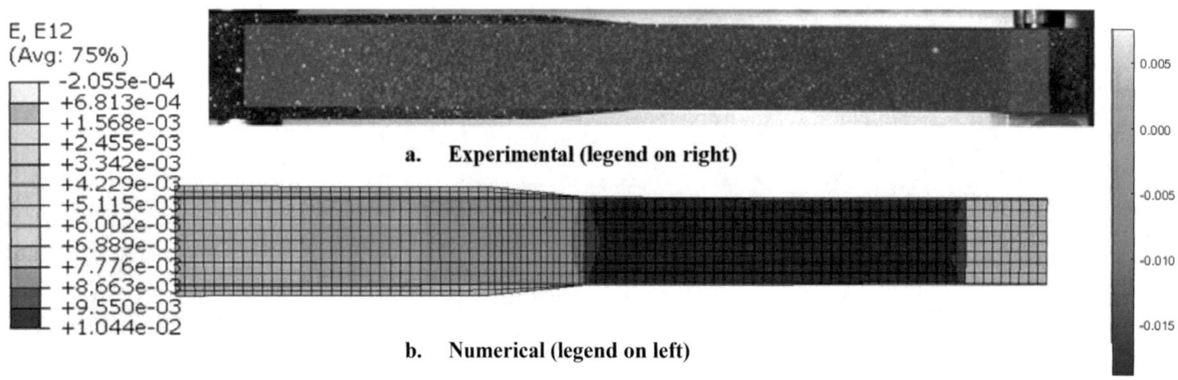

Figure 10 Contour of shear strain in a 3mm doubler with 1:10 taper

The figures show that the experimental and numerical shear strain contours follow the same trend along the length of the core. The shear strain is low in the central aluminium block, due to a much higher shear stiffness. The shear strain gradually increases in the foam core moving away from the central aluminium block.

In the region at the end of the doubler the shear strain then decreases, this decrease appears to be more significant for the specimen without any taper. At the end of the doubler the shear strain then substantially increases, remaining at this increased level until just before the outer aluminium block, where it decreases again. The shear strain is again low in the outer aluminium piece due to the higher stiffness.

The 'dip' in shear strain in the region towards the end of the doubler is most profound where there is no taper on the doubler (Figure 9). In the numerical model the dip in

The location of the end of the taper is constant between all specimens, both numerically and experimentally, other than the specimens with no doubler. It is located at a normalised position of x = 0.466. In the following graphs this location has been indicated by a vertical black dashed line. A second vertical black dashed line indicates where the doublers taper begins, for those specimens with a taper.

shear strain can be seen by the turquois section while in the experimental data it can be seen by the small amount of green at the end of the doubler. This dip in shear strain is not as profound in Figure 10 where the doubler has a 1:10 taper. This is likely due to the gradual decrease in doubler thickness with the taper. The numerical model shows a very minimal decrease in shear strain at this location, however a small section of green can clearly be seen in the experimental contour in the region of the taper. Indicating that the dip does still occur in the tapered cases, although it is smaller in magnitude.

In order to quantify the variation with respect to the position along the beam, the shear strain has been extracted along a line at the centre of the specimen in the y-direction. For the purposed of comparison, only data for the foam component has been extracted. The length in the x-direction has been normalised to 1, as seen in Figure 10.

For consistency, graphs of the results have been presented with different cases indicated by different colours. The experimental, numerical and analytical results are also indicated by different line types. A summary of the colours and line types are given in Table 3.

Table 3 Line types and colours for graphs

Case	Colour	Result type	Line type
No doubler	Purple	Experimental	Solid line
1mm doubler, no taper	Green	Numerical	Dashed line
3mm doubler, no taper	Blue	Analytical	Dotted line
3mm doubler, 1:10 taper	Orange		
3mm doubler, 1:4 taper	Black		

6.1. Experimental

Three test panels of each doubler type were tested to ensure the data obtained was consistent throughout the testing. Figure 11 shows the measured centreline shear strain within the foam core for a doubler thickness of 3 mm with no taper, with respect to the normalised x-location. Data from the three panels is shown for a 1000 N applied load. Similarly, Figure 12 shows the same data but for a doubler thickness of 3 mm and taper of 1:10.

The resulting shear strain from the each of the three test specimens can be seen to follow approximately the same trend with very similar magnitudes (typically ±10%). There is a significant amount of noise in the shear strains extract from the DIC. To make the data clearer, Figure 11 and Figure 12 have had a moving average (20 points) applied to reduce the noise. The significant local noise is likely due to the edge effects of cut cells within the foam. The deformation across each cell will vary depending on the geometry and depth of the partial cell. By taking a moving average the results, the values better represent the bulk behaviour of the foam, rather than the local material/cell behaviour.

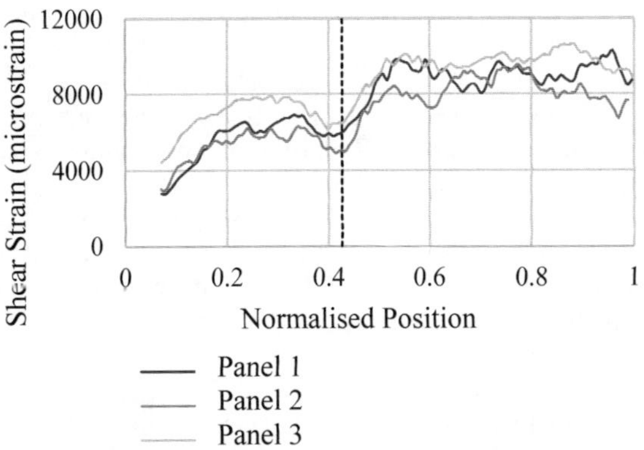

Figure 11 Experimental shear strain against distance for three panels with a 3mm doubler and no taper with a 1000 N load

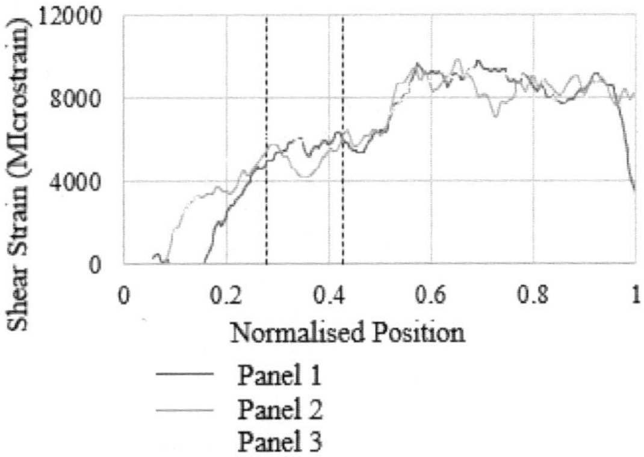

Figure 12 Experimental shear strain against distance for three panels with a 3mm doubler and taper of 1:10 with a 1000 N load

Generally, results are repeatable. In the following experimental results and subsequent comparison to models the averages of the centreline shear strains for the three specimens tested have been presented. This further reduces the spatial noise in the data.

The shear strain for each doubler type (average of the three specimens) are plotted in Figure 13, demonstrating how the different doubler geometries affect the induced shear strain.

The result from the specimen with no doubler can be treated as a baseline. It remains approximately consistent along the length of the specimen, between 8,000 and 10,000 με. For all other specimens the shear strain in the region without the doubler (x=0.466 to x=1) matches that of the specimen with no doubler.

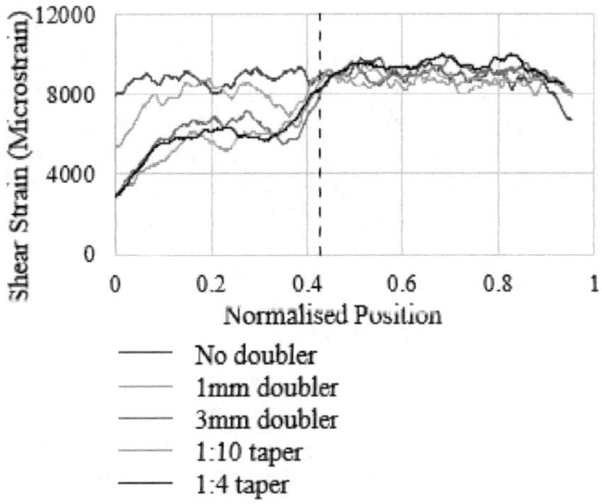

Figure 13 Average experimental shear strain for all doubler types and a load of 1000 N

The 1 mm doubler with no taper reaches approximately the same shear strain as the no doubler specimen half-way between the central aluminium block and the end of the doubler. At the central aluminium block and the end of the doubler there is a significant decrease in shear strain, 5,500 με and 7,200 με respectively down from ~ 9,000 με. This indicates two effects, firstly, an effect of increasing the flexural rigidity of the skins at their intersection with the aluminium block. This increase in flexural rigidity appears to result in a decreased local rotation, similar to 'clamped' boundary conditions, and reduces the local shear. The second effect is that of the termination of the doubler on the local transverse shear. This is of primary interest in this work.

The case with a 3 mm doubler and no taper follows a similar shape to the 1 mm doubler case. The main difference is that the shear stress in the doubler region reaches a lower peak of approximately 7,200 με. This is a result of the increased thickness of the doubler caring more shear stress. The shear strain then sharply dips at the end of the doubler to a minimum value of about 5,500 με before increasing to the no doubler strain value of approximately 9,000 με.

Specimens with a 3 mm doubler with a taper reach similar shear strains as the 3 mm doubler case with no taper. The main difference is the variation in shear strain where the doubler stops and a slight decrease in the peak shear strain within the doubler region. When there is no taper there is a significant dip at the end of the doubler, well below the peak reached within the doubler region. A dip of approximately 1,700 με. In the 1:10 taper case the shear strain gradually increases from its peak within the doubler region until it reaches the no doubler shear strain. It does not display the dip in shear strain that the no taper cases doe. The 1:4 taper shows a result somewhere in the between these two cases. It has a smaller dip (< 500 με) than the no taper case before increasing to the no doubler shear strain value. This indicates that the sharper the doubler drop of is the larger the dip in shear strain will be. There also appears to be evidence to support the conclusion that the shaper the taper the higher the peak shear stress will be within the doubler region. For 3 mm thick doublers with zero, 1:4 and 1:10 tapers the peak shear strains within this region are; 7,200, 6,300 and 6,100 με respectively.

6.2. Model Validation

The developed numerical model was compared against the experimental and analytical results. Figure 14 displays the analytical, numerical and experimental results for the two no taper cases which had doubler thicknesses of 1 and 3 mm. The numerical results are taken from a path through the centre of the specimen in the y-direction, in the same way as the experimental results were extracted from the DIC. Importantly the results were taken on the outer face of the core as this correlates with where the DIC measurements are made experimentally.

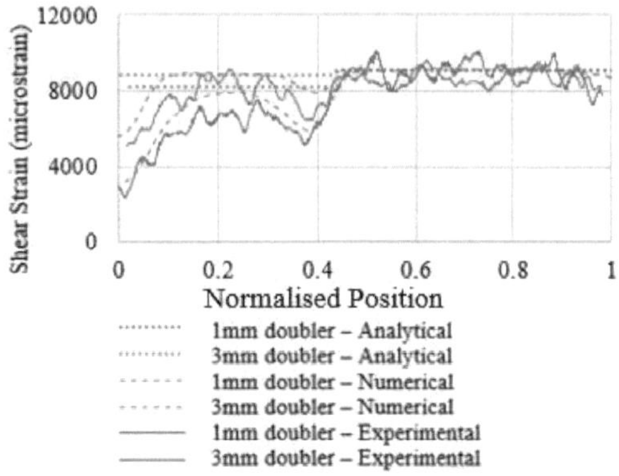

Figure 14 Numerical, experimental and analytical results for 1mm and 3mm doublers with no taper

From Figure 14 it can be seen that the analytical and numerical results reach the same shear strain from x = 0.5 to x = 0.95, where there is no doubler: 9,070 ± 5 με. The experimental results, accounting for the level of noise present, are very similar.

For the 1 mm doubler the shear strain from the numerical model matches the analytical shear strain (8,830 με) between x = 0.1 and 0.3. For the 3 mm doubler the shear strain does not quite reach the analytical prediction (8,160 με) in the doubler region, reaching a peak of 7,890 με.

Numerical results for both doubler thicknesses show a decrease in shear stress in the region of the central aluminium block and at the end of the doubler. This corresponds with the experimental shear strain distribution. With lower noise than the experiment it is clear from the numerical results that there is a difference in the rate at which the shear strain increases moving away from the central aluminium block. The 3 mm doubler takes a greater distance to reach its maximum, despite its maximum being approximately 1,000 με lower.

Overall, there is good correlation between the experimental and numerical results. Matching each other well along the length of the beam. The homogeneous approach to modelling the core material in the numerical model leads to less spatial noise in shear strain when compared with the experimental data. The key features of the distribution and magnitudes seen experimentally are well represented in the numerical results.

Of key interest is the shear strain in the region of the end of the doubler. For the 1 mm doubler the minimum shear strain in the dip for the experimental case is measured at around 6,500 με while in the numerical model it is measured at around 8,000 με. This is a reasonable difference that may potentially be attributed to the singularity nature of having no taper in the end of the doubler. It is however preferable that the model overpredicts the shear strain as the component to ensure conservatism. There is significant conservatism in this region when comparing either the numerical or experimental results with the analytical solution, which predicts no decrease in shear strain in this region. There appears to be better correlation between the numerical and experimental results in this region for the 3 mm doubler. In both cases, however, there is difficulty in determining the specific degree of validation due to the noise within the experimental results.

Figure 15 illustrates the numerical, analytical and experimental results for two 3 mm doublers, one with a taper of 1:10 and the other with a taper of 1:4. As with the specimens with no taper, the numerical and experimental results both agree well with the analytical results in the no taper region (± 0.5%). In the doubler region both the numerical and experimental results show lower shear strains than predicted by the analytical analysis. The numerical results show peaks of 7,060 με and 6,820 με for the 1:4 and 1:10 tapers respectively. These peaks are 14 % and 16 % lower than the analytically predicted value of 8,160 με.

The experimental and numerical results follow the same shape within the doubler region, increasing moving away from the central aluminium block, raising up to a peak. There is then a decrease in shear strain (except for the experimental 1:10 taper results), before a rapid increase up to the no doubler shear strain value. In the numerical results it's clear that the sharper the taper the sharper and more defined the dip in shear strain at the end of the doubler is and the higher the peak in shear strain is within the doubler region. It is clearer in the numerical results than in the experimental results, largely due to the level of noise within experimental the results. Overall, the numerical model appears to accurately replicate the shear strain shape and magnitudes seen experimentally. It appears to slightly overpredict the experimental shear strain in the doubler region. In the region with no doubler the analytical model predicts the shear strain well, however the correlation within the doubler region diminishes with increasing doubler thickness.

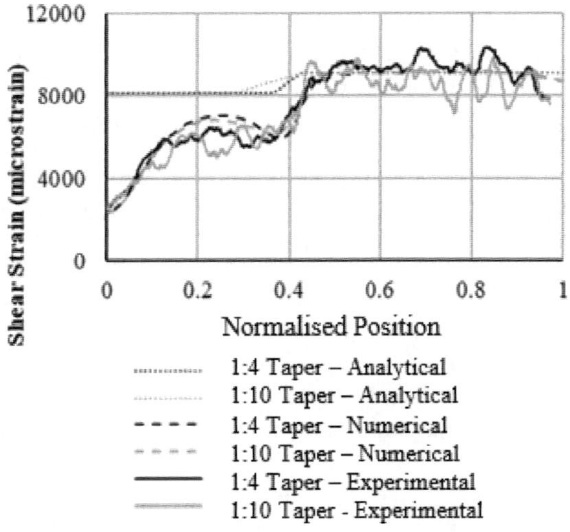

Figure 15 FEA and analytical results for 3mm doublers with a 1:10 and 1:4 taper

6.3. Transverse variation in numerical model

In order to investigate the differences between the transverse shear stress observed on the face of the specimens (which is the only location that can be measured experimentally) and that present in other sections of the beam, a transverse section (y-plane) of the numerical model has been taken. This contour, shown in Figure 16, indicates negligible variation through the width of the beam. This gives confidence that the measurements made in the experimental tests are applicable for the core in its entirety and not limited to the surface. The trend shown for this case is representative of the other cases investigated.

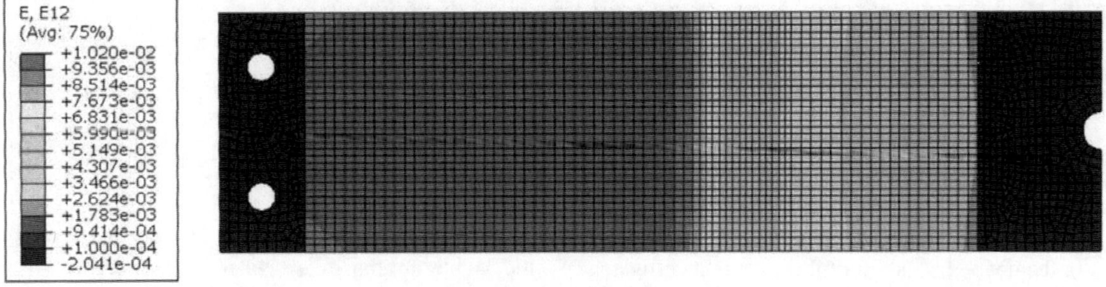

Figure 16 Shear strain cross section for a 3 mm doubler with a 1:10 tap

8. PARAMETRIC STUDY

Following validation using the five experimental cases, the numerical model has been used to undertake a parametric study. This study further investigated the relationship between shear stress, doubler thickness and doubler taper. To do this a python script was written to generate and run a series of cases. These cases had a range of thicknesses between 1 and 5 mm and taper ratios of 0:10 to 160:10. The results of the parametric study were extracted using a path through the centre of the part, as per the previous numerical results.

Figure 17 presents the shear strain distribution for all the cases run within the parametric analysis. There is little variation in the non-strengthened region beyond the doubler end. There is a clear trend of decreasing shear strain within the region of the doubler with increasing doubler thickness. The five horizontal groups of lines each represent a separate doubler thickness. In order of decreasing shear strain, 1 to 5 mm thick doublers.

It's notable that as the doubler thickness increases there is a change in the shape of the shear stress distribution in the doubler region. For the 1 mm doubler (highest strain values) there is a defined plateau with a gradual decrease before and after. This plateau is evident in the 1 mm doublers and the 2 mm doubler with no taper, for all other cases there is only a gradual increase then decrease with no plateau. It is possible that the ratio of the doubler thickness to the length has an affect and that for a longer reinforced region a plateau might have been evident for the thicker doublers.

Within the cluster of results for each doubler thickness there is a progressive change in the shape of the strain distribution. For sharp tapers (closest to no taper) there is a very definitive 'dip' in the region of the end of the doubler, this raises to a peak in the centre of the doubler region. As the taper increases, this 'dip' decreases and progressively moves to an almost linear distribution between the end of the doubler and central aluminium block. Also, of note is a small increase in strain, above the no-doubler value, at the end of the doubler region. This appears to be most significant for thicker doublers with no taper. This increase is relatively minor (4.1 %), however should be considered for significant reinforcement drop-offs.

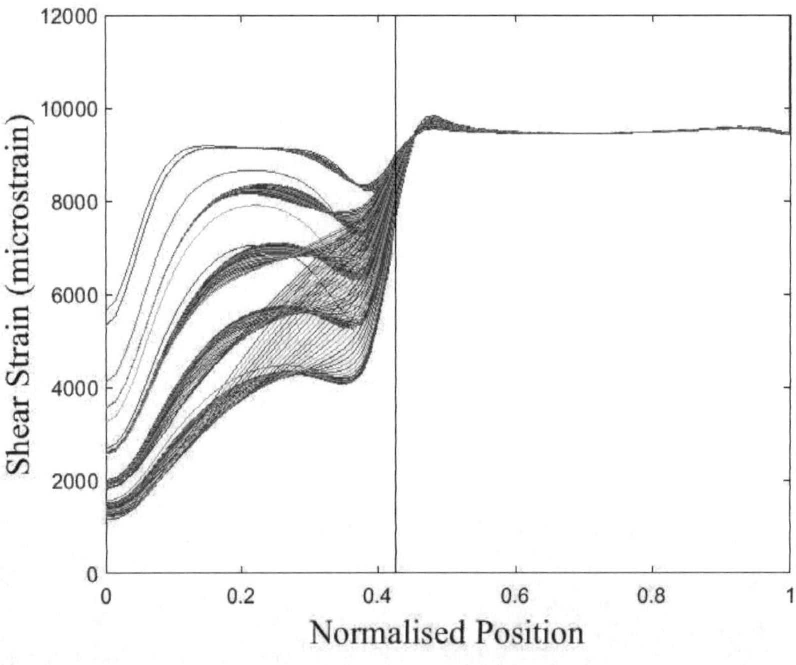

Figure 17 Shear strain through the middle of the part along the distance

Figure 18 presents the maximum transverse core shear strain in the doubler region for all the cases investigated within the parametric study. The graph has been plotted for the 5 doubler thicknesses relative to the taper ratio. This data considers the maximum between the central aluminium block and 'dip' seen in the strain distribution at the doubler end. Only those cases are plotted where there is a dip in the strain. As the taper ratio increases, this local maximum slowly decreases. It's also clear that with increased doubler thicknesses, the 'dip' in the strain is not there at smaller tapers.

Figure 19 presents the minimum shear strain at the end of the doubler for those cases where there is a 'dip'. As with Figure 18 the taper angle at which there is no longer a 'dip' decreases with increasing doubler thickness. It can also be seen that as the taper ratio increases the minimum within the 'dip' increases. This supports the evidence for a transition in distribution from a peaked shear strain in the centre of the doubler region to an approximately linear distribution between the end of the doubler and the central aluminium block.

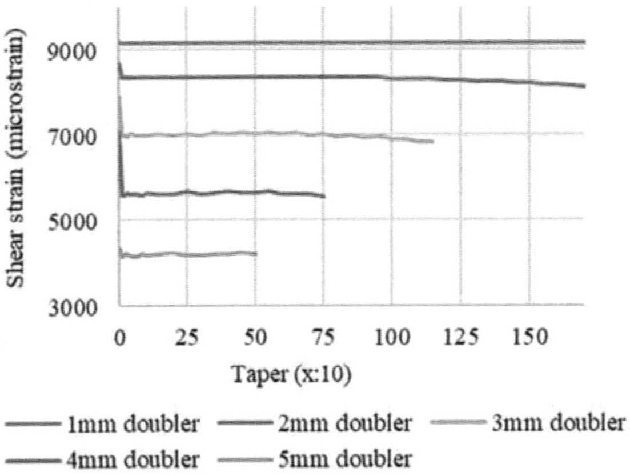

Figure 18 Maximum shear strain within the doubler region for all cases within parametric study

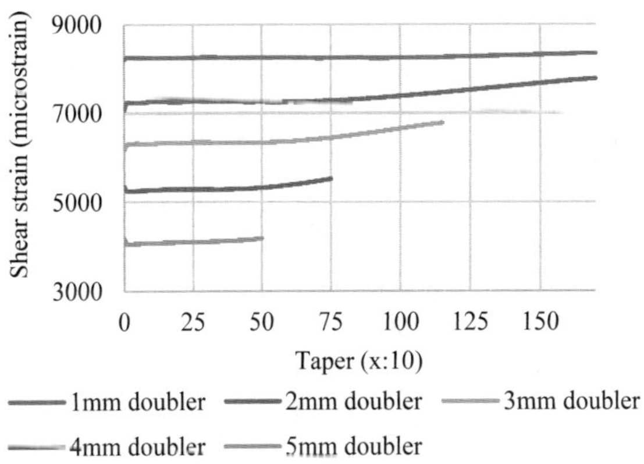

Figure 19 Minimum shear strain at the end of the doubler for all cases within parametric study

9. CONCLUSIONS

This paper describes an investigation on the effects of local reinforcement of facesheets on composite sandwich structures, specifically with respect to the transverse shear strains within the core material.

A specimen has been designed to allow for the experimental characterisation of a range of reinforcement configurations, varying both the thickness of the reinforcement and the taper at its termination. A series of experiments have been undertaken using aluminium facesheets and doublers as reinforcements on a polymeric foam core. The transverse core shear strains have been extracted using Digital Image Correlation.

Numerical modelling of the same scenarios has been undertaken and a numerical modelled validated using the experimental results. The numerical modelling and experimental results have also been compared with the analytical predictions of transverse shear strains. It has been shown that in non-reinforced areas the analytical prediction accurately matches both the experimental and numerical predictions. When a reinforcement is added to the facesheet, the analytical solution is shown to over predict the maximum core shear stress for doublers thicker than 1 mm (for a 0.45 mm thick facesheet)

It has also been shown that there is a distinctive, unpredicted, decrease in transverse shear strain in the region of the end of the reinforcement. This local 'dip' in shear strain is seen to decrease as the taper ratio on the end of the reinforcement is increased.

A numerical parametric study has been completed, looking at the effect of increasing doubler thickness and taper. A clear transition, with increasing taper ratio, from a peak of shear strain within the reinforced region to a linearly decreasing distribution has been observed. A small increase (<10 %) in core transverse shear stress has also been observed for thick reinforcements with little to no taper, at the very end of the doubler.

It has been shown, by and large, the analytical solutions provide a conservative calculation of the transverse shear strain in regions of local facesheet reinforcement and that an increased taper ratio on local reinforcement reduces local transverse shear stress

References

[1] R. Eliasson, L. Larsson, and M. Orych, *Principles of yacht design*. A&C Black, 2014.

[2] D. Zenkert, *An introduction to sandwich construction*. Engineering materials advisory services, 1995.

[3] K. He, S. V. Hoa, and R. Ganesan, "The study of tapered laminated composite structures: A review," *Compos. Sci. Technol.*, vol. 60, no. 14, pp. 2643–2657, Nov. 2000, doi: 10.1016/S0266-3538(00)00138-X.

[4] F. Mortensen and O. T. Thomsen, "A simple approach for the analysis of embedded ply drops in composite and sandwich laminates," *Compos. Sci. Technol.*, vol. 59, no. 8, pp. 1213–1226, Jun. 1999, doi: 10.1016/S0266-3538(98)00160-2.

[5] D. Roach and K. Rackow, "Development and Validation of Bonded Composite Doubler Repairs for Commercial Aircraft," in *Aircraft Sustainment and Repair*, Elsevier, 2018, pp. 545–743.

[6] ASTM, Ed., "C393 - 00 Standard Test Method for Flexural Properties of Sandwich Constructions." .

[7] Y. Chen, R. Das, and M. Battley, "Effects of cell size and cell wall thickness variations on the stiffness of closed-cell foams," *Int. J. Solids Struct.*, vol. 52, pp. 150–164, 2015, doi: https://doi.org/10.1016/j.ijsolstr.2014.09.022.

AUTHOR INDEX

Abiven, N. 76
Alas, R. 1
Allen, T. 129, 136
Arredondo-Galeana, A. 98
Aubin, N. 108
Basset, T. 129
Battley, M. 129
Bot, P. 1, 87
Braud, C. 1
Braun, J. 118
Cater, J. 108
Copeland, E. 39
D'Arco, F. 1
Dhome, U. 59
Drummond, M. 39
Gerard, R. 129
Giovannetti, L. 7, 67
Grant, E. 136
Hawthorne, E. 49
Horteborn, A. 59
Jochum, C. 76
Kuttenkeuler, J. 59
Larsson, L. 25
Leroux, J. 76
Ljungqvist, K. 67
Lundback, O. 59
Marcillat, P. 1
Minerva, L. 15
Montero, F. 15
Morvan, A. 76
Nava, S. 108
Neme, A. 76
Nicols, J. 49
Norris, S. 108
Peart, T. 108
Persson, A. 25
Pezeril, B. 1
Prabahar, N. 25
Razola, M. 59
Sabadell, I. 39
Sacher, M. 76
Simhony, T. 39
Soulier, A. 1
Souppez, J. 87
Stenius, I. 7
Viola, I. 87, 98
Voisin, D. 1
Werner, S. 59, 67